Three Shots at Prevention

Three Shots at Prevention

The HPV Vaccine and the Politics of Medicine's Simple Solutions

Edited by Keith Wailoo, Julie Livingston, Steven Epstein, and Robert Aronowitz

The Johns Hopkins University Press
Baltimore

© 2010 The Johns Hopkins University Press
All rights reserved. Published 2010
Printed in the United States of America on acid-free paper
9 8 7 6 5 4 3 2 1

The Johns Hopkins University Press
2715 North Charles Street
Baltimore, Maryland 21218-4363
www.press.jhu.edu

Library of Congress Cataloging-in-Publication Data

Three shots at prevention : the HPV vaccine and the politics of
medicine's simple solutions / edited by Keith Wailoo . . . [et al.].
    p. ; cm.
  Includes bibliographical references and index.
  ISBN-13: 978-0-8018-9671-2 (hardcover : alk. paper)
  ISBN-10: 0-8018-9671-1 (hardcover : alk. paper)
  ISBN-13: 978-0-8018-9672-9 (pbk : alk. paper)
  ISBN-10: 0-8018-9672-X (pbk : alk. paper)
   1. Papillomavirus vaccines—Social aspects.   I. Wailoo, Keith.
[DNLM: 1. Papillomavirus Vaccines—therapeutic use.
2. Uterine Cervical Neoplasms—prevention & control.   3. Policy
Making.   4. Risk Factors.   5. Sexual Behavior—psychology.
6. Vaccination—ethics. WP 480 T531 2010]
QR189.5.P36T48 2010
612.6'2—dc22   2009052693

A catalog record for this book is available from the British Library.

*Special discounts are available for bulk purchases of this book. For more
information, please contact Special Sales at 410-516-6936 or specialsales@
press.jhu.edu.*

The Johns Hopkins University Press uses environmentally friendly
book materials, including recycled text paper that is composed of at
least 30 percent post-consumer waste, whenever possible. All of our
book papers are acid-free, and our jackets and covers are printed on
paper with recycled content.

Contents

Acknowledgments

The science of immunology has spawned many dramatic developments over the years, yet its by-products, from transplantation to vaccination, are frequently shadowed by debate and controversy. In 2008, Julie Livingston, Steven Epstein, Robert Aronowitz, and I brought together a group of scholars (from sociology, history, medicine, medical ethics, pathology, psychology, communication arts, and health policy) for two meetings to reflect on one of the latest unfolding controversies: the then-new HPV vaccine, Gardasil. Because the vaccine had become a focal point for contentious debate—about pharmaceutical advertising; scientific evidence; cancer prevention and risk; the role of government in public health; girls, families, and sexuality; and global health inequalities—we envisioned this volume as a case study of the fraught intersection where the science of immunology, medical practice, and public health collide with society, culture, and politics. The resulting volume can be read both as a chronicle of the evolution of the debate and as an analysis of the perspectives, values, and policy questions unearthed by the vaccine in particular and by immunology more generally.

This multidisciplinary analysis could not have been possible without the generous support I received from the James S. McDonnell Foundation's Centennial

Fellowship in the History of Science—a grant to encourage exploration of a range of developments in the biomedical sciences in fields from cancer to genetics and from pain to immunology. This book follows on the heels of an earlier McDonnell-supported, coedited volume, *A Death Retold: Jesica Santillan, the Bungled Transplant, and the Paradoxes of Medical Citizenship*, which focused on another recent "immunological event" (an organ transplantation error) and used similar methods to examine the administrative, technical, social, political, and ethical complexities of organ donation, organ matching, medical error, and immigrant health today. Both volumes reflect one important goal of the McDonnell Fellowship: to promote innovative, cross-disciplinary scholarship and thereby advance the understanding of the biomedical sciences and their sweeping implications in the modern world. Additional support for this project came from the Rutgers University Institute for Health, Health Care Policy, and Aging Research and from the university's Center for Race and Ethnicity.

The volume is also the result of the insight, wisdom, and effort of many extraordinary people. I am particularly grateful that Julie Livingston, Steven Epstein, Robert Aronowitz, and I made such a marvelous team. Several research assistants provided important assistance by organizing our conferences and meetings and providing editorial insight at various stages. They are Isra Ali, Dana Brown, Nadia Brown, Fatimah Williams-Castro, Jeffrey Dowd, Bridget Gurtler, Shakti Jaising, Melissa Stein, Anantha Sudhakar, and Dora Vargha. In this group, Shakti Jaising deserves our special gratitude for her insightful editorial contributions. Mia Kissil provided excellent administrative and conference-planning support, as did Maureen DeKaser. Others who offered helpful comments on conference presentations, chapters, sections, or the entire volume are Mia Bay, Carol Bigman, Carlos Ulises Decena, Christine Gorka, Amy Leader, Stephen Pemberton, and Charles Rosenberg. We are particularly grateful to Elizabeth Armstrong for her strong, challenging suggestions for improving the volume, to Jacqueline Wehmueller at the Johns Hopkins University Press for her wide-ranging editorial insight, and to copyeditor Lois Crum for deft work with words, themes, and concepts. Finally, Julie, Steven, Robby, and I owe profound thanks to the essay contributors for their diligence, thoughtfulness, and brilliance in analyzing the many features of the HPV story—from the global to the local, and from science to politics and culture—and for collectively illuminating what makes the vaccines at once promising and controversial.

Keith Wailoo

Introduction

A Cancer Vaccine for Girls?
HPV, Sexuality, and the New Politics of Prevention

The first advertisements for Gardasil in the United States in 2006, part of
Merck's "One Less" campaign, featured an array of active adolescent girls—
soccer players, double-dutch rope jumpers, and other youngsters—staring de-
fiantly into the camera. Each one declared her intention to be vaccinated
against the human papillomavirus (HPV) and to be one less victim of cervical
cancer; all it would take was three injections over six months. The Spanish-
language version of the advertisement that soon followed ("Una Menos")
portrayed a Latina girl dressing for her quinceañera—her rite of passage to
adulthood—making a similar defiant declaration. With young girls framed as
the potential victims of a sexually transmitted infection and as the hoped-for
beneficiaries of the new vaccine, Merck's promising advertisements invited
much public attention, hope, and scrutiny. A few months after the first appearance of the ads, the HPV question took another turn when Texas governor
Rick Perry issued an executive order mandating vaccination of all sixth-grade
girls in the state. Other U.S. states began considering similar efforts in the name
of safeguarding girls against a sexually transmitted virus associated with cervical, anal, and other cancers. Slowly, it became clear—to physicians, public

health officials, legislators, families, citizens, and girls—that the vaccine discussion in the United States was not just about the vaccine per se but also about long-simmering cultural and political tensions, now inflamed by the arrival of Gardasil.

The emerging HPV vaccine controversy also brought public health to a challenging crossroads as elected officials began championing Merck's consumer health product as a protector of girls, a salve for parental anxiety, and a new kind of public health weapon against cancer. With the development of vaccines like Gardasil and, later, GlaxoSmithKline's Cervarix, public health questions spilled into the spotlight: Should government mandate vaccination in the name of cancer prevention? Should the choice of whether to vaccinate their daughters against the sexually transmitted virus be left to parents? Should "at-risk" groups other than young girls and women be targeted by HPV vaccination campaigns? In the United States, some parents welcomed the new cancer-prevention tool while others recoiled against government mandate as an unwarranted intrusion into private matters; and legislators and health policymakers also entered the fray. At every twist and turn in the controversy, different interests came into view: state and federal government, parents and families, pharmaceutical companies, and advocacy groups. Questions of trust, knowledge, and sexuality undergirded the controversy: some Americans distrusted the drugmaker's claims about the vaccine's effectiveness and the company's shrewd appeals to the empowerment of girls. Other critics, concerned about sexual morality and abstinence, alleged that a vaccine against a sexually transmitted infection sent the wrong message to girls and encouraged sexual activity.

One simple question for girls, families, policymakers, and states was how much faith to place in the messages and underlying knowledge claims put forth by experts in public health or by pharmaceutical companies such as Merck. Opinions varied widely on whether harnessing pharmaceutical development for the public good (through the power of a vaccine) would really maximize the health of the population and reduce harm. In truth, no one knew the answer. Even scientists and clinicians were not certain whether the new vaccine would confer lasting cancer immunity. Some governments, however, felt compelled to act by the existing evidence and the availability of public resources for the costly vaccine. In the UK, the National Health Service launched an extensive, market-savvy campaign—promising teenage girls that with three voluntary Cervarix vaccinations they would be "armed for life" against cervical cancer.[1] Like other vaccine efforts in the past, this new one would be a large-scale public health experiment, and only time, vaccine uptake, follow-up studies, and long-term surveillance of the vaccinated population would reveal its effectiveness.

In the gap between anxiety and evidence, however, many sociopolitical and cultural dynamics became visible. The vaccine's unusually high cost (approximately $350 per three-dose sequence), for example, refocused attention on the divide between the world's haves and have-nots. The high price ensured that while the citizens of wealthy nations debated the option, those in poor and developing nations would remain outsiders to this apparent marvel of prevention. Cervical cancer has long been linked not only to a virus and to sexual activity, but also to poverty and socioeconomic circumstances. The number of deaths caused by cervical cancer is highest in developing countries, while in wealthier nations this number has been dropping for years, with better access to Pap smears, routine cancer surveillance, early intervention, and prevention of sexually transmitted diseases. Arguably, the vaccine is most needed in the global South, and yet the debate has been most intense in the global North. Gardasil, with its consumerist model of vaccination, marked a kind of "pharmaceuticalization" of public health, to use a term coined by health scholar João Biehl, that had implications nationally as well as globally.[2]

The HPV controversy was not, therefore, a one-dimensional health debate, for it threaded many questions—family values, the role of government, the reliability of scientific evidence, the oversight of sexuality, global equity, and trust in drug companies—into a dense tangle of scientific claims and political assertions. At the center of the storm were young girls, with intense anxieties swirling around them about their futures, their sexuality, their health, and the world of risks confronting them. In their shadow stood young boys, whose own potential susceptibility to HPV-linked penile, anal, and oral cancers was caught up in a politics of sexuality rendered invisible in the marketing blitz and the ensuing public debate.

This project brings together scholars and clinicians from diverse perspectives to analyze and dissect the science, politics, and symbolism of the HPV vaccine debates, not only in the United States, but globally. Organized in four parts, the book examines (1) the knowns and unknowns of the HPV–cervical cancer link; (2) what it means that images of girls and girlhood have dominated and skewed the HPV debates; (3) how the politics of family and government informs HPV vaccine opinions and policies; and (4) the relationship of the American vaccine debate to debates across Europe and to cervical cancer mortality in Africa.

With its diverse voices and perspectives, the book offers—section by section—a roadmap for understanding, analyzing, and navigating the many uncertainties surrounding HPV vaccines. We analyze the underlying message of the vaccine advertisements, describe the global and regional politics that the vaccine brought into view, and explore how parents' groups and policymakers formed their positions and framed the vaccine debate. Through the essays that follow, the vaccine debate emerges not as a finished story with clear answers, but as an unfolding controversy in which both evidence and interests evolved rapidly. Because this vaccine debate involves science and sexuality as well as the interests of government and the public, business and private individuals, it provides a unique opportunity to explore how these multiple interests and underlying values intersect within the arena of public health.

Vaccines have long been prized as one of medicine's apparently simple solutions to disease, yet their use sparks contention and generates new social, political, and cultural complexities. Although many have fantasized about a vaccine to prevent cancer, the realities of vaccination are inevitably more complicated than such fantasies allow. Vaccines present a locus of controversy precisely because they offer future protection from present threats and because they depend (to some extent) on public fear to mobilize citizens and states to take action. Indeed, public health mobilization itself can foster fear. Compulsory vaccination—although validated by the courts early in the twentieth century—was seldom invoked before the 1960s because health professionals worried that compulsion would be counterproductive (see chapter 1). Wherever states acted aggressively to safeguard the health of the population, the overriding of individual rights raised questions. But driven by the ideals of the Great Society and the expansion of health opportunity, a new era of compulsion began in the 1960s and 1970s. Mobilizing science in the name of this effort connected vaccines to a broader story of politics and American governance. Even so, the history of vaccination remains defined by fundamental tensions around the limits of state power, scientific uncertainty, and trust in science and government. With Gardasil, these issues were given new cultural potency because they were organized around the sexuality of young girls.

Part One: The Known and the Unknown. What do we know about the relationship between HPV and various forms of cancer, about the effectiveness of Gardasil, and about its global impact on HPV-related disease? Scientific, clinical, and political uncertainties and disagreements abound. Historian James Colgrove observes in chapter 1 that there is disagreement even among those who oppose the vaccine. The great diversity of reasons expressed reflects an "incompletely theorized agreement" driving the opposition. Historian-physician Robert Aronowitz examines the character of scientific and clinical uncertainties, noting, for example, that even a Merck representative conceded during a regulatory hearing that knowledge about the causal connection between high-risk HPV types and cervical cancer was indirect and open to revision. Pathologist Lundy Braun and cancer researcher Ling Phoun delve deeper into the complex relationship between virus and disease process (pathogenesis) on the one hand and the erasure of that complexity in many public discussions on the other hand. Sociologist Steven Epstein points to the erasure of another complexity: the link between HPV, anal cancer, and gay men's health. Discussion of this linkage has been almost entirely absent from public debate about the safety, efficacy, and availability of Gardasil and Cervarix, making it the "great undiscussable" issue. In the final essay in the section, Doreen Ramogola-Masire, an obstetrician, gynecologist, and practicing clinician in Botswana, explores the ambiguities and challenges posed by cervical cancer in an African nation, where the death rate is higher than in the West, where Merck occupies a central philanthropic role in the health care infrastructure, where the state apparatus is impoverished and externally dependent, and where the expense of the vaccine puts it far out of reach for average patients.

Part Two: Girls at the Center of the Storm. Merck's "One Less" advertisements, the state-by-state debates over compulsory vaccination, and intense media coverage put the vaccine squarely into the vortex of debates about young women's sexual activity, responsibility, risk, and morality. As historian Heather Munro Prescott explains, however, this is not the first time that cultural anxieties about adolescent female sexuality have led to critiques of a public health initiative. Similar concerns over virginity arose decades earlier around the pelvic exam and the Pap smear; highlighting the vaccine did not create but rather reactivated a biomedical politics that often hinged on girlhood and sexuality. For sociologist Laura Mamo and her collaborators, the core of the HPV debate was the visual and narrative production of "risky girlhood" in need of protection. Those authors focus on how the Merck campaigns portrayed the vaccine as part of a "do-it-yourself biological project," highlighting Gardasil's capacity to enhance girls' power and choice by protecting them against risk—thus issuing a gendered call to action. In the aftermath of Merck's campaigns, communication arts scholar Giovanna Chesler and environmental psychologist Bree Kessler found that the prominence of fifty-something mothers protecting young girls crowded out many other, more subversive, representations. Their essay documents how they created a skeptical and irreverent participatory Web channel, Tune in HPV, a site illuminating diverse alternative narratives of girlhood, queer identity, and sexuality.

Part Three: Focus on the Family. When Texas governor Rick Perry issued an executive order mandating that all sixth-grade girls submit to vaccination, concerns about the role of state government, the governance of sexuality, potential conflict of interest, and the meaning of family values came into the foreground. The angry reaction from opponents in his own Republican Party—strong advocates for "family values"—was swift and intense, fueling parental efforts to overturn the ruling, fight the mandate, and opt out. Sociologist Jennifer Reich draws upon Colorado-based research to examine this fractious, oppositional family politics, revealing how opponents viewed HPV vaccine development and government endorsement as an unhealthy collaboration—a corruption of politics by business and an interference with parental autonomy and teenage sexuality in the name of cancer prevention. Psychologist Gretchen Chapman offers insight into another dimension of family values, showing how parents' perceptions of risk (a powerful predictor of vaccine behavior) often differ markedly from their children's perceptions and also exist in marked tension with scientific assessments of the risks of HPV and cervical cancer. Bioethicists Nancy Berlinger and Alison Jost examine the rise of opt-out clauses that appeared in HPV legislation as a result of parental opposition. These policies, they argue (often framed as matters of "conscience"), have become part of a broader, misguided pattern in public health that leads to flawed public health policy. Sociologists Steven Epstein and April N. Huff examine how Merck, faced with the characterization of the vaccine as a matter of sexuality, pursued a strategy to desexualize the vaccine. They carried out this plan with considerable success at the same time that other important health interventions (such as the Plan B for emergency contraception and the condom) attracted withering attack from the family-values Right and from policymakers in the Bush administration.

Part Four: In Search of Good Government. Given the complex swirl of issues surrounding this vaccine, how far should government—any government—go in mobilizing its public health apparatus, married to pharmaceutical innovation, to safeguard citizens? Far beyond the United States, wherever the topic of the HPV vaccine came up, it raised sweeping questions of governance, for individuals managing their own sexuality, for families and their children, and for legislators and regulators. Taking a comparative view of the United States and Africa, historians Julie Livingston, Keith Wailoo, and Barbara Cooper portray the vaccine as a political tool, opening new possibilities for control while bringing great risk for government by unleashing protest and backlash. They find that across states in America and nations in Africa, HPV vaccine skepticism reflects instabilities in the social contract, skepticism about capital, and doubt about government's motivation and role in people's lives. Looking across Europe, sociologist Andrea Stöckl asks why the HPV story unfolded so differently in England and Italy, Austria and Germany; she finds that in each nation the HPV vaccine took on a meaning shaped powerfully by antecedent vaccine and public health controversies, among them the measles, mumps, rubella vaccine controversies in the United Kingdom. Stöckl finds that the HPV vaccine, although controversial, never produced obsessions akin to the American anxiety over the vaccination as a "passport" to promiscuity. Fundamentally, Europeans worried about questions of state power and citizenship. In time, the United Kingdom did mandate the vaccine nationwide for young girls, laying the groundwork for recurring controversies about whether the vaccine was responsible for the deaths of some girls and whether the vaccine had been prematurely adopted. Historian Ilana Löwy explores the distinctive French response to the HPV vaccine. There, passionate reaction to HPV vaccination was rare, but 1980s and 1990s scandals surrounding the hepatitis B vaccine encouraged officials to avoid mandating HPV, so it was left as a voluntary measure.

Four years after the 2006 approval of Gardasil, the swirl of intersecting controversies surrounding the HPV vaccines continues. Nor will they cease anytime soon, for the interests remain vocal, numerous, and conflicted, and the economic and public health imperatives driving the vaccine, both as a private consumer product and as a public health tool, remain powerful. As we discuss in the book's conclusion, the uptake of the vaccine has been uneven worldwide, reflecting the diverse political, cultural, and epidemiological landscapes in various states and nations. In the United States alone, twenty-two state legislatures rejected proposals to mandate the vaccination; eleven states mandate only that schools or health programs provide HPV information; and only Virginia and the District of Columbia passed mandatory vaccination laws, but even there parents retained the right to opt out on behalf of their children.[3] This book cannot provide uniform recommendations or guidance to parents or HPV policymakers. Rather, our goal is to describe and analyze the turbulent HPV vaccine landscape, calling attention to how diverse controversies in politics and governance gave life to the debate and how different cultural topographies across the globe explain the skepticism, faith, and radical unevenness in the new vaccine's reception.

NOTES

1. For one example of the NHS's marketing, see the 2008 poster: www.immunisation .nhs.uk/Files/HPV-press-ad-launch-ad.pdf.

2. João Biehl, "Pharmaceutical Governance," in *Global Pharmaceuticals: Ethics, Markets, Practices*, ed. Adriana Petryna, Andrew Lakoff, and Arthur Kleinman (Durham, NC: Duke University Press, 2006), 206–239; João Biehl, "Pharmaceuticalization: AIDS Treatment and Global Health Politic," *Anthropological Quarterly* 80 (Fall 2007): 1083–1126.

3. California, Colorado, Connecticut, Florida, Georgia, Illinois, Kansas, Kentucky, Maryland, Massachusetts, Michigan, Minnesota, Mississippi, Missouri, New Mexico, New York, Ohio, Oklahoma, South Carolina, Texas, Vermont, and West Virginia rejected mandating the vaccine. Colorado, Indiana, Iowa, Louisiana, Maine, Michigan, New Jersey, North Carolina, North Dakota, Texas, and Washington require school or health departments to provide HPV information. "A Look at States' Legislation on an HPV Vaccine," Associated Press, August 30, 2009.

Vaccine Time Lines

Five selective time lines place the HPV vaccine controversies in broader context. Scientific, clinical, and policy developments appear in them alongside social, cultural, economic, and political influences on vaccine debates in general and on the HPV vaccine debate in particular.

I. Establishing the HPV-Cancer Link. Many developments over the past century made it possible for scientists to link the human papillomaviruses to cervical cancer. Among those advances were the ability to distinguish cervical cancer from endometrial cancer of the uterus, the discovery of and ability to visualize viruses, and increasingly widespread population screening. By the late twentieth century, after more than 100 years of observations that nuns had high rates of breast cancer and married women had high rates of uterine cancer, epidemiologists and others reached consensus about causal connections between breast cancer and hormones and menstrual cycles and between cervical cancer and sexual life and sexually transmitted viruses. Despite these biological and clinical milestones, there is continuing uncertainty about cervical cancer causation and the impact of social, cultural, and economic factors on the disease.

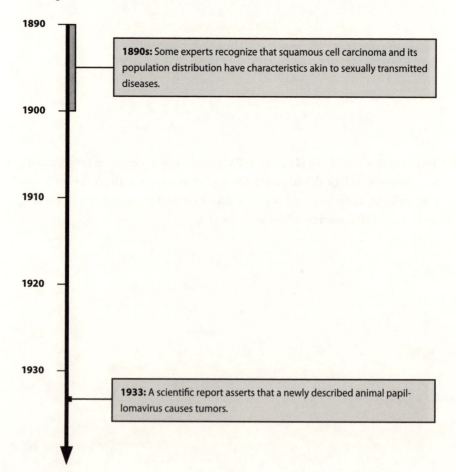

1890

1890s: Some experts recognize that squamous cell carcinoma and its population distribution have characteristics akin to sexually transmitted diseases.

1900

1910

1920

1930

1933: A scientific report asserts that a newly described animal papillomavirus causes tumors.

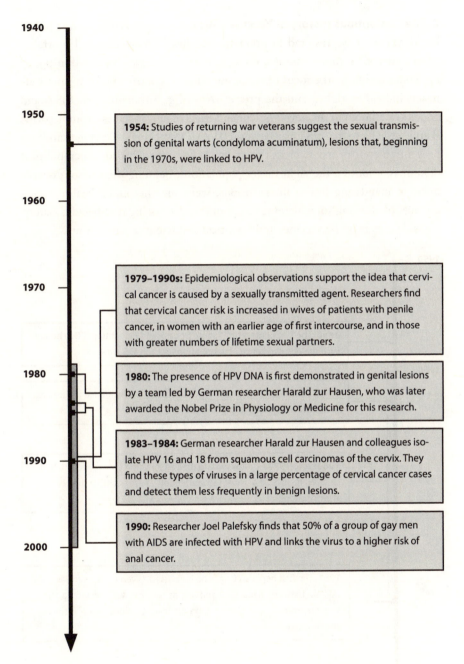

1940

1950

1954: Studies of returning war veterans suggest the sexual transmission of genital warts (condyloma acuminatum), lesions that, beginning in the 1970s, were linked to HPV.

1960

1970

1979–1990s: Epidemiological observations support the idea that cervical cancer is caused by a sexually transmitted agent. Researchers find that cervical cancer risk is increased in wives of patients with penile cancer, in women with an earlier age of first intercourse, and in those with greater numbers of lifetime sexual partners.

1980

1980: The presence of HPV DNA is first demonstrated in genital lesions by a team led by German researcher Harald zur Hausen, who was later awarded the Nobel Prize in Physiology or Medicine for this research.

1983–1984: German researcher Harald zur Hausen and colleagues isolate HPV 16 and 18 from squamous cell carcinomas of the cervix. They find these types of viruses in a large percentage of cervical cancer cases and detect them less frequently in benign lesions.

1990

1990: Researcher Joel Palefsky finds that 50% of a group of gay men with AIDS are infected with HPV and links the virus to a higher risk of anal cancer.

2000

II. The Contentious History of Vaccine Policy. Even as vaccines against public health threats have resulted in dramatic declines in morbidity and mortality from measles, pertussis, mumps, polio, and other diseases, vaccination policy has continued to be the focus of recurring political controversy over safety, efficacy, individual rights, and the prerogatives of government. The 1980s and 1990s saw a boom in the development and trials of new vaccines (from hepatitis B and varicella to the still unsuccessful HIV/AIDS vaccine), spurring numerous prevention controversies across the globe. At stake were broader sociopolitical dilemmas: choosing the right target populations, defining the role of government in mandating health interventions, weighing the public health consequences of allowing some parents to opt out, and assessing true benefits, safety, and side effects of the vaccines amid political and scientific uncertainty.

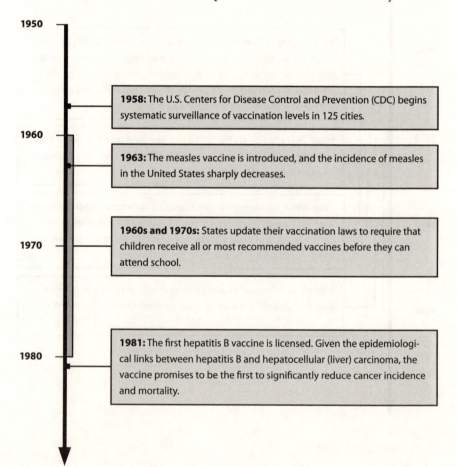

1950

1958: The U.S. Centers for Disease Control and Prevention (CDC) begins systematic surveillance of vaccination levels in 125 cities.

1960

1963: The measles vaccine is introduced, and the incidence of measles in the United States sharply decreases.

1970

1960s and 1970s: States update their vaccination laws to require that children receive all or most recommended vaccines before they can attend school.

1980

1981: The first hepatitis B vaccine is licensed. Given the epidemiological links between hepatitis B and hepatocellular (liver) carcinoma, the vaccine promises to be the first to significantly reduce cancer incidence and mortality.

1990

2000

2010

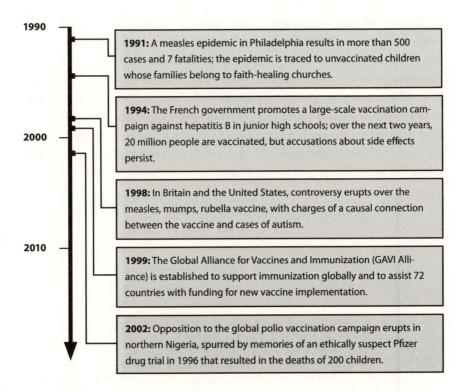

1991: A measles epidemic in Philadelphia results in more than 500 cases and 7 fatalities; the epidemic is traced to unvaccinated children whose families belong to faith-healing churches.

1994: The French government promotes a large-scale vaccination campaign against hepatitis B in junior high schools; over the next two years, 20 million people are vaccinated, but accusations about side effects persist.

1998: In Britain and the United States, controversy erupts over the measles, mumps, rubella vaccine, with charges of a causal connection between the vaccine and cases of autism.

1999: The Global Alliance for Vaccines and Immunization (GAVI Alliance) is established to support immunization globally and to assist 72 countries with funding for new vaccine implementation.

2002: Opposition to the global polio vaccination campaign erupts in northern Nigeria, spurred by memories of an ethically suspect Pfizer drug trial in 1996 that resulted in the deaths of 200 children.

III. Vaccine Policy: Agendas and Perspectives. Controversy over the HPV vaccine grew heated in 2006. Tensions over vaccine policy pitted physicians against parents and advocacy groups against vaccine manufacturers. Skeptics bemoaned the influence of business marketing and the power of the state over its citizens. The debate emerged against a backdrop of declining cervical cancer–related deaths in the developed world (e.g., cervical cancer incidence sharply declined from 3.49 per 100,000 in 1991 to 2.41 per 100,000 in 2004), rising mortality in the developing world, and lingering tensions over the power of government and the health and well-being of young girls.

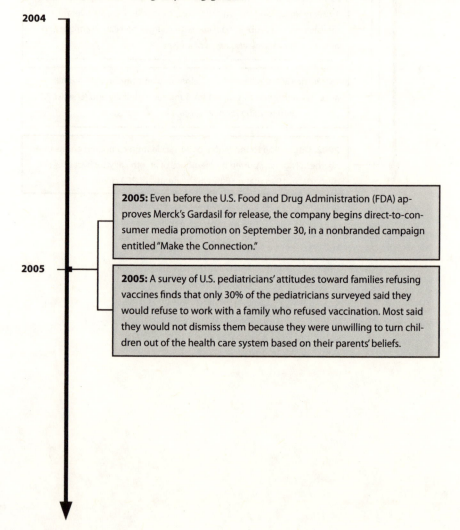

2004

2005: Even before the U.S. Food and Drug Administration (FDA) approves Merck's Gardasil for release, the company begins direct-to-consumer media promotion on September 30, in a nonbranded campaign entitled "Make the Connection."

2005

2005: A survey of U.S. pediatricians' attitudes toward families refusing vaccines finds that only 30% of the pediatricians surveyed said they would refuse to work with a family who refused vaccination. Most said they would not dismiss them because they were unwilling to turn children out of the health care system based on their parents' beliefs.

2006

2006 (February): The Italian health secretary declares Italy the first state within the European Union in which the HPV vaccine will be provided for all girls free of charge.

2006: In an indication of brewing U.S. controversy, the president of the South Dakota–based group Abstinence Clearinghouse characterizes the vaccine as an endorsement of adolescent sexual activity, declaring, "I personally object to vaccinating children . . . against a disease that is one hundred percent preventable with proper sexual behavior. Premarital sex is dangerous, even deadly. Let's not encourage it by vaccinating ten-year-olds so they think they're safe."

2006 (June): The FDA approves Merck's Gardasil, a quadrivalent vaccine to prevent anogenital infection with four HPV types: 6, 11, 16, and 18. A few weeks later, the Advisory Committee on Immunization Practices of the CDC adds the vaccine to the Vaccine for Children Program and formally recommends it for females of ages 11 or 12.

2007

2006 (mid to late): A flurry of legislative activity begins, as various U.S. states take steps to maximize the benefits to their populations; proposals for compulsory HPV vaccination for girls attending middle school are introduced in at least 24 states; conservative groups voice opposition. Merck begins its "One Less" advertising campaign for Gardasil. Meanwhile, the FDA rejects an application to make Plan B (emergency contraceptive) available over the counter despite strong support for doing so from its own advisory panel.

2006 (September): The first bill to mandate HPV vaccination is introduced in Michigan; according to a MSNBC television public opinion poll, 80% of parents of children under 15 would give their daughter the vaccine; by the end of 2006, a total of 2,151,000 doses of Gardasil vaccine have been distributed.

2008

2006–2007: Twenty-two U.S. state legislatures reject legislation mandating HPV vaccination (California, Colorado, Connecticut, Florida, Georgia, Illinois, Kansas, Kentucky, Maryland, Massachusetts, Michigan, Missouri, Minnesota, Mississippi, New Mexico, New York, Ohio, Oklahoma, South Carolina, Texas, Vermont, and West Virginia); 11 states require that schools or health programs provide HPV information (Colorado, Indiana, Iowa, Louisiana, Maine, Michigan, New Jersey, North Carolina, North Dakota, Texas, and Washington); and only the District of Columbia and Virginia pass mandatory vaccination laws, but parents retain the right to opt their daughters out.

IV. Mandating Prevention? Legislation, Sexuality, and State Power. In 2007, European nations and U.S. state legislatures took up the question of mandating HPV vaccination for young girls. In some U.S. states, the mandate was passed despite deep controversy. But conservatives (most notably in Texas, where the proposal was defeated) successfully portrayed the mandate as a government attack on the prerogatives of the family and as unwarranted endorsement of teenage sexual relations. The vaccine had become a health intervention with multiple and expanding political meanings. From the political Right and Left, advocates and opponents infused vaccination policy debates with fears over the loss of family rights, anxieties over the power of the state and the public interest, anger over drug-company financial greed, and concerns over adolescent sexuality and the protection of girls.

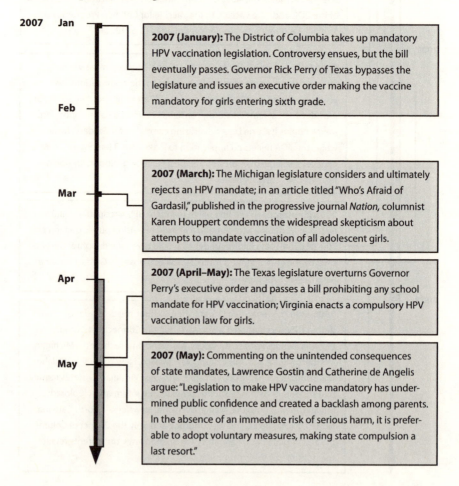

2007   Jan

2007 (January): The District of Columbia takes up mandatory HPV vaccination legislation. Controversy ensues, but the bill eventually passes. Governor Rick Perry of Texas bypasses the legislature and issues an executive order making the vaccine mandatory for girls entering sixth grade.

Feb

2007 (March): The Michigan legislature considers and ultimately rejects an HPV mandate; in an article titled "Who's Afraid of Gardasil," published in the progressive journal *Nation,* columnist Karen Houppert condemns the widespread skepticism about attempts to mandate vaccination of all adolescent girls.

Mar

2007 (April–May): The Texas legislature overturns Governor Perry's executive order and passes a bill prohibiting any school mandate for HPV vaccination; Virginia enacts a compulsory HPV vaccination law for girls.

Apr

2007 (May): Commenting on the unintended consequences of state mandates, Lawrence Gostin and Catherine de Angelis argue: "Legislation to make HPV vaccine mandatory has undermined public confidence and created a backlash among parents. In the absence of an immediate risk of serious harm, it is preferable to adopt voluntary measures, making state compulsion a last resort."

May

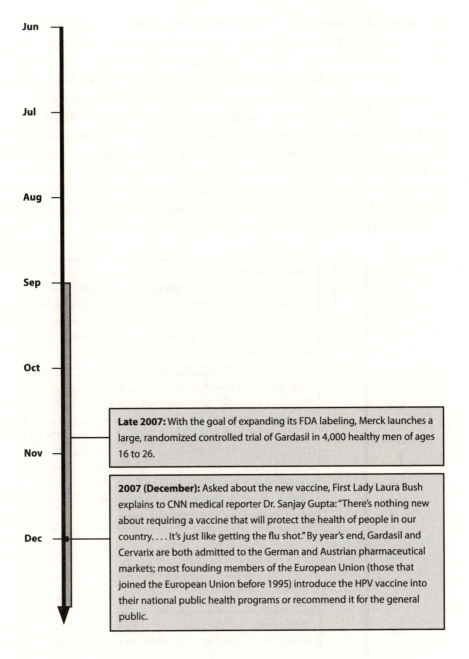

Jun

Jul

Aug

Sep

Oct

Nov

Late 2007: With the goal of expanding its FDA labeling, Merck launches a large, randomized controlled trial of Gardasil in 4,000 healthy men of ages 16 to 26.

Dec

2007 (December): Asked about the new vaccine, First Lady Laura Bush explains to CNN medical reporter Dr. Sanjay Gupta: "There's nothing new about requiring a vaccine that will protect the health of people in our country. . . . It's just like getting the flu shot." By year's end, Gardasil and Cervarix are both admitted to the German and Austrian pharmaceutical markets; most founding members of the European Union (those that joined the European Union before 1995) introduce the HPV vaccine into their national public health programs or recommend it for the general public.

V. States and Citizens at the Crossroads of Prevention. In 2008 and 2009, discussions of the HPV vaccine in the United States remained polarized, and debates swirled in multiple directions, raising profound issues of citizenship and state governance. In some locales, controversy revolved around the morality of vaccination mandates. In others, the debate focused on the adverse side effects of the vaccine itself or on the public health implications of vaccine refusal. Outside the United States, discussions also focused on how best to launch vaccination programs and on the cost-effectiveness of Gardasil versus another new HPV vaccine, Cervarix (which, unlike the Merck vaccine, targeted only the two "cancer-causing" HPV types). And for some observers, the global disparity remained striking—with high cervical cancer mortality in the developing world amid the vaccine's rapid, if uneven, uptake in the developed world.

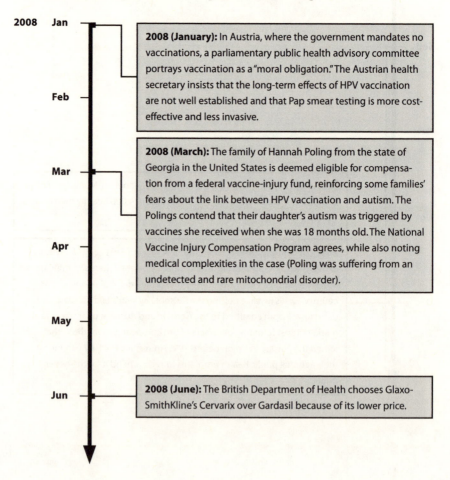

2008 **Jan**

2008 (January): In Austria, where the government mandates no vaccinations, a parliamentary public health advisory committee portrays vaccination as a "moral obligation." The Austrian health secretary insists that the long-term effects of HPV vaccination are not well established and that Pap smear testing is more cost-effective and less invasive.

Feb

2008 (March): The family of Hannah Poling from the state of Georgia in the United States is deemed eligible for compensation from a federal vaccine-injury fund, reinforcing some families' fears about the link between HPV vaccination and autism. The Polings contend that their daughter's autism was triggered by vaccines she received when she was 18 months old. The National Vaccine Injury Compensation Program agrees, while also noting medical complexities in the case (Poling was suffering from an undetected and rare mitochondrial disorder).

Mar

Apr

May

Jun

2008 (June): The British Department of Health chooses GlaxoSmithKline's Cervarix over Gardasil because of its lower price.

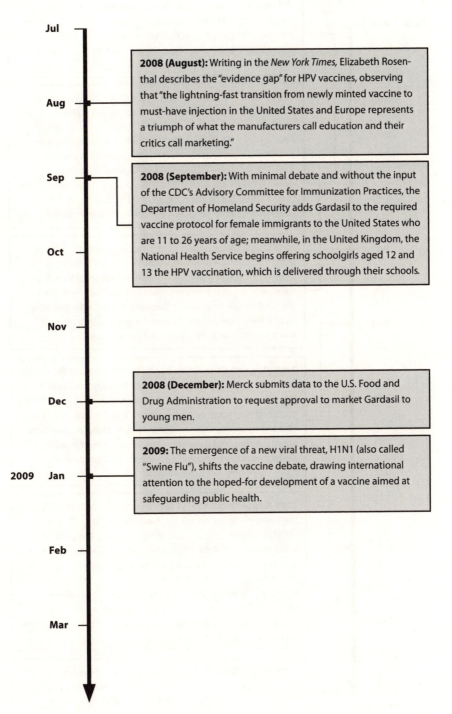

Jul

Aug

2008 (August): Writing in the *New York Times,* Elizabeth Rosenthal describes the "evidence gap" for HPV vaccines, observing that "the lightning-fast transition from newly minted vaccine to must-have injection in the United States and Europe represents a triumph of what the manufacturers call education and their critics call marketing."

Sep

2008 (September): With minimal debate and without the input of the CDC's Advisory Committee for Immunization Practices, the Department of Homeland Security adds Gardasil to the required vaccine protocol for female immigrants to the United States who are 11 to 26 years of age; meanwhile, in the United Kingdom, the National Health Service begins offering schoolgirls aged 12 and 13 the HPV vaccination, which is delivered through their schools.

Oct

Nov

Dec

2008 (December): Merck submits data to the U.S. Food and Drug Administration to request approval to market Gardasil to young men.

2009   Jan

2009: The emergence of a new viral threat, H1N1 (also called "Swine Flu"), shifts the vaccine debate, drawing international attention to the hoped-for development of a vaccine aimed at safeguarding public health.

Feb

Mar

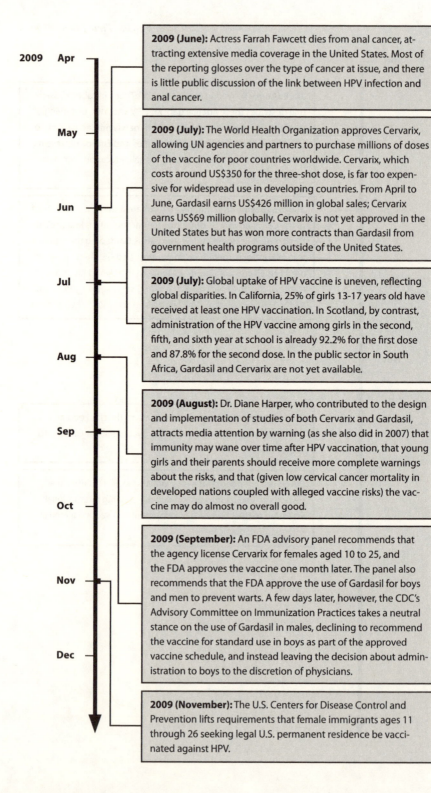

2009 Apr

2009 (June): Actress Farrah Fawcett dies from anal cancer, attracting extensive media coverage in the United States. Most of the reporting glosses over the type of cancer at issue, and there is little public discussion of the link between HPV infection and anal cancer.

May

2009 (July): The World Health Organization approves Cervarix, allowing UN agencies and partners to purchase millions of doses of the vaccine for poor countries worldwide. Cervarix, which costs around US$350 for the three-shot dose, is far too expensive for widespread use in developing countries. From April to June, Gardasil earns US$426 million in global sales; Cervarix earns US$69 million globally. Cervarix is not yet approved in the United States but has won more contracts than Gardasil from government health programs outside of the United States.

Jun

Jul

2009 (July): Global uptake of HPV vaccine is uneven, reflecting global disparities. In California, 25% of girls 13-17 years old have received at least one HPV vaccination. In Scotland, by contrast, administration of the HPV vaccine among girls in the second, fifth, and sixth year at school is already 92.2% for the first dose and 87.8% for the second dose. In the public sector in South Africa, Gardasil and Cervarix are not yet available.

Aug

2009 (August): Dr. Diane Harper, who contributed to the design and implementation of studies of both Cervarix and Gardasil, attracts media attention by warning (as she also did in 2007) that immunity may wane over time after HPV vaccination, that young girls and their parents should receive more complete warnings about the risks, and that (given low cervical cancer mortality in developed nations coupled with alleged vaccine risks) the vaccine may do almost no overall good.

Sep

Oct

2009 (September): An FDA advisory panel recommends that the agency license Cervarix for females aged 10 to 25, and the FDA approves the vaccine one month later. The panel also recommends that the FDA approve the use of Gardasil for boys and men to prevent warts. A few days later, however, the CDC's Advisory Committee on Immunization Practices takes a neutral stance on the use of Gardasil in males, declining to recommend the vaccine for standard use in boys as part of the approved vaccine schedule, and instead leaving the decision about administration to boys to the discretion of physicians.

Nov

Dec

2009 (November): The U.S. Centers for Disease Control and Prevention lifts requirements that female immigrants ages 11 through 26 seeking legal U.S. permanent residence be vaccinated against HPV.

Part I / The Known and the Unknown

Vaccination Decisions amid Risk and Uncertainty

CHAPTER ONE

The Coercive Hand, the Beneficent Hand

What the History of Compulsory Vaccination Can Tell Us about HPV Vaccine Mandates

James Colgrove

The licensure of Gardasil in June 2006 set off a flurry of legislative activity as states around the country took steps to maximize the benefits of the product among their populations. States select from a variety of policy approaches to increase uptake of a vaccine. They can sponsor educational and promotional efforts through mass media or in clinical settings; allocate public funds to pay for the vaccine; require that private insurers doing business in the state reimburse it; or require or allow schools to provide education about it as part of health curricula or other science units.[1] One of the most effective and efficient approaches, but also the most controversial, is to mandate a vaccine as a condition of attending school. Vaccination requirements, like all compulsory health measures, may be ethically troubling and politically sensitive because they represent an intrusion on individual autonomy. Weighed against these concerns is the success of mandates in achieving high levels of coverage.[2]

Mandates may be especially appropriate for adolescent vaccination because teens are in the age group least likely to have contact with a primary care provider. A recent study found that of numerous legal, financial, and promotional strategies for increasing rates of adolescent immunization, school entry mandates were the only consistently effective intervention.[3] An ancillary benefit is that a clinical visit to comply with a school vaccine requirement can present an opportunity for health care providers to offer other needed screening, prevention, and care for teens.[4]

Proposals for compulsory human papillomavirus vaccination in middle school were among the first out of the legislative chute in late 2006 as states began their deliberations. Bills to require HPV vaccination for girls attending middle school were introduced in at least twenty-four states.[5] Over the next three years, however, almost all of these bills stalled or were abandoned.

This chapter situates the deliberations about HPV vaccine mandates within the context of two centuries of legally enforced vaccination. I begin by briefly tracing the initial efforts on the part of some legislators and policymakers to enact middle school mandates after the licensure of Gardasil. I then examine attempts since the nineteenth century to increase uptake of vaccines via the law and the popular responses these efforts have produced. Many aspects of compulsory vaccination have been subject to debate: their rationales, their purposes, their ethical and legal bases, and their consequences, both intended and unintended. I conclude by considering whether there are historical precedents or analogies that might be informative for current policy development.

The Rise and Fall of Compulsory HPV Vaccination

Much of the initial concern about mandating protection against HPV centered around the sexual behavior of teens. Before the vaccine was licensed, it was dubbed a "promiscuity vaccine," and some religious conservatives expressed concern that the availability of a preventive against a sexually transmitted disease would increase the likelihood that teens receiving it would engage in sex. An early salvo was fired by the South Dakota–based group Abstinence Clearinghouse, whose president, Leslee Unruh, said, "I personally object to vaccinating children . . . against a disease that is one hundred percent preventable with proper sexual behavior. Premarital sex is dangerous, even deadly. Let's not encourage it by vaccinating ten-year-olds so they think they're safe."[6] Although high-profile conservative groups including Focus on the Family and the Family Research Council were wary of the vaccine initially, their stated positions following licensure were generally in support of making the vaccine available to parents who wanted it. When school mandates began to be placed on the table in the fall of 2006, however, conservative groups were united in their opposition. In their view, such a requirement constituted an attempt by the state to force a child to undergo an intervention that was irreconcilable with her family's religious values and beliefs.[7]

Concerns about sexuality were soon joined—and often overshadowed—by criticisms related to potential adverse health effects. The high-profile debates in recent years about the alleged connection between autism and the measles-mumps-rubella (MMR) vaccine and the preservative thimerosal have created an environment of suspicion toward all vaccines. Even though the data on Gardasil's safety were very favorable, the proposals to mandate HPV vaccination unfolded in a context in which there was considerable background noise about vaccine policy in general and about mandates in particular. Members of antivaccination groups at the national, state, and local levels were on the alert for any new application of state coercion, and organizations such as the National Vaccine Information Center emerged as prominent adversaries of proposed school-entry requirements.

The first bill to mandate HPV vaccination was introduced in September 2006 by Beverly Hammerstrom, a state legislator in Michigan. Hammerstrom was a member of Women in Government (WIG), a Washington, D.C.–based advocacy organization made up of member legislators around the country who supported bills aimed at advancing women's issues. WIG members made it a top priority to introduce HPV-related legislation, not just mandates but a variety of measures related to funding and educating about the vaccine.[8]

In late January 2007, the situation grew politically explosive. Governor Rick Perry of Texas, a state with a conservative political culture and a well-organized antivaccination movement, bypassed the legislature and issued an executive order making the vaccine mandatory for girls entering sixth grade. This was an unusual move; vaccine mandates are typically enacted either via the legislative process or through the rule-making authority delegated to state public health officials. Amid a political firestorm, critics charged that Perry had been influenced by a Merck lobbyist who was his former chief of staff. Around the same time, newspaper reports revealed that Merck had given heavily to Women in Government.[9] In response to criticism of its lobbying, Merck discontinued efforts to persuade states to make the vaccine mandatory. The Texas legislature subsequently overturned Perry's order and passed a bill prohibiting any school mandate for HPV.[10]

The revelation of this aggressive lobbying was a public-relations debacle for Merck, which was already under fire for accusations that it had concealed the risks of its pain medication Vioxx; the firm had been ordered to pay billions of dollars in damages resulting from civil litigation. The putative connection between the vaccine mandates and the company's financial woes gave rise to a wry joke: it proposed that to Merck, HPV stood for "Help Pay for Vioxx." There had already been criticism of the price of Gardasil, approximately $350 for a full course of three shots, which was much higher than the costs of other routinely given childhood and adolescent vaccines. Criticism of the price became more vocal after Merck's donations to Women in Government were made public.

Amid the fallout from the revelation of Merck's campaign, groups and individuals who were otherwise supportive of vaccine mandates took a stand against requiring HPV vaccination. These included prominent figures such as public health legal scholar Lawrence Gostin, who coauthored a commentary in the *Journal of the American Medical Association* with Catherine de Angelis, the journal's editor. "Legislation to make HPV vaccine mandatory," they warned, "has undermined public confidence and created a backlash among parents. In the absence of an immediate risk of serious harm, it is preferable to adopt voluntary measures, making state compulsion a last resort."[11] Their argument echoed the position of the Association of Immunization Managers, which argued in a 2006 position paper that mandates "must be used sparingly, approached cautiously, and considered only after an appropriate vaccine implementation period. . . . Inappropriate application of mandates risks loss of support for immunization programs and reversal of policy and program gains."[12]

During the state-level debates in 2007 and 2008, diverse voices from across the political spectrum coalesced in opposition to compulsory HPV vaccination. This broad-based opposition was an example of what the legal scholar Cass Sunstein has called incompletely theorized agreement, in which parties to a debate agree on a particular case or policy but disagree at greater levels of abstraction; that is, they arrive at the same position from differing principled bases.[13] Religious conservatives resisted Gardasil mandates because they believed such laws interfered with the guidance they wished to provide their children regarding sexuality; antivaccinationists were opposed because they believed vaccines were dangerous; and public health advocates turned away because they believed that it was too early in the life of the vaccine for such laws and that adding new vaccine requirements could undermine support for the better-established vaccines already covered under state laws.

As of this writing, only one state, Virginia, has enacted a compulsory HPV vaccination law. The bill passed the legislature in April 2007 and was signed into law by Governor Tim Kaine.[14] Although Kaine was a democrat, both houses of the Virginia legislature were controlled by Republicans, and the bill passed the house by a wide margin and the senate unanimously, before the fallout from Texas Governor Perry's action and the revelation of Merck's lobbying rendered mandates politically poisonous. Significantly, the bill contained a very broad exemption clause that allowed parents to opt out of the requirement for any reason. This provision differed from the state's policy toward all other mandated vaccines, which was to allow exemptions only for medical or religious reasons.[15]

By mid-2009, Virginia looked like a policy outlier. The District of Columbia had also enacted a middle school mandate, but no other state seemed poised to follow a similar course. Once the controversy over Merck's lobbying has receded from public memory, however, and more conclusive data are available on the long-term safety and efficacy of Gardasil, it is not inconceivable that attempts to make HPV vaccination mandatory in middle schools may be made once again. Should this occur, policymakers will be forced to return to underlying questions about the ethical and political acceptability of such mandates. Both supporters and opponents have used historical analogies to claim that the proposed mandates were or were not comparable to prior vaccination laws. It is therefore worth looking closely at how a middle school requirement for HPV vaccination fits within the broader history of compulsory vaccination in the United States.

The First Generation of Compulsory Vaccination Laws

It is useful to distinguish between what one might call the first generation and the second generation of compulsory vaccination laws. The first generation of laws arose in the nineteenth century in response to smallpox and typically applied to the general population, adults and children alike, since the vaccine's protection lasted only seven to ten years and periodic revaccination was necessary. These laws imposed a variety of penalties on those refusing, including fines and imprisonment.[16] With the spread of public education, the schoolhouse emerged as a hotspot for the spread of contagion, and states and localities began to pass laws requiring children to be protected against smallpox before they could attend school. Massachusetts, a pioneer in the use of public health law and regulation, passed the first compulsory vaccination law for the general population in 1809 and became the first to tie school attendance to vaccination in 1855.[17] Many other states and localities followed suit.

Vaccination laws eventually became a victim of their success: as they contributed to the steady decline in the incidence of smallpox over many decades, people lacked firsthand experience of the terrors of the disease and became reluctant to undergo a procedure they considered unpleasant and unnecessary. With complacency came resistance in the form of court challenges over the use of compulsion. Vaccination requirements provoked dozens of lawsuits in state courts around the country. These cases produced conflicting decisions, with some judges upholding the laws and others declaring them unconstitutional. (Laws pertaining to schools withstood challenge better than those covering the entire population.) Opponents of compulsory vaccination pressed their case not just in courthouses but in statehouses: in the second half of the nineteenth century many legislatures repealed or modified their laws in response to pressure from antivaccination activists.[18]

The backdrop for these controversies was the rise in the second half of the nineteenth century of a robust network of laws and regulations designed to protect the public health—measures such as isolation and quarantine and limitations on where and how "noxious trades" such as slaughterhouses could operate.[19] Such intervention by the state into private and commercial behavior often triggered resistance, and vaccination provoked an especially vociferous response. Sanitary measures to advance communal well-being typically required that people *refrain from* an action or behavior: discharging sewage into common waterways, for example, or disembarking from a ship while suffering from a communicable disease. In contrast, vaccination laws required people to *submit to* a procedure, one that involved discomfort and whose safety and efficacy were widely challenged.[20] Bodily invasiveness remains a significant factor that makes vaccination an especially contentious area of public health law.

Both supporters and antagonists of compulsory vaccination found a basis for their arguments in the harm principle of utilitarian philosopher John Stuart Mill, who argued that the only justification for the use of coercive state power was to prevent imminent harm to others; a person's own good was insufficient reason. Health officials claimed that those who refused vaccination posed a danger to other members of the community, thereby justifying state intervention, while resisters argued that the laws were an impermissible violation of individual liberty. (Mill, notably, believed that children were fit subjects of paternalistic action, and as a member of the British parliament he did not oppose compulsory vaccination legislation in 1867.)[21]

After decades of conflicting legal rulings, the U.S. Supreme Court affirmed the constitutionality of compulsory vaccination laws in the famous 1905 case of *Jacobson v. Massachusetts*, in which a Lutheran minister in Cambridge, Henning Jacobson, contested the five-dollar fine he was forced to pay after refusing to be vaccinated. In a seven-to-two ruling, the justices determined that the laws were a legitimate exercise of states' "police powers" to protect the health, welfare, safety, and morals of citizens. The ruling did not directly address the question of vaccination as a school attendance requirement, and in 1922 the Supreme Court dispensed with that question in a case arising from Texas. In *Zucht v. King*, a unanimous decision relying primarily on the opinion in *Jacobson*, the court determined that no constitutional right was abridged by excluding unvaccinated children from school.[22] Justice Louis Brandeis, who wrote the opinion, later claimed that he did not believe the court should have taken the case because it presented no new constitutional issues.[23]

One striking aspect of this history is that after the rulings in *Jacobson* and *Zucht*, compulsory vaccination laws largely fell into disuse: Health officials rarely invoked the power to which the Supreme Court had given its constitutional imprimatur. Laws mandating smallpox vaccination were not rigorously enforced, and as three new vaccines (against diphtheria, pertussis, and tetanus) became routine between around 1920 and 1950, only a handful of states made them compulsory. This was partly a consequence of bitter public conflicts that followed the ruling in *Jacobson*. Antivaccination activism thrived during the increased popular democracy of the Progressive Era, as activists used the new ballot processes of initiative and referendum to challenge their states' laws. Washington state voters repealed their mandatory school vaccination law in 1919, for example, and Wisconsin residents did the same the following year. Utah and North Dakota both enacted laws expressly forbidding the use of compulsory vaccination.[24]

In addition to growing out of new forms of democratic action, these moves were a result of antagonism toward what was perceived as overreaching and paternalism on the part of medical professionals. The pharmaceutical industry grew rapidly during the first decades of the century and introduced an array of new products for preventing and treating illness. These advances provoked an antimodernist backlash against the potentially coercive uses to which scientific medicine might be put. The brief that Henning Jacobson, the plaintiff in

Jacobson v. Massachusetts, filed with the Supreme Court articulated an argument against what we might today call public health "mission creep":

> The present tendency of medical science is toward the treatment of contagious
> diseases by the use of serums, and it is entirely possible that public authorities and
> physicians may be encouraged to extend the vaccination scheme to all other con-
> tagious diseases and set up a general compulsory medical regime, which will sub-
> ject a healthy community to attack by boards of health under compulsory laws.
> If it be justifiable to compel the inoculation of a citizen for one disease, then by
> a parity of reasoning it is for the public interest that every citizen should be in-
> oculated to render him immune against all possible contagions which may men-
> ace the community.[25]

The public battles over smallpox vaccination that raged for decades left an indelible mark on the consciousness of health professionals in later eras, who worried that the use of compulsion would be counterproductive and would give rise to new agitation against vaccines. They had other pragmatic concerns, such as the fear that linking vaccination to school entry would encourage parents to postpone vaccines that were supposed to be given in the first two years of life. Finally, many health care providers felt it was ethically wrong to compel parents to do something they didn't want to do, even if it meant that children would go unprotected.[26] Instead of legal coercion, vaccine proponents sought to encourage high levels of uptake through a range of persuasive strategies: advertising and mass-media promotional campaigns, printed materials such as pamphlets and brochures with expert guidelines, and individual education and counseling by clinicians.

How effective were noncoercive approaches in achieving widespread use of these vaccines? We lack data on levels of protection against diphtheria, pertussis, and tetanus in the middle decades of the century. Health officials at the local and state levels typically had only the most general idea what percentage of their population had been vaccinated against a given disease, and there was no systematic nationwide surveillance of vaccination levels until 1958, when the Centers for Disease Control and Prevention (CDC) began regular monitoring of vaccination in 125 cities.[27] But anecdotal reports from health officials suggested that persuasion, though preferable to coercion for both ethical and political reasons, had clear limits.

In the case of diphtheria immunization, which was introduced to a generally favorable public response in the 1920s, health officials found that coverage levels reached a plateau within a few years and that achieving the degree of herd immunity needed to keep outbreaks under control required constant reinforcement that was time-consuming and expensive. In frustration, a few experts argued that legal coercion could be justified on grounds of efficiency: mandates would enable health officials to redirect elsewhere the considerable energies they spent educating the public about the need for vaccination.[28]

The limits of persuasion were starkly revealed in the years following the licensing of Jonas Salk's polio vaccine in 1955. The acclaim that greeted the Salk vaccine represented the twentieth-century apotheosis of public faith in medical science and vaccination in particular. But even after a celebrated medical breakthrough against a highly feared disease, health officials found that achieving high levels of vaccine coverage was a painstaking uphill climb, especially among people of lower socioeconomic status. Surveys found a consistent gradient in vaccine uptake along the lines of income and education, with those at the bottom of the scale much less likely to be protected than those at the top. When use of the Salk polio vaccine began to lag within only a few years of its licensure, a few states turned to the law to increase coverage levels, and by 1963 eight states had enacted laws requiring polio vaccination for students.[29] But compulsory vaccination in schools remained the exception, not the norm.

The Second Generation of Laws

The legal landscape around vaccination was transformed in the 1960s and 1970s, when the second generation of state laws—those now in place—were enacted. These laws grew out of a national measles-eradication campaign (ultimately unsuccessful) that the CDC launched in 1967. One of the chief motivations driving this campaign was evidence of wide gaps in vaccine coverage between children of higher and lower socioeconomic status. After the vaccine was licensed in 1963, measles became disproportionately a problem of poor black and Latino children in inner-city ghettoes. Epidemics continued to break out in public housing projects in Chicago, New York, Los Angeles, and Houston at a time when the disease was declining among middle-class white children.[30]

To support its eradication campaign, the CDC urged states to enact laws requiring that students be vaccinated against measles; the agency further recommended that the new requirements extend to the other vaccine-preventable diseases. The number of states with school immunization requirements quickly grew. In 1968, just half the states had a law requiring one or more vaccinations prior to school entry, and most of these were antiquated smallpox laws that had not been updated or enforced in decades. By 1974, forty states had laws that covered all or most of the recommended childhood shots, including the recently licensed vaccines against mumps and rubella.[31] By 1981, Idaho, Iowa, and Wyoming became the last states to fall in line with the national trend and make all or most vaccines compulsory for school.

These laws were very much a product of the Great Society era, which was the high-water mark of activist government in the twentieth century. During Lyndon Johnson's administration, Congress passed approximately fifty pieces of legislation related to health, providing funds that flowed to states and localities through units of the federal government including the Office of Economic Opportunity, the Children's Bureau, and the U.S. Public Health Service. Medicaid and Medicare, enacted in 1965, were the signature products of this period; also significant was the Early and Periodic Screening, Detection, and Treatment (EPSDT) program, passed in 1967, a Medicaid benefit to ensure that poor children received preventive care.[32]

It was not coincidental that the CDC began pushing for school vaccination mandates and that politicians around the country took up this charge at the same time that Medicaid, Medicare, EPSDT, and other pieces of health-related legislation were pouring forth from legislative chambers in Washington, D.C., and state capitols. All of these initiatives represented a common impulse: an attempt to remedy health disparities using the tools of the administrative state. The supporters of mandatory vaccination laws saw them as a uniquely effective and efficient way to ensure that all children, rich and poor alike, derived protection from vaccines. Mandates were not only about controlling the spread of contagions in schools, although that was one of their functions. They were also about bringing medical benefits to children who would not receive them otherwise.

There was another, less well publicized impetus that drove the enactment of these laws. Measles-related encephalitis, a sequela in about one in one thousand cases of the disease, was a major cause of mental retardation, and the Joseph P. Kennedy Foundation, a charity based in Washington, D.C., and concerned with mental retardation, took great interest in the licensure of the measles vaccine. The organization had significant political connections (its president was Massachusetts senator Edward Kennedy, and its executive vice president was Eunice Kennedy Shriver, sister of the senator and of former president John F. Kennedy and wife of Sargent Shriver, head of the federal Office of Economic Opportunity), and it encouraged legislators and health officials around the country to mount aggressive measles-vaccination programs even before the eradication program was officially announced. In early 1968 the Kennedy Foundation began a full-court press to get states to place compulsory measles-vaccination laws on the books. Eunice Kennedy Shriver sent letters to state governors and congressional delegations urging them to act; each letter cited the state's measles incidence and warned of the number of cases of measles-related mental retardation and deaths that might be expected in the coming year and how much the state would have to pay for medical and rehabilitative care for stricken children.[33] The actions of the Kennedy Foundation reveal the extent to which controlling contagion was not the sole, or even primary, motivation for the laws. Reducing incidence of a disease's long-term sequelae and the resulting health care costs were also explicit rationales.

These laws came into place with strikingly little public debate or controversy. The laws did provoke scattered legal challenges, however, centered on the issue of exemptions. Unlike mandates in the nineteenth century, almost all of the new laws contained exemptions for children whose parents had religious scruples against the practice. These clauses were included in response to the lobbying of Christian Scientists.[34] Legislators in some states wrote their exemptions narrowly, allowing them only for members of "recognized" or "established" religious denominations whose tenets specifically prohibit vaccination. But limiting exemptions in this way left the laws on shaky constitutional ground, and since 1968 state courts have heard more than a dozen challenges to the opt-out provisions of compulsory vaccination laws. Most courts have found that providing exemptions only to members of certain faiths violates the First Amendment proscription against state establishment of religion, and as a result many states' laws also allow for secular "philosophical" or "personal belief" exemptions.[35]

Given the presence of exemptions, one could argue that, at least in some states, vaccination is not truly compulsory in the strict sense of the word. Indeed, proponents of school requirements framed the laws in terms of their persuasive function and characterized them as helpful prompts to action rather than tools of coercion. "Some additional stimulus is often needed to provoke action on the part of a basically interested person who has many other concerns competing for attention," argued one CDC official in the 1970s. In this view, the laws were essentially hortatory, serving as a "means of bringing to individuals' attention the continuing publicly perceived need for immunization."[36]

Assisted by the legal scaffolding that was erected during the 1970s, public health officials and lawmakers have added mandates as new vaccines have been licensed. Between 1985 and 2006, the number of vaccines recommended by the CDC for children and adolescents doubled from seven to fourteen, and states have made most of these newly recommended vaccines school requirements.

The dramatic expansion in the vaccine schedule was a critical factor in the emergence of the modern-day antivaccination movement (or, to use the activists' preferred term, vaccine safety movement). These activists, who make up a loosely organized but highly vocal social movement that questions the safety of childhood immunization, alleged connections between vaccines and a wide range of illnesses in children, including autism, diabetes, sudden infant death syndrome, and multiple sclerosis. The claims spread via the Internet and attracted increasing attention in the mainstream media, to the extent that health officials at the highest level of the federal government made investigating and responding to such theories a top priority.[37]

Are There "Lessons" from History?

The implications of history for current policy are rarely straightforward, and the history of legally enforced vaccination is sufficiently long and complex that one might draw a variety of conclusions, some mutually contradictory, about the most appropriate course of action for the HPV vaccine. Rather than suggest "lessons" from history, I would like to make two claims that I believe aspects of this history reinforce. The first is related to why compulsory HPV vaccination should be considered ethically defensible; the second has to do with why such a mandate is politically unwise and is likely to remain so for the foreseeable future.

The Ethics of Mandates

Vaccination is unique among public health interventions in that it conveys a dual benefit: to the individual who undergoes it and to the surrounding community members who enjoy the protection afforded by herd immunity. Put another way, vaccines protect against both an individual and a communal risk. Because HPV is sexually rather than casually transmitted, however, and because the effect on herd immunity of a vaccine given only to girls remains unclear, many commentators have argued that HPV does not fit in the category of diseases for which vaccine mandates are appropriate. According to this rationale, mandates should be used only to prevent imminent danger to vulnerable third parties in the community from diseases that spread easily. This principle derives from John Stuart Mill's harm principle and was the rationale behind the first generation of laws requiring smallpox vaccination.

More recent history suggests a counterargument. An additional ethical basis for mandates—and a major premise behind the second generation of laws—is that they foster the equitable distribution of the benefits of vaccines, especially among children whose life circumstances make them less likely to be fully immunized. Seen in this way, mandates embody not just the coercive hand of the government but also the beneficent or therapeutic hand. While concerns about potential violations of parental autonomy have dominated the debates around HPV vaccination, the events of the 1960s call our attention to the significance that considerations of equity and justice have played in vaccination laws. These principles are especially relevant in light of the disproportionate epidemiological burden of cervical cancer in poor women of color, who are less likely to receive regular Pap smears and follow-up care after abnormal results.[38]

Furthermore, the presence of exemptions, which have been a standard component of vaccine mandates since the 1960s, attenuates arguments rooted in individual liberty and the harm principle. As of this writing, forty-eight states allow parents who have religious objections to vaccination to excuse their children from requirements, and twenty of those states also allow exemptions for those who have secular or philosophical concerns. Most of the compulsory HPV vaccination bills introduced in 2006 and 2007 contained broad exemptions. Permissive opt-out provisions raise their own set of ethical concerns; Nancy Berlinger and Alison Jost in chapter 11 persuasively lay out some of the problems that exemption clauses may present related to fairness and the undermining of civic responsibility. These issues aside, the exemption clauses that would likely be part of any HPV vaccine mandates make it difficult to argue that the laws would be an unacceptable intrusion on parental autonomy.

There is precedent for mandating vaccines against diseases that are not casually transmissible. Most states require that students be protected against hepatitis B, which is primarily transmitted through sexual contact and injection drug use, and tetanus, which is not communicable. These laws can be justified on the grounds that the benefits of lower rates of hepatitis and tetanus—in terms both of reduced human suffering and of costs to the health care system—outweigh the potential infringement on parental autonomy. The laws also express the principle that the government has a justifiably paternalistic concern for the welfare of minor children.

A caveat to an argument grounded in beneficence is that several important aspects of Gardasil remain unknown: the duration of its protection, its potential effects on the prevalence and severity of other HPV strains, its long-term impact on the incidence of cervical and other cancers, and—a function of all these factors—its ultimate cost-effectiveness.[39] If the vaccine's effect on cervical cancer and other HPV-related illnesses is less than initial data suggest, then both the population-level and the individual-level benefits will be fewer, and a beneficence-based argument for the use of government compulsion will become concomitantly weaker.

Moreover, compulsory vaccination laws must be judged not only by their intent but by their effects as well. A critical question as the debates over HPV vaccine policy proceed is whether states' approaches embody the principles of justice and beneficence—that is, whether mandates (if mandates are indeed enacted) are accompanied by policies and laws that represent the government's providing hand. For example, will lawmakers provide public funding for the vaccine or require private insurers doing business in the state to reimburse its cost? Over the past forty years, compulsory vaccination has often failed to fulfill its ethical promise, as states have done a poor job of enabling or helping parents to comply with the laws they have passed. This is a critical consideration in the case of HPV vaccination given the high cost of Merck's Gardasil relative to other routinely given vaccines.

The Politics of Mandates

The history I have presented in this chapter demonstrates that the danger of popular resistance to vaccine mandates should not be underestimated. Public health gains, after all, are not irreversible. The United States saw an extended period, beginning in the mid-nineteenth century and lasting through the early twentieth century, of widespread repeal or weakening of compulsory vaccination laws, and there is no reason to suppose that the same thing could not occur again, especially since the libertarian rhetoric and cultural hostility toward elites that animated earlier challenges remains strong today. This is one reason many prominent figures in public health such as Lawrence Gostin, who are usually sympathetic to the robust use of law, have come out in opposition to HPV vaccine mandates.

One can find cautionary tales about this danger in other realms of public health law. Three decades ago almost all states required that motorcyclists wear helmets while riding; today, only twenty states have such laws. A vocal and politically astute group of citizen-activists has used powerfully resonant arguments based in individual liberty and choice to trump a large body of epidemiological evidence about the benefits of helmets.[40] As helmet laws have been rolled back, the toll of highway injury and death has risen steadily. Helmet laws have proved vulnerable to challenge because they protect against an individual risk rather than a communal one; a key rhetorical strategy of the laws' proponents has been to recast the risk as a shared one because of the costs to the health care system that motorcycle injuries impose.

As Jennifer Reich's essay (chapter 9) demonstrates, parental resistance to HPV vaccination rests on a variety of foundations. The diversity of reasons that have been put forth against mandating HPV immunization is typical of the opposition to vaccine mandates throughout U.S. history. The current network of laws may be vulnerable, because people can and do oppose them for a wide range of reasons. This is a very real threat, but it should not be overstated; there is also a great deal of incompletely theorized agreement about why we should have the laws. But one of the benefits of looking at history is recognizing that laws that may seem permanent and enduring may not really be so. While the concept of compulsory vaccination may be fairly elastic in both its rationale and its purposes, it may not be elastic enough to encompass a mandate for an HPV vaccine.

Strikingly, for all the grassroots agitation in recent years related to vaccine safety and the alleged risks of autism and neurological harm, the United States has not seen the kind of dramatic declines in vaccine coverage that have occurred in other industrialized democracies. In the United Kingdom, for example, concerns about safety caused uptake of the pertussis vaccine to plummet in the 1970s, and in the 1990s the public turned away in droves from the MMR vaccine.[41] In both episodes, vaccine refusal led to deadly resurgences of disease. The contrasting experience of the United States is at least in part due to the extensive system of compulsory vaccination laws that protect coverage levels from fluctuations in public trust.[42] Should resistance to HPV mandates provide the impetus to roll back vaccination laws more generally, the consequences could be severe. Another international example, from Japan, is instructive. Following the abandonment there of mandatory vaccination in favor of "strong recommendations" in 1994, the incidence of several childhood illnesses spiked sharply upward. Japan now has the highest rate of measles among economically developed nations and is the leading "exporter" of the disease to the United States through travelers.[43]

With new recommendations for vaccines to be delivered in early adolescence, middle schools have begun adopting requirements similar to those of elementary schools. In the past decade, most states have enacted laws requiring middle school enrollees to be vaccinated against hepatitis B and varicella and to receive booster doses of the combined tetanus-diphtheria-pertussis vaccine. These laws have been effective in increasing vaccine uptake and lowering disease incidence.[44] An argument based in utility and beneficence could be made for adding protection against HPV to the growing roster of required middle school vaccines.

From a pragmatic standpoint, however, this step must be balanced against the potential disutilities of inflaming parental resistance. Should people become convinced that lawmakers and health officials, in advocating for HPV vaccine mandates, are overreaching, acting in ways contrary to the public interest, or beholden to pharmaceutical industry influence—and anecdotal evidence suggests that such views were widespread during the first round of debates over mandating Gardasil—they could use legislative and electoral processes to weaken or repeal a seemingly well-established network of vaccination laws. Sensitivity to social and political context, as much as adherence to ethical principles, is essential to good public health practice.

NOTES

1. Peter A. Briss, Lance E. Rodewald, Alan R. Hinman, et al., "Reviews of Evidence Regarding Interventions to Improve Vaccination Coverage in Children, Adolescents, and Adults," *American Journal of Preventive Medicine* 18 (2000): S97–S140.

2. James Colgrove, "The Ethics of Vaccination," in *The Penn Center Guide to Bioethics*, ed. Vardit Ravitsky, Arthur Caplan, and Autumn Feister (New York: Springer, 2008).

3. Elyse Olshen, Barbara E. Mahon, Shuang Wang, et al., "The Impact of State Policies on Vaccine Coverage by Age 13 in an Insured Population," *Journal of Adolescent Health* 40 (2007): 405–411.

4. Richard Rupp, Susan L. Rosenthal, and Amy B. Middleman, "Vaccination: An Opportunity to Enhance Early Adolescent Preventative Services," *Journal of Adolescent Health* 39 (2006): 461–464.

5. National Conference of State Legislatures, "HPV Vaccine," available at www .ncsl.org/programs/health/HPVvaccine.htm, accessed Nov. 1, 2008.

6. Michael Specter, "Political Science," *New Yorker*, March 13, 2006, 58–69.

7. R. Alta Charo, "Politics, Parents, and Prophylaxis—Mandating HPV Vaccination in the United States," *New England Journal of Medicine* 356 (2007): 1905–1908.

8. James Colgrove, "The Ethics and Politics of Compulsory HPV Vaccination," *New England Journal of Medicine* 355 (2006): 2389–2391.

9. Liz Austin Peterson, "Merck Lobbies States to Require Cervical Cancer Vaccine for Schoolgirls," Associated Press, Jan. 30, 2007.

10. Ralph Blumenthal, "Texas Legislators Block Shots for Girls against Cancer Virus," *New York Times*, April 26, 2007, A5.

11. Lawrence Gostin and Catherine de Angelis, "Mandatory HPV Vaccination: Public Health vs. Private Wealth," *Journal of the American Medical Association* 297 (2007): 1921–1923.

12. Association for Immunization Managers, "Position Statement: School and Child Care Immunization Requirements," www.immunizationmanagers.org/pdfs/SchoolrequirementsFINAL.pdf.

13. Cass Sunstein, *Legal Reasoning and Political Conflict* (New York: Oxford University Press, 1996).

14. National Conference of State Legislatures, "HPV Vaccine," available at www.ncsl.org/programs/health/HPVvaccine.htm, accessed Nov. 1, 2008.

15. Dena Potter, "Bill Would Push Back Mandatory HPV Vaccinations till 2010," *Associated Press State and Local Wire*, Jan. 15, 2008.

16. William Fowler, "Principal Provisions of Smallpox Vaccination Laws and Regulations in the United States," *Public Health Reports* 56, no. 5 (1941): 167–173; Charles L. Jackson, "State Laws on Compulsory Immunization in the United States," *Public Health Reports* 84, no. 9 (1969): 787–795.

17. John Duffy, "School Vaccination: The Precursor to School Medical Inspection," *Journal of the History of Medicine and Allied Sciences* 3 (1978): 344–355.

18. Martin Kaufman, "The American Anti-Vaccinationists and Their Arguments," *Bulletin of the History of Medicine* 41 (1967): 463–467.

19. William Novak, *The People's Welfare: Law and Regulation in Nineteenth-Century America* (Chapel Hill: University of North Carolina Press, 1996).

20. Jackson, "State Laws on Compulsory Immunization," 787–795.

21. Nadja Durbach, *Bodily Matters: The Anti-Vaccination Movement in England, 1853–1907* (Durham, NC: Duke University Press, 2005).

22. *Zucht v. King*, 260 U.S. 174 (1922).

23. Louis Brandeis to Felix Frankfurter, Dec. 17, 1924, quoted in *"Half Brother, Half Son": The Letters of Louis D. Brandeis to Felix Frankfurter*, ed. Melvin Urosky and David W. Levy (Norman: University of Oklahoma Press, 1991).

24. James Colgrove, *State of Immunity: The Politics of Vaccination in Twentieth-Century America* (Berkeley: University of California Press, 2006).

25. *Jacobson v. Massachusetts*, 197 U.S. 11 (1905).

26. Colgrove, *State of Immunity*.

27. Robert E. Serfling, R. G. Cornell, and Ida L. Sherman, "The CDC Quota Sampling Technic with Results of 1959 Poliomyelitis Vaccination Surveys," *American Journal of Public Health* 50 (1960): 1847–1857.

28. Colgrove, *State of Immunity*.

29. Adelaide M. Hunter, Robert Ortiz, and Joe Martinez, "Compulsory and Voluntary School Immunization Programs in the United States," *Journal of School Health* 33 (1963): 98–102.

30. Colgrove, *State of Immunity*.

31. "Measles—United States," *Morbidity and Mortality Weekly Report* 26 (1977): 109–111.

32. Robert Stevens and Rosemary Stevens, *Welfare Medicine in America: A Case Study of Medicaid* (New York: Free Press, 1974).

33. Colgrove, *State of Immunity*.

34. Rita Swan, "On Statutes Depriving a Class of Children of Rights to Medical Care: Can This Discrimination Be Litigated?" *Quinnipiac Health Law Journal* 2 (1988–1999): 73–95; William J. Curran, "Smallpox Vaccination and Organized Religion," *American Journal of Public Health* 61 (1971): 2127–2128; Rennie B. Schoepflin, *Christian Science on Trial: Religious Healing in America* (Baltimore: Johns Hopkins University Press, 2003), 196–199.

35. James G. Hodge and Lawrence O. Gostin, "School Vaccination Requirements: Historical, Social, and Legal Perspectives," *Kentucky Law Review* 90 (2001–2002): 831–890.

36. Alan Hinman, "Position Paper," *Pediatric Research* 13 (1979): 689–696, quotations on 695.

37. Robert Johnston, "Contemporary Anti-Vaccination Movements in Historical Perspective," in *The Politics of Healing*, ed. Robert Johnston (New York: Routledge, 2004).

38. Jennifer S. Smith, "Ethnic Disparities in Cervical Cancer Illness Burden and Subsequent Care: A Prospective View," *American Journal of Managed Care* 14 (2008): S193–S199.

39. George F. Sawaya and Karen Smith-McCune, "HPV Vaccination: More Answers, More Questions," *New England Journal of Medicine* 356 (2007): 1991–1993.

40. Marian Moser Jones and Ronald Bayer, "Paternalism and Its Discontents: Motorcycle Helmet Laws, Libertarian Values, and Public Health," *American Journal of Public Health* 97 (2007): 208–217.

41. Jeffrey Baker, "The Pertussis Vaccine Controversy in Great Britain, 1974–1986," *Vaccine* 21 (2003): 4003–4010.

42. James Colgrove and Ronald Bayer, "Could It Happen Here? Vaccine Safety Controversies and the Specter of Derailment," *Health Affairs* 24 (2005): 729–739.

43. Margie C. Andreae, Gary L. Freed, and Samuel L. Katz, "Safety Concerns Regarding Combination Vaccines: The Experience in Japan," *Vaccine* 22 (2004): 3911–3916.

44. Thad R. Wilson, Daniel B. Fishbein, Peggy A. Ellis, et al., "The Impact of a School Entry Law on Adolescent Immunization Rates," *Journal of Adolescent Health* 37 (2005): 511–516; Francisco Averhoff, Leslie Linton, K. Michael Peddecord, et al., "A Middle School Immunization Law Rapidly and Substantially Increases Immunization Coverage among Adolescents," *American Journal of Public Health* 94 (2004): 978–984.

Gardasil

A Vaccine against Cancer and a Drug to Reduce Risk

Robert Aronowitz

Merck began its "One Less" advertising campaign for Gardasil, its vaccine against human papillomavirus, in 2006, not long after Gardasil received U.S. Food and Drug Administration approval. In a widely broadcast television advertisement, girls and young women of different ethnicities and apparently different socioeconomic backgrounds chant "I want to be one less . . . one less statistic." One woman says, "Gardasil is the only vaccine that *may* help protect you against the four types of HPV that may cause 70% of cervical cancer," and another (older) woman warns of side effects.[1] Viewers of these advertisements might have wondered why, if Gardasil was "the only cervical cancer vaccine," its name had to be endlessly repeated. Some might also have pondered the uncertain efficacy introduced by "may" and just how likely they were to be one of the "thousands of women" who develop cervical cancer each year.

People looking at a print advertisement for Gardasil that appeared around the same time were offered a different rationale than becoming one less cancer statistic. This advertisement suggested that Gardasil allowed the consumer to evade the *risk* of contracting HPV. "She won't have to tell him she had HPV . . . because she doesn't."[2]

This chapter explores the details and implications of Gardasil's co-construction as a vaccine against cancer and as a proprietary drug that promises reduction and control of individual risk. One might immediately question whether Gardasil is any different from some vaccines introduced in the past, which had elements of public goods as well as of commodities and which aimed to benefit populations as well as individual vaccine consumers. While these continuities exist, Gardasil has entered American medicine and society at a time when intervening in individual risk has become central to the health care economy, medical and lay understandings of efficacy, and the experience of health and disease. These changed features, as they relate to Gardasil and generally to health risk interventions, are sometimes difficult to appreciate because researchers, public health officials, drug manufacturers, and clinicians have blurred the border between risk and disease and have appropriated older rationales and the language of traditional clinical interventions and public health. They have often done so intentionally and for narrow, self-interested purposes.

So although Gardasil is not a completely novel vaccine and the controversies it has sparked are not unprecedented, it has appeared under new circumstances in American medicine and society that have shaped every aspect of its development, architecture, marketing, and reception. These circumstances, described below from the medical literature, media accounts, advertisements, and policy deliberations available to the public, can help explain some of the widespread ambivalence Gardasil has evoked, especially the gap between the promise of effectively preventing cancer and many Americans' weariness with the introduction of yet another risk-reducing commodity.[3]

Gardasil's meaning has also been influenced by an epidemiological reality far different from the one in which many of the most effective vaccines against epidemic and endemic childhood diseases (e.g., polio and measles) were introduced. Because of the success of earlier vaccine programs and other developments, the potential targets for new vaccines are generally less prevalent conditions. Many new potential target diseases have other means of prevention and are treatable if contracted. This diminishing potential population benefit constitutes a void that can be filled by a greater focus on individual risk.

My argument is not that Gardasil is more a drug against risk than a vaccine. It is both of these things. The class of immunological interventions we call vaccines is a very heterogeneous category, whose prototype and namesake, the smallpox vaccine, is no longer in active service. The prototypical vaccine is directed against a specific, prevalent, serious, and communicable infectious disease. It reproduces the lifelong immunity given by natural infection. Because of its high efficacy in individuals, the prototypical vaccine contributes to herd immunity: it interrupts the spread of disease by reducing the number of susceptible individuals.[4] The prototypical vaccine thus serves population goals, not just individual ones, and thereby provides a major rationale for public funding, coercive rules such as those requiring vaccination for entry into school, and calls for universal administration.

In practice, many vaccines now and in the past have strayed from this prototype. Typhoid vaccine's immunity is so short-lived that it is mostly used to protect tourists. Pneumococcal vaccine targets many but not all of the infection-causing subtypes. Tetanus vaccine is aimed at a noncommunicable disease. Influenza vaccine is designed anew each year to respond to antigenic drift and the perceived danger of different influenza strains. *Bacillus Calmette-Guérin* (BCG) vaccine offers only partial immunity for a disease that does not normally induce immunity. To say that Gardasil is not a prototypical vaccine makes it *typical* of most vaccines.

The "Risky" Nature of HPV and Cervical Cancer and Their Causal Links

The "risky" nature of HPV and cervical cancer and the causal links between the two begins with the probabilistic meaning of HPV types and extends to the association with cancer and other clinical outcomes. HPV types are defined quantitatively and probabilistically as having a less than 10 percent difference in nucleate assays. They have been named (in this case, numbered) by their order of discovery. As a result, the ontology of any particular HPV type is necessarily problematic, as are causal claims based upon them, for example, the claim that HPV types 16 and 18 are *the cause* of 70 percent of cervical cancer.

Infection by one of these HPV types, however necessary a condition in the causal chain, does not by itself lead to cancer. HPV infection is common; cervical cancer is rare. Other *risk factors* that have emerged from epidemiological studies include smoking, having multiple HPV type infections, being older, and being coinfected with other sexually transmitted diseases such as chlamydia and genital herpes. These risk factors explain little of the population variance but serve as a reminder that we do not understand precisely the etiology of cervical cancer at either the individual or the population level (see chapter 3 for a more complete discussion of these uncertainties).

Although the mechanisms by which HPV infection leads to cancer have been more directly observed than for almost any other cancer, the carcinogenic role of particular HPV types is nevertheless dependent on the probabilities that particular regions of HPV DNA will survive host immune reaction, integrate with cervical epithelial DNA, survive DNA repair, and lead to clonal autonomy and immortality. Chance and poorly understood ecological factors (immune or otherwise) not only influence these processes but may also keep a particular cancer clone, fully capable of growth and metastasis, at bay or not, which is yet another reason that prediction of the clinical course of patients from their cervical biopsies has long been difficult.[5]

During a regulatory hearing for Gardasil, a Merck representative conceded that our knowledge about the causal connection between high-risk HPV types and cervical cancer was necessarily indirect and open to revision. "There was— there isn't any marker that says, you know, okay, here's an HPV 16. It is glomming right onto the cell and causing it to be malignant, if you know what I mean. Just the strong associations between these things and the fact that persistent HPV 16 infection is highly likely to cause disease and the association with 16 is particularly relevant for cervical cancer, 18 for adenocarcinoma and so on."[6]

Despite these uncertainties, the evidence for the causal link between HPV and cervical cancer is tighter than almost any other causal association in cancer.[7] HPV infection has been directly observed and visualized within various precursor lesions and pathological specimens of full-blown, invasive cervical cancer. There is also very precise knowledge of which HPV DNA regions seem to be most transformative and get incorporated in clones destined to become cancerous. Additionally, the role HPV is understood to play in cancer transformation makes adaptive sense when looked at from the perspective of the virus (viruses that are successful at integrating into the cervical cell genome and at inducing immortality will produce more progeny).[8]

"Scientific" Efficacy of Cervical Cancer Vaccines Is Uncertain

Despite the highly plausible observations linking HPV to cervical cancer, not only does causality remain contested but the strength of the causal association does not necessarily predict what will happen when a causal factor is removed. On ethical and statistical grounds (i.e., the numbers of subjects and the time needed to measure a significant change in cancer mortality), there will never be a randomized, placebo-controlled trial of HPV immunization with cervical cancer mortality as an endpoint. This is an old problem. For similar reasons, there has never been a randomized controlled trial of Pap smears with this endpoint. Our belief that the widespread use of the Pap smear has led to the decline in cervical cancer mortality rests on a set of assumptions about the natural history of precancerous cervical abnormalities and the effectiveness of their "early" removal.[9] Using the declining cervical cancer mortality to support the efficacy of Pap smears is problematic, if for no other reason than that this decline preceded the introduction of widespread Pap screening.[10]

So instead of cervical cancer incidence and mortality, clinical trials of HPV vaccine efficacy have had to make do with using HPV-type-specific cytological abnormalities as endpoints. The principal evidence establishing Gardasil's efficacy has been the reduction of different cervical abnormalities and putative cancer "precursors" that contain HPV 16 and 18 infections within them.[11] Using this measure, Gardasil is nearly 100 percent effective.

It remains unclear, however, whether reducing these types of abnormalities will translate into reduced cervical cancer incidence and mortality. To understand why extrapolating from such limited endpoints may ultimately be misleading, consider one scenario. Reduction in the incidence of high-risk HPV types 16 and 18 through vaccination may result in a higher incidence of other HPV types.[12] We do not know whether types 16 and 18 are intrinsically more virulent than other types or are more successful at infecting individuals relative to other types. This is a nontrivial issue. To the degree that type-specific associations with cancer are due to their greater success at infection in competition with other types rather than an inherent capacity for harm, the theoretical concern about the dangers of type replacement mentioned during regulation hearings and elsewhere is very real. Will vaccines in effect select for different dominant HPV types, which may turn out to have a much higher pathogenicity than previously appreciated? Such a phenomenon may be occurring with currently used multivalent pneumococcal vaccines.[13]

Complicating any determination of the possible impact of Gardasil on cervical cancer are the possible effects of Gardasil on the detection and meaning of cervical cancer and even on sexual behavior. Will the vaccine lead to more or less cervical cancer screening and be efficacious or not by this indirect route? Will there be a shift in cases from squamous cell carcinoma to adenocarcinoma of the cervix?[14] Will coincident HPV infection or genomic integration become part of the case definition of cervical cancer, leading to new types of cases that are necessarily influenced by HPV screening and Gardasil administration?

But my aim is not to simply fan the skeptical fire. In the case of Gardasil, there are many reasons to hope that there will be significant, positive population benefits. Yet our current ignorance and the general unpredictable nature of ecological and evolutionary developments is real. The effects of massive immunization campaigns on populations are complex and difficult to predict. Net long-term (as in decades, intergenerational) effects are even more unclear and have provided fodder for vaccine skeptics. These uncertainties underscore what is apparent from the history of other vaccination programs (see note 4). They are massive population experiments and should be treated as such. This uncertain population impact is one of the main reasons that promoters of Gardasil have appealed to additional types of efficacy besides received notions of objective and material improvement in the morbidity and mortality of individuals and populations.

Efficacy as Risk Reduction and Control

Gardasil has been constructed and marketed to have an impact on an individual's experience of risk. The vaccine promises to control fears and bring some relief from feelings of randomness, shame, and stigma. This rationale is not trumpeted. Highlighting too boldly individual risk control can potentially undermine the more scientific and idealized views of efficacy through which Gardasil and many other products are rationalized and by which they are legitimated and approved within the scientific and regulatory worlds.

Let's begin with a fantasy of Gardasil's initial development and marketing. This fantasy cannot easily be checked against reality because the negotiations behind the launching of new drugs are shrouded in privacy. Someone at Merck, GlaxoSmithKline, or elsewhere perhaps woke up one morning and realized that the firm's prestigious, effective, morally upright but only marginally profitable vaccine business was looking strangely like their most profitable, drugs-for-individualized-risk division. This fantasy depends on a "risky" vision of developments in health care and society that needs some explication.

I have elsewhere outlined some general features of a new "at risk" paradigm that is by no means confined to the pharmaceutical industry. Shifts in body metaphors, clinical practice, and perceptions of efficacy have resulted in and interacted with profound changes in lay demand for risk interventions and what individuals consume and do with their bodies.[15]

Others have also traced this history. Jeremy Greene tells this emergence-of-risk story within the pharmaceutical industry, emphasizing the industry's skillful use of clinical trials to simultaneously establish efficacy and promote new types of risky targets for drugs. Greene begins his narrative at a 1957 meeting of the American Drug Manufacturer's Association in which an industry representative reflected on the paradoxical market impact of the industry's most obvious success—the production and marketing of antibiotics. Antibiotics' success at curing infectious disease had the effect of limiting the industry's market. No profitable industry's business model would want to create products that completely consumed demand for them. The representative called for a new paradigm for drug development and marketing. In 1957 this representative could only imagine the outlines of another paradigm in which drugs would grow rather than shrink their market. In following years, the pharmaceutical industry developed and marketed risk-reducing drugs and a new probabilistic concept of efficacy as risk reduction.[16]

Unlike the antibiotics market, the market for risk-reducing drugs is potentially the whole population, and the duration of use could span a consumer's entire life. The evidence for this new "at-risk" monetary and moral economy is all around us yet is often framed in ways that obscures what is happening. While policy debates raged around Medicare financing of prescription drugs at the end of the twentieth century, few analysts were discussing that the elderly, like everyone else, were largely taking "risk-reducing" drugs.[17] Observers of Merck's Vioxx debacle, retold as a morality and liability story in various courtroom, media, and policy forums, usually pay only passing attention to the fact that Merck definitively proved or stumbled onto the up-to-then debated cardiovascular risk in the midst of a clinical trial to see if Vioxx might reduce colon cancer risk, the ultimate rationale for a mass-marketed risk-reducing pharmaceutical. It is intrinsic, not accidental, that the attempt to find a mass risk-reducing identity for drugs can backfire. The Vioxx debacle illustrates a characteristic "easy come, easy go" problem when efficacy is constructed as individualized risk reduction. Risk-reducing drugs are especially vulnerable to data suggesting that they carry their own risks. How can one aggressively sell peace of mind, insurance, and freedom from fear when the very same product shows evidence of producing further risks? Others have pointed out the American cultural aversion to imposed risk.[18]

Risk-reducing pharmaceuticals promise to eliminate or control the fears, discomfort, and hassles associated with risk. It is the *experience* of risk, not only the objective, specific danger, that is the object of elimination, control, and reduced hassle.

All of these general features of our new "risk reduction as efficacy" paradigm are operative in contemporary attempts to control and contain cervical cancer risk. A large fraction of the adult female population gets annual Pap screening.[19] Close to 3 million of these women annually will be told they have an abnormality. Some will only have repeat studies, but others will undergo procedures such as colposcopy and biopsy and perhaps live for a time with the suspicion or frank diagnosis of cervical cancer or one of the many associated precancerous diagnoses. If tested and found positive for HPV (as part of the Pap test, following a Pap abnormality, or because of a wart), they will be told—depending on whether typical testing was done and what the results were—that they are capable of transmitting this cancer- and/or wart-causing virus to their sexual partners and that even condoms might not be fully protective.

It is against this intervened-in and experienced risk state that Gardasil's efficacy for individuals must be understood. The vaccine may result in one less cervical cancer victim, but it promises every consumer the chance to evade or reduce the risk of infection with cancer- or wart-causing HPV strains. This risk state is made more palpable by being communicable and embodied. In this latter quality it is like so many other noninfectious but worrisome precancerous entities.[20]

I referred earlier to the message that appeared in a print advertisement, "She won't have to tell him she had HPV . . . because she doesn't." This message more overtly sells Gardasil's efficacy at evading or controlling the risk experience, which in the case of HPV is the stigma and worry attached to harboring a sexually transmissible disease that can lead to warts and cancer. I have had numerous patients and students over the years express their worry, shame, and confusion about what to do after being told that they had an HPV infection. They were understandably worried about infecting others with a cancer-causing virus and feared uncomfortable discussions and negotiations at extremely awkward moments.[21]

I know parents who have urged their daughters to get the HPV vaccine less out of fear that they would develop cervical cancer than to prevent or minimize the anxiety, shame, and guilt associated with HPV risk and to avoid the tests and interventions that follow an abnormal Pap smear.[22] In sum, the desire to avoid the experience of embodied risk may be a significant driver of the initial Gardasil demand, especially among the more affluent. It also helps explain why and how the vaccine/drug was marketed to middle-class women, priced high, and sold as a consumer choice, by advertisements similar in tone and content to the ones for many risk-reducing drugs such as statins. Although at low risk of cervical cancer, more affluent individuals generally have a great deal of knowledge and experience with abnormal screening tests and their psychological and other consequences.

Merck has not pushed this "risk/fear control" rationale very explicitly in print and television commercials. Fear and avoidance of shame work better as subliminal messages. The appeal of control needs little emphasis once fear and other negative consequences of a risk state have been effectively communicated. Made too explicit, such messages are easily exposed and ridiculed as fear mongering. However, evading the sexual aspects of HPV transmission and the resulting stigma can itself be open to criticism and invite parody.

Cancer risk reduction is perhaps too serious a target for parody. But another direct-to-consumer-marketed product, Valtrex, which similarly promises to protect people from sexual transmission of a disease in a highly probabilistic fashion, has invited many parodies. In one Valtrex advertisement, seen frequently in our home, different (presumably sexual) partners of various ages and ethnicity (but not same-sex partners) face the screen. One says "I have herpes." The other says, "And I don't." Voice-overs and different actors then give a litany of side effects and warnings of limited efficacy.[23] This advertisement has spawned parodies that make fun of the sanitary and inexplicit way the stigma and fear of sexual transmission and impurity are presented as well as the many hedges about the vaccine's safety and efficacy.[24] In a typical parody, the affected partner straightforwardly announces that she or he has herpes, just as in the real advertisement. But there is no smiling and understanding unaffected partner. Upon hearing that his or her partner has herpes, the unaffected partner hurls curses ("you whore"), and violence sometimes ensues. Behind the black humor is a warning to Merck and other drug manufacturers to keep the control of sexually transmitted disease at a distance from products that can be sold in a more highbrow way.[25]

Vaccine Composition, Architecture, and Cost: Strategies for Maximizing Risk and Benefit

HPV vaccines are composed of ingenious virus-like particles (VLPs). Type-specific viral proteins of the L1 class produced by recombinant genetic technology organize themselves into VLPs that strongly resemble the wild-type HPV virus but which contain no DNA. Gardasil consists of four of these VLPs, each derived from different HPV serotypes. Two of the types (16 and 18) are associated with the development of cervical cancer, the other two with warts.[26] In what sense is this combination VLP vaccine a drug against individual risk?

The rationale and function of this particular mix of VLPs is similar to the way a stock portfolio or mutual fund works to manage risk and benefit. This combination product has been designed to reach a reasonable trade-off between the probable benefits of immunity to multiple HPV types and the risk that including noncarcinogenic targets might dilute the immunity to the cancer-causing ones as well as adding costs and side effects. Like a stock portfolio or a bundled set of risky investments, Gardasil diversifies and reduces risk and appeals to multiple market segments (people who want cancer protection and people who don't want warts), making the drug more palatable and giving it a large market. In particular, the addition of the wart-causing capsids expands the drug's target market to boys and men, building on the less compelling (in terms of numbers effected) rationales for the male market: (1) preventing HPV-associated male cancers (see chapter 4) and (2) contributing to herd immunity against cervical-cancer-associated types. Overall, this composite identity increased the benefit side of the risk/benefit equation for individuals taking the vaccine and thereby helped to ease the drug's passage through regulatory bottlenecks in which cost-effectiveness considerations from the individual consumer's perspective were paramount.

At more than three hundred dollars for the three immunizations, Gardasil is currently priced out of the ballpark of cost-effectiveness associated with mandated, universally recommended, and often subsidized vaccines. It is similarly too expensive for mass use in the developing world. Merck is presumably trying to recover its development costs and make a profit by selling the vaccine at a high price to a limited segment of affluent and well-insured, worried Americans, especially while it remained the sole HPV vaccine. Later, when pressures from competitors and price reductions are sought by government mass purchase, it might be a better strategy to lower the price and widen the market, as has been done with other vaccines.

Priced in this way, however, Merck's initial marketing gives preference to Gardasil's impact on individuals rather than the population. A similar logic lies behind the decision to initially seek approval to give the vaccine only to girls. Cost-effectiveness analysis has shown a much lower marginal cost-effectiveness when boys are added to the population mix. This conclusion is more easily reached when the cost of the vaccine is high and by the use of assumptions and econometric techniques that do not measure herd immunity.

Manufactured as a Risk-Reducing Pill

As a recombinant DNA product, Gardasil's material reality and its production more closely resemble pharmaceuticals such as Humulin and many new cancer and immunological therapies such Herceptin, Interferon, and Remicaid than traditional vaccines. Vaccines as technologies and products have historically emerged from the companies or divisions of pharmaceutical companies that make biologicals. In the early twentieth century, these firms often made serums and antitoxins using live animals. The production of many of the vaccines that replaced serums and toxins similarly involved live whole organisms, eggs, tissue cultures, and so forth. Innovation and production involved a great deal of knowledge about managing and manipulating living organisms, often on a mass scale—hundreds of thousands of eggs, chickens, and other living organisms. The emergence of recombinant DNA technology may mean that this era of biologicals is coming to an end.[27]

One consequence of this new recombinant DNA production is that Gardasil and other similarly produced vaccines may find it much easier to ensure patent protection—and thus higher profits for patent holders—than the older biologicals. In the era of biologicals, there were often many different and difficult-to-specify ways to isolate immunogenic material from infectious organisms and to attenuate them while preserving immunogenicity. Once the possibility of developing a vaccine had been demonstrated, different biological techniques could be used to develop similarly effective products. These differences in production made patent protection for vaccines-as-biologicals more difficult than for the vaccines currently produced by recombinant DNA technology.

Furthermore, the government formerly—because of differences in regulatory politics and the more severe nature of many of the diseases that were the targets of older vaccines—had a heavier hand in fixing or negotiating low prices or demanding multiple suppliers for vaccines. Vaccines as a class were therefore not necessarily big moneymakers, nor did they predictably give a large return on investment. In contrast, Gardasil's recombinant manufacture, along with changes in the regulatory environment, has made this vaccine much more like a highly profitable risk-reducing pharmaceutical product than the *biological* vaccine, which was more highly valued as a public good but less profitable.

Gardasil's Efficacy in Reducing Existing Risk Uncertainties

In terms of predictable economic and human costs, Gardasil's major impact is likely to be a reduction in the number of abnormal Pap smears, a direct result of the reduced incidence of infections from the vaccine-targeted HPV types. It has been estimated that less than 10 percent of the total economic costs associated with HPV ($4 billion annually) is direct medical costs of cervical cancer treatment.[28] Instead, the greater share of HPV's economic impact is due to the costs of screening and the work-up of the close to 3 million abnormal Pap smears in the United States yearly, especially the finding that has been termed atypical squamous cells of undetermined significance. If the vaccine ends up dramatically reducing the number of these abnormalities, as is expected, the cost savings might be immense. So would be the savings in hassle, fear, and wasted time of patients and doctors. This kind of efficacy has not been widely touted, certainly relative to avoiding cervical cancer, because it is essentially the reduction of iatrogenic costs associated with screening rather than improvements in the actual or objective health or well-being of individuals. This benefit would not accrue to populations where Pap screening is not currently done. But it is typical of a prominent type of efficacy within our risky medical paradigm—reducing the harm and costs of existing risk interventions.

But the fact that there is something convoluted and self-referential about this benefit makes it no less real. It is a central and predictable consequence of our highly medicalized risk interventions. Looked at from this perspective, the initial construction and marketing of Gardasil as a drug against individual risk for Western markets is not an example of mistargeting but one that is right on target: Gardasil promises to reduce this noise in the expensive and wasteful system.

It is, of course, difficult to predict how much economic saving will actually occur. Gardasil may contribute to other trends that may increase medical costs. By increasing clinical and consumer interest in the HPV-cancer continuum, for example, Gardasil may lead to more HPV testing and its associated expense. There is already a direct-to-consumer advertising (DTCA) campaign aimed at increasing HPV testing. "Take the test not the risk" is the slogan of one such DTCA HPV test.[29]

Yet Gardasil is not simply a risk-reducing drug for individuals. If it were, then a more individualized test and vaccination strategy might make more sense than the current recommendations of universal administration to women at different ages. From a purely individual perspective, there is little reason to be vaccinated against an HPV type to which a person has already been exposed and which has already provoked an immune response. From this individual perspective, it would make more sense, in terms of avoiding whatever vaccine-related risks exist, to first test for HPV infection and then get vaccinated only if negative. This would apply to sexually active girls and women, who have a high likelihood of HPV type exposure in their first two years of coital life. But from a population perspective, this strategy would greatly increase the total costs associated with a Gardasil vaccine campaign.

Population Goals One Individual at a Time

As mentioned earlier, vaccines have long held elements of individualized preventive treatment and population-level effects. But current vaccines are being researched, produced, and marketed in Western societies and medical systems in which risk identification and intervention at the level of the individual consumer has become central. So it is not surprising that mass immunization can strongly resemble mass-marketed risk-reducing products and procedures. In effect, vaccines, by means other than herd immunity or other contextual means, are being promoted as a one-person-at-a-time ("One Less") population-level intervention.

This scaling up of individual treatment to accomplish population health goals has become widespread. It is partly a response to criticism of the "high-risk" prevention strategy offered by proponents of population-level interventions. While high-risk prevention strategies can be meaningful and efficacious for the individual screened and treated, they often have only minimum population impact because only the high-risk tail of the population distribution is impacted.[30] In contrast, population strategies aim at moving the mean of the entire distribution "to the left." A population strategy for preventing heart disease might include changes in national food policy and zoning regulations that aim to influence the entire population rather than screening and then treating individuals with high blood pressure and lipid levels. One retort to the criticism of high-risk prevention strategies has been to advocate for the mass use of preventive interventions that hitherto were used only for selected, "high-risk" individuals. This is the last stop of the logic of risk intervention, made possible by high demand for risk reduction *and* population impact and permitted by the presumed safety and acceptable cost of some interventions.

This new paradigm in disease prevention has emerged following different paths. One path, the one followed by coronary heart disease prevention through the identification and treatment of individuals with high blood pressure, has been the gradual reformulation of the meaning of risk categories and their extension to huge swaths of the population. First, medicines were developed to treat hypertensive crises, then asymptomatic hypertension above a certain threshold was identified as a risk factor to be screened and treated, then this threshold was gradually lowered, and finally a new disorder, prehypertension—being at high risk for hypertension—was defined and promoted in such a way that a very large segment of the population could labeled and treated.

The other path, today still more hypothetical than actual, is to put nearly everyone in the population on a hypertensive medication. Serious proposals for giving everyone a *polypill*—a concoction of low doses of safe, cheap, and efficacious medicines hitherto used to lower blood pressure, lipid levels, thrombus formation (aspirin), and hyperglycemic states—aim to move the population mean "leftward" one patient at a time.[31] Mass marketing of vaccines to individual consumers to become "One Less" cancer statistic shares many features with this brave new world of individual-by-individual population-level intervention.

Implications

It is possible that when vaccines are co-constructed as proprietary drugs against individual risk, they are in danger of losing their appeal as public goods. The proprietary character potentially undermines the belief that taking the vaccine is a civic responsibility rather than a consumer choice. As such, it may be a much harder sell to require them as part of school entry requirements or even to convince the public to voluntarily comply. Elements of this scenario already occurred with the recent failure of the Lyme disease vaccine. Backlash has become routine in response to medications that have been advertised directly to consumers and for which demand has been manipulated in other heavy-handed ways. Witness the severe public outcry against Merck's funding of women's legislators around the country and the backlash against the Texas governor—whose aide was formerly a Merck lobbyist—who attempted to mandate Gardasil as a vaccine required for school entry. Fueling this backlash is the apparent hypocrisy of the joint construction: dressing the vaccine as a public good but selling it and profiting by it as a consumer product.[32]

Many people have worried that Gardasil will be too expensive to be used in developing countries, where the global burden of cervical cancer is greatest.[33] The high cost is one manifestation of a product conceived of and developed for the individual consumer who can afford it rather than the populations who most need it. This is perhaps the latest installment in a long history of the way different aspects of the economic and structural realities of American medical practice have interfered with effective public health practices. Yet despite the evident mismatch of initial market and greatest need, Gardasil—as a tangible and visible commodity—may ultimately prove to be salutary. It may increase the visibility of the disparity between cervical cancer rates in the developed and developing world, thereby constituting an imperative to intervene. In this and other instances, Gardasil's societal and population impact is an experiment-in-progress.

ACKNOWLEDGMENTS

The research for this chapter was funded in part from an Investigator Award in Health Policy from the Robert Wood Johnson Foundation and from grant 1G13 LM009587-01A1 from the National Library of Medicine, the National Institutes of Health, and the Department of Health and Human Services. The views expressed in any written publication or other media do not necessarily reflect the official policies of the Department of Health and Human Services; nor does mention by trade names, commercial practices, or organizations imply endorsement by the U.S. government.

NOTES

1. YouTube, www.youtube.com/watch?v=hJ8x3KR75fA, accessed Jan. 11, 2008.
2. The advertisement is available at http://drflisser.com/flisserhpvcolposcopy.html.
3. See chapter 8, where some of this weariness is captured; see especially Chesler and Kessler's hpvboredom.com Web site. Vaccines have long been at the forefront of

popular opposition to medicalization. They have also repeatedly ignited controversy over how benefit and risk are distributed (see James Colgrove, *State of Immunity: The Politics of Vaccination in Twentieth-Century America* [Berkeley: University of California Press, 2006]; and chapter 1 in this volume). In particular, vaccines have generally been given to healthy people at no special risk, often to children. Sometimes groups have borne risk only for the benefit of others (as in rubella vaccination for boys).

4. While vaccines have had positive effects on population health, mostly by herd immunity, they may also have deleterious population-level effects. The abandonment of smallpox vaccination, a consequence of the global eradication campaign's success, has led to population vulnerability to bioterrorism. Varicella vaccination, while reducing morbidity in childhood, is less effective than wild-type infection in producing immunity in the entire population. As a result, widespread immunization might lead to more widespread susceptibility in adults, who are more likely to suffer serious disease. Other possible types of negative population impact include hyperreactivity (e.g., asthma) that may result from delayed exposure to infectious diseases because of mass vaccination and other phenomenon (the so-called hygiene hypothesis) and the ecological changes in the type and virulence of infectious agents induced by vaccination's potential selection effects on wild-type infectious disease (considered in the section "Scientific" Efficacy of Cervical Cancer Vaccines Is Uncertain).

5. P. Vineis and M. Berwick, "The Population Dynamics of Cancer: A Darwinian Perspective," *International Journal of Epidemiology* 35, no.5 (2006): 1151–1159.

6. Testimony of Dr. Barr, minutes of FDA Center for Biologics Evaluation and Research, Vaccines and Related Biological Products Advisory Committee, May 18, 2006, p. 86.

7. Testimony of Laura Koutsky, who said that "we probably know more about the way HPV 16 and 18 cause cervical cancer than we know about how other agents cause other cancers," ibid., p. 125.

8. See the chapter "The Epidemiology of Cancer," in *Matters of Life and Death: Perspectives on Public Health, Molecular Biology, Cancer, and the Prospects for the Human Race*, by J. Cairns (Princeton, NJ: Princeton University Press, 1997), esp. 188–190.

9. Although since the introduction of the polio vaccine in the 1950s we have had clinical trials of vaccines, at no time has the problem of evaluating efficacy and safety been easy. Clinical trials have not extinguished uncertainty or controversy. Even when done well, such trials must use limited endpoints and short time horizons. By chance alone, trials may produce putative evidence of benefit and harm. Efficacy has been especially hard to evaluate at the population level, since many diseases are already declining for apparently nonspecific reasons related to socioeconomic advancement. It has also not helped contain controversy that so many "clinical trials" of vaccines were ad hoc experiments done on vulnerable populations such as prison inmates and institutionalized children.

10. While I find it highly likely that Pap smear screening has been effective at reducing cervical cancer mortality, other explanations for the larger secular trend are possible, e.g., rising hysterectomy rates in the post–World War II period.

11. Chapter 10 shows that this near 100% efficacy in reducing cervical abnormalities associated with targeted HPV serotypes can potentially contribute to a "pseudocertainty" effect in which the vaccine's overall efficacy against cancer is overestimated.

12. The inherent probabilistic nature of HPV infection and its natural history also follow from how cervical and vaginal specimens have been collected. This is not simply a question of possible selection bias in the population of women tested for HPV; cytological screening, colposcopy, and biopsy do not yield a random sample of the histology of the individual sampled. Instead, particular parts of the cervix are sampled and subjected to particular assays.

13. C. Munoz-Almagro, I. Jordan, A. Gene, et al., "Emergence of Invasive Pneumococcal Disease Caused by Nonvaccine Serotypes in the Era of 7-valent Conjugate Vaccine," *Clinical Infection Disease* 46, no. 2 (Jan. 15, 2008): 174–182.

14. We have the earlier precedent of the leftward migration of colon cancers that resulted from the negative selection of right-sided lesions and precursors identified by screening.

15. R. A. Aronowitz, "Situating Health Risks," in *American Health Care History and Policy: Putting the Past Back In*, ed. R. Burns, R. Stevens, and C. Rosenberg (New Brunswick, N.J.: Rutgers University Press, 2006), 153–165.

16. J. Greene, *Prescribing by Numbers: Drugs and the Definition of Disease* (Baltimore: Johns Hopkins University Press, 2006). This argument was presented in very similar form in R. A. Aronowitz, "Framing Disease: An Underappreciated Mechanism for the Social Patterning of Health," *Social Science and Medicine* 67, no. 1 (2008): 1–9.

17. Aronowitz, "Situating Health Risks."

18. A. Brandt, "Blow Some My Way," *Clio Medicine* 46 (1998): 164–191.

19. Herschel Lawson, presenter, minutes of CDC National Immunization Program, Advisory Committee on Immunization Practices, June 29, 2005, p. 69, noted that an astounding 82% of U.S. women have had a Pap test in the three years prior to 2000, according to the National Health Interview Survey and the Behavioral Risk Factor Surveillance System.

20. See my discussion of lobular-carcinoma-in-situ and other embodied risks in R. Aronowitz, *Unnatural History: Breast Cancer and American Society* (Cambridge: Cambridge University Press, 2008), 273–275. See also chapter 15 in this volume.

21. Given these fears and concerns, I wondered why HPV risk was not given more medical or popular attention in recent years (until Merck's "tell someone" campaign), especially since wide media attention was given to genital herpes and HIV risk. One explanation, but one that itself needs some analysis, is the minor cultural visibility of cervical cancer and other HPV-related health problems, as evidenced and constituted by the lack of prominent cervical cancer lay advocacy groups.

22. See chapter 9 for an insightful and textured analysis of the decision making with regard to cervical cancer vaccines. In contrast to what is commonly said about parental concerns, for example, no parents in this study worried that the vaccine would encourage promiscuity or early sexual activity.

23. YouTube, www.youtube.com/watch?v=grVqRnDgS8w, accessed Jan. 11, 2008.

24. YouTube, www.youtube.com/watch?v=nsRoPHTQzFE; accessed Jan. 11, 2008.

25. See A. Leader, J. Weiner, C. Bigman, R. Hornik, and J. Cappella. "The Effects of Information Framing on Intentions to Vaccinate against Human Papilloma Virus" (abstract), Fifth AACR International Conference on Frontiers in Cancer Prevention Research, Nov. 12–15, 2006, for some experimental evidence for lesser demand for a vaccine aimed at an STD and cancer rather than cancer alone.

26. A competing bivalent vaccine (16, 18) has recently been introduced in the American market. The makers of this vaccine have made a different bet—that there might be fewer side effects and greater immunity to the two cancer-associated HPV types without the VLPs of other HPV types, as well as a different (AS04) adjuvant. I suspect that efficacy in reducing invasive cervical cancer incidence or mortality will never be adequately and directly studied, and it will be difficult to claim greater efficacy on serological or other indirect evidence alone.

27. Galambos and Sewell observed that varicella vaccine was the last attenuated live-virus vaccine and hepatitis A vaccine the last killed-virus vaccine. Louis Galambos and Jane Eliot Sewell, *Networks of Innovation: Vaccine Development at Merck, Sharp and Dohme and Mulford, 1895–1995* (New York: Cambridge University Press, 1995), 235.

28. Lawson, in minutes of CDC, National Immunization Program, p. 70.

29. See The Digene HPV Test, at www.thehpvtest.com, accessed May 2, 2008.

30. G. Rose, "Sick Individuals and Sick Populations," *International Journal of Epidemiology* 14 (1985): 32–38.

31. N. J. Wald and M. R. Law, "A Strategy to Reduce Cardiovascular Disease by More Than 80%," *British Medical Journal* 326 (2003): 1419–1424.

32. Advertising and aggressive marketing of vaccines are not new phenomena. There was considerable social marketing of diphtheria vaccine. Merck was very active in marketing its measles vaccine. Nor is the backlash from different quarters about a drug company's overreaching in its manipulation of demand and the regulatory process in itself new to Gardasil. Colgrove documented the way excessive actions of Dow's promotion of its rubella vaccine led to a backlash against it (see his *State of Immunity*).

33. See chapter 5, which highlights the very different disease that any HPV vaccine will have to target in Africa and other resource-poor places. Not only is there a different prevalence of cancer-causing HPV serotypes, but coinfection with HIV (leading to greater numbers of affected women with more aggressive disease), nearly absent cervical cancer screening, and little attention to women's health complicate the picture.

CHAPTER THREE

HPV Vaccination Campaigns

Masking Uncertainty, Erasing Complexity

Lundy Braun and Ling Phoun

On June 8, 2006, the U.S. Food and Drug Administration (FDA) approved Gardasil, a quadrivalent vaccine marketed by Merck & Company to prevent anogenital infection with human papillomavirus types 6, 11, 16, and 18. A bivalent vaccine produced by GlaxoSmithKline, which contains only HPV 16 and HPV 18 virus-like particles (VLPs), was recently approved by the FDA. As a vaccine against virus-induced cancers that are ubiquitous throughout the world, the HPV vaccine has been hailed by physicians, public health practitioners, and the popular press as one of the most important public health advances in recent years.

Such claims are, in many ways, well deserved. HPVs play a central role in the development of a variety of human cancers, such as cervical cancer. Cervical cancer is the second leading cause of cancer-related death in women worldwide and the leading cause of cancer deaths in many developing countries. Eighty-three percent of cases of cervical cancers occur in the developing world (Parkin and Bray 2006). Moreover, concerns about HPV infection extend beyond just cervical cancer. Some head and neck cancers, for example, contain HPV (Gillison et al. 2000), and HPV-induced anal cancers cause substantial morbidity among women and among men who have sex with men. Precancerous cervical lesions (variably referred to as cervical dysplasia, cervical intraepithelial neoplasia, or squamous intraepithelial lesions),[1] though easily treated, trigger considerable anxiety among women and increase health care costs. In the United States, the lifetime probability of infection with heterosexually transmitted anogenital HPV is 75–80 percent. Infection is, in other words, common; however, only a small proportion of women who acquire infection develop detectable lesions, and an even smaller proportion of precancerous lesions become cancerous. As Steven Epstein argues in chapter 4, much less is known about the course of disease (pathogenesis) of anal papillomavirus infection in men.

Based on synthetic VLPs rather than killed or weakened virus, the vaccine represents a stunning accomplishment after years of labor-intensive and technically difficult research. Given claims for 100 percent efficacy, it certainly cannot be dismissed. Yet, as chapters 1, 9, 12, and 13 in this volume describe, approval of the vaccine triggered an almost immediate controversy centered on whether vaccination should be mandatory and whether it would encourage promiscuity. Some raised larger issues, such as the ethics of vaccinating only girls for an infection that was sexually transmitted; the high cost of the vaccine and its availability in industrialized as compared to developing countries; Merck's marketing strategies; and the safety of vaccines in general. With minimal debate and without the input of the Centers for Disease Control and Prevention's (CDC's) Advisory Committee on Immunization Practices (ACIP), the Department of Homeland Security added Gardasil to the required vaccine protocol for female immigrants to the United States of ages 11–26 (Magor 2008). Yet, many other technical and social issues remain unresolved. For example, there is little data on the duration of immunity or the potential effects of the vaccine on the utilization of Pap smear screening.

Despite the intensity of the controversy, acceptance, albeit tentative, for what many journalists and scientists term a "medical breakthrough" appears to be widespread in the United States, and some states have moved to make the vaccine mandatory for girls. An MSNBC poll indicated that 80 percent of parents of children under 15 would give their daughter the vaccine (Gibbs 2006). The results of this poll pose the question, Why does the vaccine appear to have been accepted so widely and so quickly in the midst of controversy? Reasons for acceptance of this cancer vaccine defy simple explanation, but, as shown in this volume, interdisciplinary perspectives can provide valuable insights into the issue. What is emerging from the discussion of the HPV vaccine and associated campaigns is that for the vaccine to be accepted, cervical cancer had to be constructed as a visible, tangible, and yet solvable problem for women—one that was easily preventable with a simple and culturally acceptable technical intervention.

Making cervical cancer a problem solvable through a technical solution is simultaneously a political, sociocultural, and scientific process, which is situated in a long history of erasing the social context and the biological complexity of this cancer, although the particular forms of erasure have shifted over time. In the nineteenth century, for example, the solution to contradictory epidemiologic evidence on what groups of women were considered to be at higher risk involved constructing women's sexuality as the cause of disease. In the contemporary moment, framing Gardasil as a cancer vaccine, not as a vaccine against a sexually transmitted disease (STD), made women as sexual agents invisible—an invisibility that was perhaps considered necessary in order to market the vaccine and make it palatable to parents of preteen and teenage girls in the United States.

In addition to masking female sexuality, the Gardasil campaign elided the many scientific uncertainties about the pathogenesis of HPV infection and the generally successful management of invasive cervical cancer through Pap smear screening for those with access to high-quality health care in the United States. One of the consequences of framing Gardasil as a simple technical solution to cervical cancer, then, is to deflect attention away from the structural barriers to screening for low-income communities in the United States and elsewhere in the world where rates of cervical cancer are high. Successful management of cervical cancer requires *access* to high-quality diagnosis and treatment of precursor lesions, something the U.S. health care system does not provide on an equitable basis. More than half of all cervical cancers in the United States occur in unscreened women (Steinbrook 2006). Indeed, the Gardasil campaign is informative for what it highlights and what it silences.

This chapter takes an interdisciplinary perspective to explore how biological uncertainties about the pathogenesis of HPVs, as well as the social context of prevention, have been masked in public discourse around the vaccine in the U.S. context.[2] We begin with a discussion of what is known about the complex pathogenesis of anogenital infection with HPV. We then highlight some of the continuities and discontinuities between contemporary ideas about prevention through vaccination and a longer history of etiologic theories of cervical cancer. Drawing on articles in the popular press, pharmaceutical industry publications, and two newsletters targeted to physicians, we explore some of the ways in which erasure of complexity, or what Deborah Lupton (1995) has called "taming uncertainty," is being accomplished. We end with a consideration of the effects and implications of the vaccination campaign's erasure of the biological complexity as well as the sociocultural and political economic context of disease prevention in the contemporary United States.

The Basic Biology of HPV Infection of the Genital Tract

Despite more than two decades of elegant scientific work, much about how HPV infection of the anogenital and the oropharyngeal tract causes disease remains to be elucidated. These scientific uncertainties, operating at multiple levels, are important to keep in mind when evaluating technical features of the vaccine and claims for its significance as a mechanism of individual risk reduction and as a broad public health measure.

HPVs are small DNA viruses with a simple genome structure, but they engage in highly complex molecular interactions with host cells. There are more than one hundred types of HPVs distributed throughout the globe, although the biological significance of some of those variants is unclear (de Villiers et al. 2004). Unlike HIV, the genome of HPVs evolves slowly, thus facilitating vaccine development. Most HPVs are associated with benign warts or are undetectable and asymptomatic in people with intact immune responses. Only a subset is associated with cancer. There have been periodic reports of the presence of HPV DNA sequences in other tumor types, such as tumors of the prostate and the colon, but the evidence remains weak for a broader role of HPVs beyond the anogenital and oropharyngeal tracts in humans.

HPVs are highly species- and tissue-specific, infecting basal cells in tissues that line body surfaces and cavities, specifically mucosal and keratinizing (squamous) epithelium. Access of the virus to basal epithelial cells, which are the only cells in squamous epithelium with the capacity to divide, most likely occurs through some sort of microtrauma to the epithelium. For reasons related to the molecular and biochemical functioning of this epithelial surface, complete virus particle production, a step in the virus life cycle that is required for transmission, takes place only in the most superficial layers, where squamous epithelial cells acquire their distinctive functional characteristics (Stanley, Pett, and Coleman 2007). This feature of HPVs has made it very difficult to grow the virus in tissue culture, creating a major impediment to acquisition of the large amounts of virus necessary for vaccine production. Consequently, beginning in the early 1990s, researchers worked on developing methods for producing virus-like particles by expressing structural proteins of the virus capsid in vector or carrier systems.[3] Incredibly, such proteins self-assembled, mimicking a naturally occurring virus that would activate a protective immune response in vaccinated individuals (zur Hausen 2006). That such an approach, which involved nearly two decades of imaginative and labor-intensive research, has been successful is technically remarkable.

Of the 100 distinct HPVs, approximately 40 infect the cervix, the vagina, the vulva, the penis, and the anus, and approximately 15 have oncogenic potential (Muñoz et al. 2003; de Villiers et al. 2004).[4] Some types, such as HPV 6 and HPV 11, produce benign warts that can be persistent and troublesome but do not progress to cancer. In addition, under rare circumstances, babies born to mothers with genital warts can develop recurrent papillomas of the laryngeal tract with airway obstruction that can be life-threatening but is not cancerous. A majority of the various histological subtypes of anal cancers, which occur twice as frequently in women as in men and occur at high rates in men who have sex with men, contain HPV sequences (Parkin and Bray 2006). Some types, such as HPV 16 and HPV 18, are found both in benign lesions and in cancers. Despite their characterization as "high-risk" HPVs, the vast majority of infections with HPV 16 and HPV 18 are transient and generally regress spontaneously without treatment within about two years (Schiffman et al. 2007). By middle age, 75–80 percent of women (and presumably of men) will have acquired infection with one or more types of anogenital HPV.[5] Much less is understood about the biological interactions between host cells and the less frequently occurring anogenital HPVs.

The extent to which HPV can remain in the genital tract without symptoms is still unknown, as are the other contributing factors in disease progression. Postulated cofactors include smoking, oral contraceptive use, persistent HPV infection,[6] and immune suppression, but precise evidence is lacking as to how such cofactors might work mechanistically. Thus, in biomedical parlance, persistent infection with a "high-risk" HPV, such as HPV 16, HPV 18, or other less common types is considered a necessary but not a sufficient cause for cervical cancer.

Males and females are infected at similar rates, but the biological manifestation of infection differs, for reasons that are not clear. Penile cancers are rare relative to cervical cancers, comprising less than 0.5 percent of cancers in men globally. Moreover, given that only about 40–50 percent are associated with HPV (Parkin and Bray 2006), there are most likely HPV-independent pathways to penile cancer. Whether hormonal differences, differences in the nature of the respective epithelial surfaces or differentiating pathways, or as yet unidentified factors account for variable rates of progression in HPV-positive penile as compared to cervical cancer is unknown. The differential gender effects of infection, however, are significant for knowledge-making claims. In a very concrete way, it is more challenging to study absence than presence of disease.

Efforts to develop a therapeutic vaccine for premalignant and malignant tumors that could interfere with the progression of disease have been unsuccessful but remain a high priority among researchers (zur Hausen 2006). This approach will require a better understanding of the mechanisms by which HPVs interact with host cells, the mutual relationship between viral oncogenes and cellular genes, and the environmental context, all of which are important in transformation to malignancy. To complicate matters even more, a large body of evidence indicates that concurrent or sequential infection with different HPV types is common and that there are type-specific interactions. But whether the various virus types compete or are synergistic in vivo is unknown. According to researchers, "the possibility of genotype replacement cannot be excluded" (Woodman, Collins, and Young 2007, 12). Moreover, there is already evidence, although not statistically significant, from clinical trials that HPV types other than HPV 16 and HPV 18 are causing an increasing number of precancerous lesions (Haug 2008).

The Epidemiology of HPV-Induced Cervical Disease

In the 1930s, uterine cancer (some databases still combine cervical and endometrial cancer into one category of uterine cancer) was the leading cause of cancer-related deaths in the United States (Jemal et al. 2008), responsible for just slightly more deaths than those caused by stomach cancer. For unknown reasons, death rates declined precipitously (in parallel with those from stomach cancer) *before* the introduction of mass screening of exfoliated cervical cells pioneered by George Papanicolaou in the third and fourth decades of the twentieth century. Designated the now familiar Pap smear, Papanicolaou's method detected cellular abnormalities indicative of precursors to cancer, which were easily removable. After Pap smears became available on a widespread basis, rates of uterine cancer continued to fall, although there were never any randomized clinical trials definitively demonstrating the efficacy of Pap smears in reducing mortality from cervical cancer. Nonetheless, despite its problematic history in becoming a widespread screening technology or its disciplinary function and invasive nature (Casper and Clarke 1998; Lupton 1995, 97), trend analyses indicate that Pap smear screening has been effective in managing squamous cell carcinoma of the cervix, the most common subtype of cervical cancer in the United States. Adenocarcinoma of the cervix, which accounts for about 20 percent of cervical cancers, is less easily detectable through screening programs and may be increasing in incidence.[7] Currently, there are approximately 50,000 new cases of carcinoma in situ (cancer localized to the epithelium) and 11,070 new cases and an estimated 3,870 deaths from invasive cervical cancer per year in the United States (Jemal et al. 2008). Most, but unfortunately not all, cases of invasive cervical cancer are preceded by histologically well-defined precursor lesions.

Since the early 1990s, the death rate from cervical cancer has continued its sharp decline from 3.49 per 100,000 in 1991 to 2.41 per 100,000 in 2004, a 31 percent decrease (Jemal et al. 2008, 84). As noted above, the overall success in managing cervical cancer in the United States, however, obscures poor access to high-quality health care for low-income women throughout the country. Cervical cancer incidence and mortality are unequally distributed in the United States, with significant variability among women, especially African American women in the Deep South, white women in Appalachia, Latinas along the Mexico-Texas border, Native Americans of the northern plains, Alaska Natives, and Vietnamese women (Freeman and Wingrove 2005, 5). Thus, access to Pap smear screening and high-quality treatment in the context of improved primary health care would presumably lower cervical cancer rates to diminishing numbers without vaccination.[8]

When we are dealing with a disease that can lead to death in younger women (the peak incidence of invasive cervical cancer is at approximately 45 years [Kumar, Abbas, Fausto, and Aster 2010, 1021]) but is theoretically preventable, it is difficult to argue with the public health goal of "One Less." Why not just get vaccinated, after all? At the same time, we suggest that making this case requires an exaggeration or at least a serious overstatement—such as the Gardasil campaign made—of the public health burden of cervical cancer in the United States relative to other diseases and an erasure of the many uncertainties and complexities related to the biology of HPVs, their mode of transmission, and their role in the development of cancer.

Uncertainty, Women's Bodies, and Etiologic Theories of Cervical Cancer

In many respects, as argued in chapters 6 and 7, some of the issues highlighted by the HPV vaccine, most notably those related to gender, are reminiscent of earlier understandings of cervical cancer causality. From the early nineteenth century, when the epidemiologic patterns of uterine tumors were first described,[9] etiologic theories centered on what now seem to be spectacular and moralistic notions of women's bodies, their deviant sexual desires, and their aberrant behaviors. Similar moralizing narratives dominated medical discourse in the early nineteenth century in Germany, although Nolte (2008) argues that cellular theories had replaced moralism by the end of the century.

The first systematic scientific studies of uterine cancer causality were those of Domenico Antonio Rigoni-Stern, provincial surgeon of Verona, who published his "Statistical Facts about Cancers on Which Doctor Rigoni-Stern Based His Contribution to the Surgeons' Subgroup of the IV Congress of Italian Scientists" in 1842 (Scotto and Bailar 1969; de Stavola 1987). In his review of mortality records from the town of Verona between 1760 and 1839, Rigoni-Stern noted higher death rates of uterine cancer in married women and lower rates in nuns. Uncertain as to the causes, he speculated that "decrease of vital powers," "temporary execution of the natural functions . . . or the periodicity" of the uterus, or "mechanical damage" (Scotto and Bailar 1969, 71) might account for the propensity of married women to develop these cancers.

In the late nineteenth and early twentieth centuries, etiologic theories in the United States ranged from "irritations" (Watkins and DeLee 1923, 157) to natural functions of this puzzling organ, numbers of pregnancies, the trauma of childbirth, "perverted nutrition" (Emmet 1884, 509), "constitutional predisposition" (Galabin 1879, 205), and "immoderate coitus and excessive sexual excitation" of the woman "who yields herself to a husband she loves" (F. W. Scanzoni, quoted in Ricci 1945, 215; see also Graves 1929). Interestingly, Emmet and others considered uterine cancer to be a disease of "the better classes" (508), occurring primarily in white women. The role of heredity was controversial for decades (Graves 1929), and the ways in which race and social class were incorporated into theories of disease causality changed over time.

In the nineteenth century, the development of benign genital warts and cancer of the body of the uterus or of the cervix were often considered a part of the same biological process, even as they were being distinguished histologically. By the first decades of the twentieth century, more careful distinctions were made. The legacy of such theories, however, masked the fundamental features of cervical cancer—that is, its sexual transmission and its many biological uncertainties.

No mention of the possible contribution of men to this disease appears in the early literature. Women remained the center of causal narratives for most of the twentieth century. Males were sometimes present in these narratives as "donors," but explanatory frameworks and research investigations continued to focus on women, their vulnerable epithelium, and their unsanitary genital tracts. By the 1970s, etiological concepts were more specific, encompassing infectious and non-infectious transmissible agents, such as smegma from uncircumcised males, syphilis, and, in the 1960s and 1970s, the herpes simplex type 2 (HSV2) virus (Rotkin 1973; Kessler 1981), or spermatozoa themselves (Jeffcoate 1975).

By this time, sexual transmission was clearly established. Even in the 1970s, however, mechanisms through which transmissible agents worked, while establishing a role for men, were nonetheless deeply intertwined with assumptions about the sexual practices, especially the risky behavior, of low-income women. As articulated by Rotkin in an influential review (1973), for example, "the act of coitus effectively launched risk, and an agent of some kind was passed from male to female during this time of life when cervical epithelium was most readily available for transformation. There were and are a number of candidate agents." Whether "Spanish, black, Puerto Rican or East Indian" or "white," all affected women shared "the cultural commonality of low socioeconomic status . . . and exposure to a plurality of sexual consorts, each with a discrete probability of conveying a carcinogenic influence to her. . . . The speculation now is that one of these carcinogenic influences may be herpesvirus type 2" (1354–1355). At the same time, gynecologic textbooks highlighted race as well as social class, linking disease in "Africans" to "conditions of squalor" and in the lower classes to their promiscuity (Jeffcoate 1975; Benson 1982). While the role of male partners in transmission was acknowledged, it was still possible to write in the mid-1980s (without defining the term) that "*promiscuity* obviously predisposes to every STD thus producing strong but not necessarily causal association with cervical neoplasia" (Francheschi, La Vecchia, and Decarli 1986, 65, emphasis added). It was women's sexual behaviors alone that mattered.

In a seminal paper published in *Cancer Research* in 1976, Harald zur Hausen offered an important challenge to the dominant view that HSV2 was the causative agent of cervical cancer.[10] Drawing on more than a half century of parallel and generally not overlapping basic science research on animal papillomas, oncogenic viruses, animal models of skin carcinogenesis, anogenital tract cancers, and rapid changes in technologies for analyzing DNA sequences in tumor tissue, zur Hausen proposed a role for the "condyloma agent" in anogenital cancers (zur Hausen 1976).[11] As it turned out, he did not get the specifics quite right at that time, and the virus he referred to as the "condyloma agent" did not turn out to be oncogenic, or only rarely so. Nonetheless, his intervention shifted causal thinking about uterine cancer of the cervix sharply away from HSV2 to the human papillomaviruses, an entirely different family of viruses.

Subsequent studies demonstrated that a particular type of previously unrecognized "wart," termed *flat wart*, occurred on the cervix, suggesting that flat warts, not classic papillomatous genital warts, were precursors to cancer (Meisels and Fortin 1976; Purola and Savia 1977). While immunolocalization studies showed that these lesions contained HPV (Woodruff et al. 1980; Morin et al. 1981), molecular hybridization techniques proved the most definitive in identifying specific HPV types and in establishing a causal relationship between HPVs and cervical cancer (Durst et al. 1983). Nonetheless, whether HPVs were etiologically related to anogenital cancers remained controversial until the mid-1980s, when the publication of *Viral Etiology of Cervical Cancer* signaled scientific consensus, thereby closing the debate over etiology (Peto and zur Hausen 1986). In 1995, the International Agency for Research on Cancer declared HPV 16 and HPV 18 to be human carcinogens (Muñoz et al. 2003). Many aspects of the pathogenesis of infection remained confusing, however, and continued to be topics of intensive investigation.

The Invisibility of HPV Prior to Vaccination Campaigns

Although anogenital infection with HPV is common and scientific research on the complex biology of HPVs and their relationship to anogenital cancers had expanded dramatically during the mid-1980s, HPV was still largely ignored in the popular press and in health education programs. Within a short time, however, HPV has come to be seen as easily controllable with a simple technological intervention, and despite some controversy, many parents are accepting HPV vaccination for their young daughters. Interesting questions related to the HPV vaccination campaign include why there has been such an abrupt shift in the discourse around HPV and cervical cancer and what role the vaccine as a technological solution has played in the sudden visibility of HPV.

As a researcher studying the mechanisms by which HPVs led to cancer, in the mid-1990s I (Braun) became interested in an apparent disconnect between the sophisticated and exciting molecular research on HPV infection and what informal discussions with college students revealed to be minimal awareness of the infection, its transmission, and its pathogenesis among young people. Since previous work showing little knowledge of HPV infection among college students focused exclusively on women (Vail-Smith and White 1992; Ramirez et al. 1997), I initiated a survey of awareness of HPV among both male and female first-year college students to determine whether understandings of infection varied by gender (Baer, Allen, and Braun 2000). We found that a large proportion (95%) of both males and females had heard of genital warts, but few knew there was an etiological relationship between HPV and genital warts. Only about one-third had heard of HPV infection of the cervix or the penis. Not surprisingly, there were distinct gender differences in views regarding transmission, with most men (82.6%) and 45.6 percent of women indicating they did not know how HPV was transmitted. Most likely confusing what they had learned about HIV infection, many students reported that HPV was transmitted through the exchange of bodily fluids. Other studies conducted in the same period reported similar findings (Anhang, Wright, et al. 2004). According to a survey conducted by the Kaiser Family Foundation in 2000, 70 percent of a sample of adults over age 18 had not heard of HPV and 89 percent had never discussed the infection with health care providers (Kaiser Family Foundation 2000).

The reasons for this lack of knowledge in the late 1990s and early 2000s are complex and have not yet been fully explored. However, the reliance of sex education programs on simple messages as they negotiate the political and cultural minefield of sexuality in the United States points to one possible explanation. Young adults in our sample reported receiving most of their education about STDs in health education classes, which at this time were focused on HIV/ AIDS.[12] Focus group participants in our study dismissed the educational impact of such programs. "I don't really know what [genital warts] are, I've just heard nasty jokes about them . . . this was on TV sitcoms I think." "I had two years of health class in high school, but I don't remember anything I learned from that class." "I remember in health class I learned about a bunch of STDs, and I don't remember any of that now." While the majority of students considered themselves at low "risk" of HPV, they were very much aware of and concerned about STDs more generally, particularly HIV/AIDS. HPV was last in a list of eight common sexually transmitted diseases in our survey. According to this group of students, herpes virus and HIV were the most common sexually transmitted diseases.

Many have written thoughtfully of the limitations and consequences of AIDS-prevention policies (see Odets 1995; Lupton 1995; Patton 1996; Epstein 1996, for example). What is important for the purposes of this discussion of HPV is that during the period prior to the introduction of the vaccine, there was no simple message of prevention that could be disseminated. Because HPVs are distributed widely in infected tissue, whether in the anogenital or the oropharyngeal tract, condom use and other components of "safe sex" reduce to a certain extent but do not eliminate virus transmission.[13] Since abstinence or absolute monogamy of both partners is the only sure means of preventing infection, liberal educators were confronted with a difficult dilemma of how to develop sex education programs centered on HPV that did not play into the hands of clearly ineffective, abstinence-only models promulgated by conservatives and the religious Right during the 1990s. As a consequence, during this period the response to our questionnaire and focus groups suggests that sex education programs simply ignored HPVs or cast the infection in such stigmatizing terms that programming had little learning value.

For a large proportion of the female students in our study, magazines were also important sources of information about sexually transmitted diseases. Again, with no simple or culturally acceptable message of prevention, health coverage in women's magazines faced a similar dilemma to that of school-based curricula. Whether in major newspapers, women's health magazines, or health education programs, accounts of HPV centered on women's reproductive organs, triggering fear and anxiety. With graphic visual depictions of the infections themselves and of humiliating, painful medical treatments, the potential threat of HPV to women's reproductive capacity was a prominent theme in the 1980s and 1990s. Men, in general, were either left out of the cycle of transmission or represented as marginal to it. There were no articles during this period on HPV in men's magazines (Baer 1997).

Making Cervical Cancer a Problem: Erasing Complexity

A successful vaccination campaign could only be launched if HPV were made visible and cervical cancer rendered a problem for middle-class U.S. women. HPV had to be seen as so serious that it required intervention in childhood before the onset of sexual activity but nonetheless as easily manageable.[14] One of the most striking aspects of the vaccination campaign to date is the way cervical cancer has been detached from historical associations with unrestrained female sexuality or "promiscuous" behavior. Indeed, de-linking HPV infection from the idea of promiscuity is important to many scientists. According to Anna Giuliano of the H. Lee Moffit Cancer Center and Research Institute in Tampa, Florida, "It's not about promiscuity. The more we can get that out of people's minds, the faster we'll be able to get prevention efforts out there.[15]

As important as it is to contest the long-standing association of female sexuality with "promiscuity," submerging or ignoring completely the sexually transmitted character of cervical cancer as well as the biological complexity of HPV infection is misleading.[16] Gardasil advertisements and commercials take this approach, for example in the *US Weekly*, April 28, 2008. Commercials portray preteen girls and young women as icons of purity and innocence, lightheartedly jumping rope, dancing, or playing the drums while advocating for "One Less." On Web sites promoting Gardasil, proud mothers bond with their daughters. According to Merck's Web site, "unlike other cancers, cervical cancer is not passed down through family genes. Cervical cancer is caused by certain types of a virus—human papillomavirus or HPV."[17] Gone is the moralism (at least momentarily) that historically characterized research initiatives and mass campaigns to eliminate sexually transmitted diseases (Brandt 1985; Ziporyn 1988). These young women are hardly the licentious beings of the past. Just like the common cold, young women can "get" HPV. How the virus is transmitted is left unstated. The Gardasil campaign, then, raises many questions. How, in such a short period of time, did cervical cancer become framed as a serious public health threat, stripped of its relation to sex? What role have the news media played in reducing the complex biology of HPV infection of the anogenital tract and its sexual transmission to a simple admonition to get vaccinated? How do professional guidelines intersect with the popular press and market imperatives?

Scholars have shown that the popular press, often working in close collaboration with scientists, plays a critical role in shaping the lay public's knowledge and understanding of scientific and health findings by emphasizing certain points and omitting others (Nelkin 1995; Lupton 1995). For example, in their analysis of the information value and "accuracy" of news articles on HPV testing and screening published between 1995 and 2002, before the period when clinical trials demonstrated the vaccines' efficacy, Anhang, Stryker, and colleagues (2004) found that a majority (79%) of the stories indicated that HPV is sexually transmitted and that the infection is common (50%), and 30 percent of the articles focused on cervical cancer. At the same time, many articles failed to report the limitations of condom use or to emphasize the key point that in the vast majority of cases HPV is an asymptomatic, transient infection that resolves spontaneously.

Soon after FDA approval for Gardasil was granted, the press began featuring articles that made grand claims for the HPV vaccine but continued to ignore the biological complexity of the infection. In the opening paragraphs of one article in *Time* magazine, Gardasil was declared an "almost universally hailed . . . medical triumph . . . the first ever designed to prevent cancer." According to the *Time* journalist (Gibbs 2006), cervical cancer was "the second most common cancer among women, and the third most deadly around the world." Her framing of the statistics gave the distinct impression that cervical cancer was a leading cause of cancer in the United States, a claim that is simply wrong.

In a content analysis of seventy-five news articles published in the *New York Times*, the *Washington Post, USA Today, GP, Doctor, Business Week,* and *Business World* from May 2003 to May 2008,[18] we found that the type of general information (the transmission, symptoms, and prevalence of HPV) articulated in the press continued to be limited. In particular, statements highlighting the complexity of the infection and its clinical manifestations were cited infrequently. Less than one-third (31%) of articles mentioned that HPV is common and ubiquitous. Importantly, only 37 percent of the articles indicated that the viruses are sexually transmitted, whereas 20 percent of the articles stated that HPV caused genital warts, and 9 percent suggested that HPVs caused other types of cancer than cervical cancer. A particularly important theme in this literature is the absence of information regarding the frequency and uncertainty of disease progression. A small proportion (11%) of the articles indicated that most HPV infections are asymptomatic. Significantly, only three (4%) of the articles pointed out that most women with HPV infections will not develop cervical cancer.

Thus, the clear message conveyed by the media through this masking of uncertainty about the pathogenesis of HPV infection is one of an exaggerated probability of developing cervical cancer in the United States and Western Europe. That the media would take up such a message is understandable. The language of "high risk" is deeply entrenched in the biomedical literature. To most readers, the notion of high risk communicates an individualized, substantial, and frightening probability of disease.[19] That high-risk HPVs are mostly asymptomatic is counterintuitive to readers. Combined with industry marketing strategies on TV and in the print media, news articles thus cast cervical cancer as a problem—but one that is easily controllable by taking the simple and socially conscious step of vaccination.

Almost entirely absent from news articles on HPV is the intriguing and completely uncertain scientific question of biological niche. Given that coinfection with different HPV types is common, there is, as discussed above, the biologically plausible possibility that HPV types (oncogenic and nononcogenic) not present in the vaccine will expand to occupy the niche vacated by HPV 6, 11, 16, and 18, the types in the vaccine. It is also possible that vaccination with one type could inhibit the viral life cycle of a type not included in the vaccine. Of all the articles examined, only three out of seventy-five (4%) noted the potential for other HPV strains not in the vaccine to take over the biological niche formerly occupied by types included in the vaccine, and only one mentioned that coinfection with different HPV strains is a frequent occurrence. Given that a recent scientific review in a prestigious journal devotes considerable attention to these uncertainties (Woodman, Collins, and Young 2007), the overall absence of discussion of this question in the popular press is notable. If an expansion of activity of other HPV strains does occur, and if it coincides with a decline in the frequency of Pap smear screening, the net effect could be to increase rates of cervical cancer. As is the case in earlier literature on disease causality, men have been absent from the contemporary discourse on cervical cancer, in terms both of their possibility of developing cancer and of their role in the cycle of transmission. (This trend now seems to be changing.)

After the introduction of the vaccine, the number of articles about HPV increased twofold. Overall, the press focused even less on general information and, while the number of articles in our analysis is small, tended to ignore biological complexity for more straightforward claims of vaccine efficacy. Statements such as "HPV is common and ubiquitous" and "most HPV infections resolve on their own" appeared less frequently after June 2006.

The guidelines and recommendations issued by professional associations, usually developed by a panel of experts, shape public opinion and acceptance of a new medical technology and set the terms by which physicians must practice. At the moment, guidelines of most of the relevant organizations (the American Academy of Family Physicians, the American College of Obstetricians and Gynecologists, the American Academy of Pediatrics, and the American Cancer Society) are based on the CDC's ACIP. The primary disagreement rests on ages at which to immunize, not the biological uncertainties associated with the concept of the vaccine. Thus, it is likely that clinicians, who often must rely on fact sheets rather than the full report of the ACIP, are not well acquainted with the nuances of this discussion.

In public pronouncements, some scientists are taking a cautious approach to the vaccine. For example, Christopher Crum, the director of women's and peri-natal pathology at Boston's Brigham and Women's Hospital, voiced concern about the duration of protection. "It's not surprising that there's still a strong immune response four years out," notes Crum. "The real question is whether that will be maintained decades later" (Rubin 2006). Similarly, Dartmouth physician Diane Harper cautioned that "the vaccine is not a silver bullet, is not a shield against cancer" (Rubin 2007). In the meantime, while simplistic news captures the public's attention and works to mobilize support for vaccination, many scientists continue to publish careful review articles and editorials, high-lighting the ambiguities and complexity of infection with this family of viruses (Roden and Wu 2006; Woodman, Collins, and Young 2007; Haug 2008).

Conclusion

In this chapter we track representations of cervical cancer and how in early theories of disease causality they were inextricably intertwined with contem-porary understandings of female sexuality. In current discussions of HPV vac-cines, cervical cancer etiology, while still centered on women, is stripped of all reference to sexual activity. The simplification of health prevention messages that results from this framing is not a grand conspiracy among public health policymakers, physicians, the press, and the pharmaceutical industry, although market interests have profoundly shaped the HPV vaccination campaign. Rep-resenting complexity in the context of a major public health measure is inher-ently a challenging undertaking. In casting the vaccine as a cancer vaccine, rather than a vaccine to prevent a sexually transmitted infection, however, Merck has made this very costly public health intervention palatable to privileged sectors of the United States. The sheer intensity of the advertising campaign for Gardasil signifies the urgency with which Merck attempted to recoup its investment before new vaccines by other manufacturers that immunize against more HPV types came to market.

Beyond market considerations, it deserves emphasis that well-intentioned, simplistic public health messages that erase complexity and ignore the messy sociocultural-scientific context of lived experience can have serious health consequences. Jessica Gregg's (2003) study in the economically deprived region of Recife, Brazil, illustrates this point in stark terms. Because a highly publicized cervical cancer campaign focused exclusively on Pap smear screening for women disconnected from sexual activity, women interpreted the lack of a Pap smear, rather than their sexual practices or those of their partners, as the major risk for cervical cancer. The Pap smear was controllable, but sex was necessary to their survival and therefore not controllable. Thus, despite the campaign, women in Recife continued to be infected with HPV and to develop invasive cervical cancer at alarmingly high rates.

HPVs are unlike smallpox in that eradication of HPVs is most likely not feasible. From a public health perspective, then, the problem to be addressed is how best to manage this biologically complicated and dynamic sexually transmitted infection. One important conclusion to be drawn from this chapter is that as we go forward we should conceptualize the messy and uncertain sociocultural and political-economic context of the vaccine *together with*, not disconnected from, the similarly messy and uncertain scientific dimensions of infection with HPV. We need to ask what uncertainties the debate over HPV vaccination obscures as well as what certainties it highlights. Not only will such an approach clarify the debate; it will also provide a framework for developing more effective public health policies and medical care for women and men worldwide.

ACKNOWLEDGMENTS

We would like to thank Dr. Marcie Richardson for her helpful comments, Heather Baer and Michael Kim for their undergraduate theses on this topic, and relatives and friends for rich conversations and good humor in entertaining our sometimes peculiar questions.

NOTES

1. Intraepithelial neoplasia (or dysplasia) refers to the microscopic changes in cells and tissues that occur after HPV infection. Intraepithelial neoplasia is graded histologically based on the degree of nuclear and cellular changes and disruption of the normal architecture of the tissue.

2. Our analysis here is specific to the U.S. context and cannot be mapped onto other contexts, either in the global North or the global South. In many countries of the global South, as demonstrated in chapters 5 and 13, the public health infrastructure for

gynecologic care is severely constrained. See chapters 14 and 15 for discussions of the particularities of the European context.

3. The virus capsid refers to the outer protein shell that surrounds the DNA and RNA of the virus. The capsid structure is the basis of the classification of viruses. Human papillomaviruses have a simple capsid structure.

4. Less common anogenital HPVs considered to be carcinogenic include HPV 31, 33, 35, 39, 45, 52, 56, 58, 59, 68, 73, and 82. Those classified as "probably carcinogenic" include HPV 23, 53, and 66.

5. An important point to keep in mind when evaluating the vaccine is that a large number of people worldwide are presently infected. There is no evidence that vaccination modifies the course of infection.

6. While persistent infection is generally accepted as important in disease progression, its definition and distinction from latency is a topic of active scientific discussion. Indeed, the relationship between viral load and disease progression is uncertain.

7. Whether this failure is due to accessibility of cells or the association of adenocarcinoma with HPV 18, which produces cytological abnormalities less frequently than HPV 16 does, is unclear (Woodman, Collins, and Young 2007).

8. Pap smears have a high false-negative rate but if they are conducted frequently in a laboratory with excellent quality control, most changes will be detected before invasion occurs.

9. Physicians did not make distinctions between cancer of the body of the uterus and cancer of the cervix at this time.

10. Zur Hausen was awarded the 2008 Nobel Prize for Physiology or Medicine for establishing the link between HPV and cervical cancer.

11. *Condyloma* is the term given to benign warts in the anogenital tract, which can be raised or flat. *Condyloma acuminata* refers to lesions with a wartlike raised morphology, which are generally caused by HPV 6 or 11.

12. Even as the complexity of the biology of HIV infection was becoming more and more clear, HIV/AIDS prevention policy sharpened its focus on a simple message of "safe sex."

13. Steven Epstein and April Huff (chapter 12) show how the religious Right exploited this point.

14. The visibility of HPV was enhanced previously by the advent of HPV testing, marketed by Digene. We thank Marcie Richardson for this insight.

15. *New York Times*, March 6, 2007.

16. See chapter 8 for an imaginative approach to reclaiming sexuality in the context of education about HPV, and see chapter 7 for a discussion of the construction of risky girlhoods.

17. Gardasil, "What Is Cervical Cancer," www.gardasil.com/hpv/human-papillo mavirus/cervical-cancer/?WT.srch=1&WT.mc_id=GL047, accessed May 1, 2008.

18. The sample for this study consisted of 75 articles from three major news papers in the United States: the *New York Times*, the *Washington Post*, and *USA Today;* 2 magazines directed toward physicians, *GP* and *Doctor;* and 2 business magazines, *Business Week* and *Business World*. All were published between May 2003 and May 2008. We searched the Lexis-Nexis database (Lexis-Nexis, Dayton, OH), using the search terms *human papilloma virus* or *HPV* and obtained the full text of each article. We selected 2003

as the earliest year in order to capture articles published during the development stage of the HPV vaccine and prior to its licensing and to extend Anhang's work, which ended in 2002. We then developed a coding instrument with thirty-five items organized into four different categories for comparison of frequency of mentions by year and news outlet.

The following four categories guided our analysis: (1) general information about HPV and cervical cancer; (2) methods and strategies for prevention; (3) the complexity and natural history of HPVs; and (4) the efficacy of HPV vaccines. For the category of general knowledge, we included information on prevalence, symptoms, transmission, and diseases associated with HPV. For the category of prevention, we included mentions of condom use and Pap smear screening as tools for cervical cancer prevention. For the category of complexity, we examined whether articles mentioned the following: the biological distinctions between low-risk and high-risk HPV types and the types of disease they cause; the generally benign course of high-risk HPVs; the role of persistent infection in the development of cervical cancer; the frequency of coinfection with different HPV types; type-specific protection and cross-protection of different types; and the question of biological niche (the growth of new strains as the vaccine-specific strains are eliminated). The final category included information on the effectiveness of the vaccine, for which groups vaccination is recommended according to clinical trial data and guidelines by professional agencies, duration of protection, and potential adverse events related to vaccination.

19. See chapter 2 for a fuller discussion of the notion of individual risk.

REFERENCES

Anhang, Rebecca, Jo Ellen Stryker, Thomas C. Wright, and Sue J. Goldie. 2004. "News Media Coverage of Human Papillomavirus." *Cancer* 100:308–314.

Anhang, Rebecca, Thomas C. Wright, Laura Smock, and Sue J. Goldie. 2004. "Women's Desired Information about Human Papillomavirus." *Cancer* 100:315–320.

Baer, Heather. 1997. "Perceptions and Representations of Human Papillomavirus: An Exploration of Popular Attitude and Literature." Undergraduate thesis, Brown University.

Baer, Heather, Susan Allen, and Lundy Braun. 2000. "Knowledge of Human Papillomavirus Infection among Young Adult Men and Women: Implications for Heath Education and Research." *Journal of Community Health* 25:67–78.

Benson, R. 1982. *Current Obstetric and Gynecologic Diagnosis and Treatment.* 4th ed. Los Altos, CA: Lange Medical Publications.

Brandt, Allan M. 1985. *No Magic Bullet: A Social History of Venereal Disease in the United States since 1880.* Oxford: Oxford University Press.

Casper, Monica J., and Adele E. Clarke. 1998. "Making the Pap Smear into the 'Right Tool' for the Job: Cervical Cancer Screening in the USA, circa 1940–95." *Social Studies of Science* 28, no. 2: 255–290.

De Stavola, Bianca. 1987. "Statistical Facts about Cancers on Which Doctor Rigoni-Stern Based His Contribution to the Surgeon's Subgroup of the IV Congress of the

Italian Scientists on 23 September 1842." *Statistics in Medicine* 6:881–884. A translation of Rigoni-Stern's original paper.

De Villiers, E. M., C. Fauquet, T. R. Broker, H. U. Bernard, and H. zur Hausen. 2004. "Classification of Papillomavirus." *Virology* 324:17–27.

Durst, M., L. Gissman, H. Ikenberg, and H. zur Hausen. 1983. "A Papillomavirus DNA from a Cervical Carcinoma and Its Prevalence in Cancer Biopsies from Different Geographic Regions." *Proceedings of the National Academy of Sciences of the United States of America* 60:3812.

Emmet, T. 1884. *The Principles and Practices of Gynecology*. Philadelphia: Henry C. Lea Son.

Epstein, Steven. 1996. *Impure Science: AIDS, Activism, and the Politics of Science*. Berkeley: University of California Press.

Francheschi, Silvia, Carlo La Vecchia, and Adriano Decarli. 1986. "Relation of Cervical Neoplasia with Sexual Factors, Including Specific Venereal Diseases." In Peto and zur Hausen 1986.

Freeman, H. P., and B. K. Wingrove, eds. 2005. *Excess Cervical Cancer Mortality: A Marker for Low Access to Health Care in Poor Communities*. NIH Pub. No. 05-5282. Rockville, MD: National Cancer Institute, Center to Reduce Cancer Health Disparities, May. Available from http://crchd.cni.nih.gov.

Galabin, A. 1879. *The Student's Guide to the Diseases of Women*. London: J. and A. Churchill.

Gibbs, Nancy. 2006. "Defusing the War over the 'Promiscuity' Vaccine." *Time CNN*, June 21.

Gillison, M. L., W. M. Koch, R. B. Capone, et al. 2000. "Evidence for a Causal Association between Human Papillomavirus and a Subset of Head and Neck Cancers." *Journal of the National Cancer Institute* 92:709–720.

Graves, W. 1929. *Gynecology*. Philadelphia: W. B. Saunders.

Gregg, Jessica L. 2003. *Virtually Virgins: Sexual Strategies and Cervical Cancer in Recife, Brazil*. Stanford, CA: Stanford University Press.

Haug, Charlotte J. 2008. "Human Papillomavirus Vaccination—Reasons for Caution." *New England Journal of Medicine* 359:861–862.

Jeffcoate, Thomas N. 1975. *Principles of Gynaecology*. 4th ed. Boston: Butterworths.

Jemal, Ahmedin, Rebecca Siegel, Elizabeth Ward, et al. 2008. "Cancer Statistics, 2008." *CA: A Cancer Journal for Clinicians* 58:71–96.

Kaiser Family Foundation. 2000. "National Survey of Public Knowledge of HPV, the Human Papillomavirus." Feb. Available from www.kff.org.

Kessler, Irving I. 1981. "Etiological Concepts in Cervical Carcinogenesis." *Gynecologic Oncology* 12:S7–S24.

Kumar, Vinay, Abul Abbas, Nelson Fausto, and Jon C. Aster. 2010. *Robbins and Cotran Pathologic Basis of Disease*. 8th ed. Philadelphia: Saunders Elsevier.

Lupton, Deborah. 1995. *The Imperative of Health: Public Health and the Regulated Body*. London: Sage.

Magor, Gavin. 2008. "US Requiring Controversial Vaccine." The Street.com, Sept. 18, www.thestreet.com/story/10438114/1/us-requiring-controversial-vaccine.html, accessed Nov. 24, 2008.

Meisels, A., and R. Fortin. 1976. "Condylomata Lesions of Cervix and Vagina. I. Cytologic Patterns." *Acta Cytologica* 20:505.

Morin, L., L. Braun, M. Casas-Cordero, et al. 1981. "Confirmation of the Papilloma-virus Etiology of the Condylomatous Lesions of the Cervix by the Perioxidase-Antiperoxidase Technique." *Journal of the National Cancer Institute* 66:831–835.

Muñoz, Nubia, F. Xavier Bosch, Silvia de Sanjosé, et al. 2003. "Epidemiological Classification of Human Papillomavirus Types Associated with Cervical Cancer." *New England Journal of Medicine* 348:518–527.

Nelkin, Dorothy. 1995. *Selling Science. How the Press Covers Science and Technology*. New York: Freeman.

Nolte, Karen. 2008. "Carcinoma Uteri and 'Sexual Debauchery'—Morality, Cancer, and Gender in the Nineteenth Century." *Social History of Medicine* 21:31–46.

Odets, Walt. 1995. *In the Shadow of the Epidemic: Being HIV-Negative in the Age of AIDS*. Durham, NC: Duke University Press.

Parkin, D. Maxwell, and Freddie Bray. 2006. "The Burden of HPV-Related Cancers." *Vaccine* 24S3:S3/11–S3/25.

Patton, Cindy. 1996. *Fatal Advice: How Safe-Sex Education Went Wrong*. Durham. NC: Duke University Press.

Peto, R., and H. zur Hausen. 1986. *Viral Etiology of Cervical Cancer*. Cold Spring Harbor, NY: Cold Spring Harbor Laboratory.

Purola, E., and E. Savia. 1977. "Cytology of Gynecologic Condyloma Acuminatum." *Acta Cytologica* 21:26.

Ramirez, J. E., D. M. Ramos, L. Clayton, et al. 1997. "Genital Human Papillomavirus Infections: Knowledge, Perception of Risk, and Actual Risk in a Nonclinic Population of Young Women." *Journal of Women's Health* 6:113–121.

Ricci, James V. 1945. *One Hundred Years of Gynaecology, 1800–1900*. Philadelphia: Blakiston.

Roden, R., and T. C. Wu. 2006. "How Will HPV Vaccines Affect Cervical Cancer?" *Nature Reviews Cancer* 6:753–763.

Rotkin, D. 1973. "A Comparison Review of Key Epidemiological Studies in Cervical Cancer Related to Current Searches for Transmissible Agents." *Cancer Research* 33:1353–1367.

Rubin, Rita. 2006. "HPV Vaccines Show Progress, but Need Booster in Question." *USA Today*, April 6, 7D.

———. 2007. "Mandate or Choice? That's the Question after Texas Governor Issues Order Requiring HPV Immunization." *USA Today*, Feb. 8, 6D.

Schiffman, Mark, Philip Castle, Jose Jeronimo, Ana Rodriguez, and Sholom Wacholder. 2007. "Human Papillomaviruses and Cervical Cancer." *Lancet* 370:890–907.

Scotto, Joseph, and John C. Bailar III. 1969. "Rigoni-Stern and Medical Statistics: A Nineteenth Century Approach to Cancer Research." *Journal of the History of Medicine and Allied Sciences* 24:65–75. With translation of original article.

Stanley, M. A., M. R. Pett, and N. Coleman. 2007. "HPV: From Infection to Cancer." *Biochemical Society Transactions* 35:1456–1460.

Steinbrook, Robert. 2006. "The Potential of Human Papillomavirus Vaccines." *New England Journal of Medicine* 354:1109–1112.

Vail-Smith, K., and D. M. White. 1992. "Risk Level, Knowledge, and Preventive Behavior for Human Papillomaviruses among Sexually Active College Women." *College Health* 40:227–230.

Watkins, T., and J. DeLee. 1923. *Gynecology, Obstetrics.* Series 1922. Chicago: Year Book.

Woodman, Ciaran B. J., S. I. Collins, and L. S. Young. 2007. "The Natural History of Cervical HPV Infection: Unresolved Issues." *Nature Reviews Cancer* 7:11–22.

Woodruff, J. D., L. Braun, R. Cavalieri, P. Gupta, F. Pass, and K. V. Shah. 1980. "Immunological Identification of Papillomavirus Antigen in Condyloma Tissue from the Female Genital Tract." *Obstetrics and Gynecology* 56:727–732.

Ziporyn, Terra. 1988. *Disease in the Popular American Press: The Case of Diphtheria, Typhoid Fever, and Syphilis, 1870–1920.* Westport, CT: Greenwood.

Zur Hausen, Harald. 1976. "Condylomata Acuminata and Human Genital Cancer." *Cancer Research* 36:794.

———. 2006. "Perspectives of Contemporary Papillomavirus Research." *Vaccine* 24S3:S3/iii–S3/iv.

CHAPTER FOUR

The Great Undiscussable

Anal Cancer, HPV, and Gay Men's Health

Steven Epstein

Local news broadcast, San Francisco, Channel 5 (CBS), circa April 5, 2007:

Ken Bastida: "Well, you've probably heard of women and girls getting the HPV vaccine to protect themselves from a virus that can cause cervical cancer. Well, now some men are taking it. Manuel Ramos is in the Castro District to explain why."

Manuel Ramos: "Well, Kenny, gay and bisexual men are saying that Gardasil isn't just for girls anymore. You see, men in England and Australia are using it, and now they're getting it here, in San Francisco."

In the United States, the ubiquitous direct-to-consumer advertising by the pharmaceutical company Merck has offered a public face for the vaccine Gardasil: young, active, multiethnic, well-informed, self-assertive, and—needless to say— female. After all, the vaccine is marketed and conventionally understood as a vaccine against cervical cancer. But recently, to a very limited extent, a new set of social actors has emerged to demand attention to their human papillomavirus– related illnesses. Advocates for gay men's health, including a small group of medical researchers and practitioners, have joined the public discussion about Gardasil as part of an attempt to highlight the threat of anal cancer, a little-known disease that is causally linked to HPV. These advocates seek to incorporate Gardasil within a larger regimen of prevention and treatment that, they argue, should include the routine screening of men who have sex with men (MSM), by means of anal Pap smears.[1] Thus, in order for the promise of medicine to reach this constituency, a largely invisible disease must be brought into the open, and medical technologies that have been associated with women's bodies must be recoded as usable in men.

To the surprise of perhaps no one, discussions of gay men's health have been almost entirely (if not completely) absent from public debate in the United States about the safety, efficacy, and availability of HPV vaccines. Gay men's interest in the prevention of anal cancer therefore serves to mark one side of the dialectic of inclusion and exclusion that often characterizes new medical technologies. On one hand, the strong push toward universal vaccination of girls in countries like the United States has raised concerns in some quarters about the pervasiveness of biomedicine and public health in the lives of those whom it grasps within its embrace—concerns about over-hasty assessments of safety and efficacy,[2] or about overly aggressive pharmaceutical marketing (chapters 2 and 12), or about the potentially problematic ways in which girls are called upon to manage their bodies, health, and sexuality (chapters 6 and 7). On the other hand, the story of HPV vaccines is also a story about those who are left out—including whole swaths of the global South, for whom the vaccine may simply be unaffordable despite the higher incidence of cervical cancer there,[3] but also distinct groups within biomedically privileged countries—and how they may fight to gain access to the fruits of biomedical progress.[4]

Just as the story of women's access to Gardasil has involved women's health advocates, medical experts, and the pharmaceutical industry, so, too, the framing of HPV as a gay health issue can be analyzed with reference to multiple players, including the organizing work of health activists in the gay community, the concerted efforts of a small network of researchers and physicians, and some cautious steps on the part of Merck. Yet insofar as advocates have found themselves waging an uphill battle for recognition of this issue, community organizing and mobilization have been central to the campaign. The interest of gay health advocates in fighting HPV infection and anal cancer—and in contesting their exclusion from the health polity—can be read as a demand for full citizenship, on both medical and sexual grounds. That is, the distinctive character of the fight against marginalization in the case of gay men's health is that it is waged simultaneously on two fronts—as one of many present-day struggles against health inequalities, and as one of many present-day struggles for full equality on the basis of sexual identity.[5]

These concerns point to the complicated politics of the submerged debate over HPV infection among gay men. At the same time, gay men's push for recognition of their medical needs is an *epistemic* issue. As gay health advocates have sought to change medical practice in relation to HPV, they have run up against the limits of medical knowledge. Repeatedly, they have encountered the response that the policy changes they promote simply cannot yet be justified by the existing evidence base. This dilemma immediately points to the politics of knowledge production or, better put, its nonproduction: How do we explain "undone science," as anthropologist of science David Hess has called it?[6] Under which social conditions do certain things remain unknown and certain topics unstudied? As historian of science Robert Proctor has argued, answering such questions demands that *epistemology* (the study of how we know) be supplemented by *agnotology* (the study of what we don't know and why we don't come to know it).[7] In this case, an agnotology of anal cancer points to the ideological linkages that have been forged in recent decades between a medically marginalized group, a socially invisible organ (the anus), and a socially stigmatized practice (anal sex). The result is to make anal cancer the "great undiscussable" in public discourse about HPV vaccines.

To advance this argument about the limits of citizenship and the nonproduction of knowledge, I explore the place of gay men's health within debates about HPV vaccines in the United States and the place of HPV vaccines within U.S. debates about gay men's health. I begin with a focus on the biomedical understanding of anal cancer, currently diagnosed in about 3,050 women and 2,020 men each year in the United States, resulting in about 680 deaths annually.[8] I trace the emergence of anal cancer as a gay health concern and its intertwining with the HIV/AIDS epidemic in the 1980s and 1990s. I then analyze the emergent and ongoing efforts to establish the anal Pap smear as a standard of care for populations at risk. Next, I consider the ways in which the advent of HPV vaccines has affected the discourse surrounding anal cancer, describe the status of vaccine research with men (including MSM), and analyze the obstacles that continue to stand in the way of attempts by gay health advocates to address anal cancer successfully. Finally, I examine the situation of adolescent gay (or "pregay") men as a limiting case in the struggle for biopolitical and sexual citizenship.[9] While these are the gay men who stand potentially to benefit most from an HPV vaccine, they are highly unlikely to receive it—unless, as may well eventually happen, HPV vaccination becomes standard for all young men, regardless of sexual identity.

Emergence

Before the 1980s, anal cancer had been recognized as a distinct health threat that affected a relatively small number of people, most of them older, and more often female than male. Looking back at the history from the vantage point of the year 2000, the authors of a review article in the *New England Journal of Medicine* observed: "Thirty years ago, anal cancer was believed to be caused by chronic, local inflammation of the perianal area and was treated with an abdominoperineal resection, necessitating a permanent colostomy." In the intervening decades, however, epidemiological and laboratory evidence had made clear that "anal cancer is associated with infection by human papillomavirus, which is usually sexually transmitted." (In this regard, anal cancer differs from the more familiar colorectal cancer, which is not linked to HPV infection.) Moreover, "in the majority of patients, the condition can be cured by concurrent chemotherapy and radiation therapy without the need for surgery."[10] This narrative of medical progress is complicated by the apparent rising incidence of the disease in recent decades, both overall and within distinct social groups.

The conception of anal cancer as a gay health threat dates to 1982, when an article in the *Journal of the American Medical Association* titled "Correlates of Homosexual Behavior and the Incidence of Anal Cancer" marked a turning point in medical discussion of that disease. The evidence was indirect and involved the analysis of public health databases: J. R. Daling and coauthors found that when they compared state cancer diagnoses from thirteen counties in western Washington state against the state syphilis registry, a much higher percentage of men with anal cancer also had syphilis, compared to men or women with other forms of cancer. In addition, this time using a national cancer database, the authors found that nearly a quarter of men with anal cancer had never married, compared to less than 8 percent of men with colorectal cancer.[11] Because, they argued, having been treated for syphilis and not having married are "correlates of homosexual behavior," and because "the majority of homosexual men practice anal intercourse," it seemed plausible to conclude "that anal intercourse is a risk factor for the development of anal cancer."[12]

Despite the limitations of the evidence, this article succeeded in creating a new conceptual framework for thinking about the issue of anal cancer—a framework that included a certain measure of confusion, which lingered, about how to define the population at risk. Daling and coauthors slid relatively seamlessly between discussions of "homosexual behavior" (or various presumed "correlates" of it), "homosexual men," and "homosexual men who practice anal intercourse." Was this, then, a "gay health threat"—one that adhered to "gay identity" and was the property of a "gay community"? Or was it a threat related to specific sexual practices, which might often be performed by men who did not adopt a gay identity and which might not be performed by some men who did identify as gay? In more recent years, and as a spillover effect of much discussion in AIDS-prevention circles, researchers concerned with anal cancer have tended to speak of "men who have sex with men" (MSM), rather than gay men, when describing those at risk.[13] Yet in part because gay-community spokespersons have had to mobilize to bring attention to the issue—in effect, to "claim" anal cancer—the disease has come to be viewed as a gay concern.[14]

Although the analysis by Daling and coauthors prefigured the present-day conception of anal cancer as a gay health threat, the article differed from subsequent research in an important respect: it preceded any recognition of a causal link between anal cancer and HPV. However, over the next decade, a small number of medical reports forged a durable set of cognitive links between anal cancer, receptive anal intercourse, HPV infection, anal intraepithelial neoplasia (AIN) (also known as anal dysplasia and considered a potential precursor of anal cancer), HIV seropositivity, and markers of immune suppression.[15] In this regard, the history of the conception of anal cancer as a gay health risk has been tightly intertwined with the history of the AIDS epidemic, which first emerged into public consciousness in 1981.

For example, in 1990, Joel Palefsky, who later became the foremost authority on HPV-related diseases of the anus, reported in the *Journal of the American Medical Association* on a study of ninety-seven gay men with AIDS. Currently a professor of medicine at the University of California, San Francisco (UCSF), Palefsky first stumbled upon the topic of HPV while completing an infectious-disease fellowship at Stanford. He became interested in whether the immunosuppression that characterized HIV/AIDS would translate into higher levels of anal HPV and anal warts, and once he joined the UCSF faculty, he began to focus on the HIV/HPV interaction as well as on anal HPV.[16] In the 1990 study, Palefsky reported that half of the gay men were found to be infected with HPV, and 15 percent of them had AIN. The men with cytological abnormalities had lower CD4 counts than those without abnormal findings, leading Palefsky and his colleagues to conclude: "Immunosuppressed male homosexuals have a high prevalence of anal human papillomavirus infection and anal intraepithelial neoplasia, and this population may be at significant risk for the development of anal cancer."[17]

Just how substantial was the threat of anal cancer, particularly for men who were *not* HIV positive? After all, most people who are anally infected with HPV "clear" the infection and do not develop major health problems, and anal cancer rates in the overall population are low compared to rates of other cancers.

Were anal cancer rates among gay men something to be worried about? Ironically, Daling and coauthors had concluded just the opposite. In the final paragraph, the authors observed that even if the link between anal sex and anal
cancer was borne out by subsequent research, "the practical consequences of
our findings . . . should not be great," because, even in gay men, "the incidence
of anal cancer is not high." Specifically, they estimated that the anal cancer risk
for homosexual men was 25 to 50 times that for heterosexual men. Based on the
assumption that homosexual men constituted between 1 and 5 percent of the
male population, they then estimated the annual incidence of anal cancer in
gay men as being between 12.5 and 36.9 per 100,000.[18]

But Palefsky had a different reading. Adopting the upper limit in Daling and
coauthors' estimate, Palefsky noted that if the incidence of anal cancer among
gay men was indeed about 35 per 100,000, then it was similar to the incidence
of cervical cancer among women in the United States before routine cytology
screening (the Pap smear) was introduced.[19] By implication, gay men merited
preventive screening on a par with what had become routine for U.S. women.
This was a powerful and dramatic way of framing a health disparity. And it has
been successful in the sense that the "35 per 100,000" figure now often appears
in medical and public health literature, often without attribution (and certainly
without indicating that this was the upper limit of a range estimated, in one
early study, from incomplete public health databases).[20] To be sure, all attempts
to quantify the health threat were bound to be somewhat problematic, not only
because of the confusion already described between behavior categories and
identity categories, but also because of the problems with determining the actual size of the group in question. Whether the group is taken to be gay men,
MSM, or "men who have receptive anal sex with men," it is difficult to provide
reliable incidence rates when the "denominator"—the total number of such men
in the overall population—is unknown.[21]

By 1997, Palefsky and other researchers, using sensitive polymerase chain
reaction testing, had found alarmingly high rates of anal HPV infection among
both HIV-negative and HIV-positive gay men (though much higher among
the latter). A prospective cohort study of anal HPV infection and anal dysplasia
in about six hundred gay men revealed that 60 percent of the HIV-negative
men and 93 percent of the HIV-positive men were anally infected with HPV.
Moreover, the most common of the many HPV subtypes detected was type 16,
a type strongly associated with the development of cancers.[22] With the advent
of highly active antiretroviral therapy, HIV-positive men began to experience

lower incidence rates of opportunistic infections as a result of improved immunological functioning. Yet by contrast, anal cancer rates were, if anything, on the increase in this group.[23]

Over time, it has become increasingly apparent that anal HPV infection is quite widespread in the population as a whole—indeed, in women, rates of anal HPV may be higher than rates of cervical HPV.[24] Moreover, anal HPV can be acquired relatively easily and very possibly without engaging in anal sex (a little-known fact that has failed to diminish the stigma surrounding the condition).[25] For example, recent evidence suggests that men with penile HPV infection and women with vulvar infection can spread HPV to their own anal regions by means of fingers or sex toys.[26] Fortunately, relatively few of the persons anally infected with HPV go on to develop cancer, though immunocompromised individuals are at particular risk of doing so. In addition, there is some evidence that rates of anal cancer may be higher, and prognoses poorer, among black gay and bisexual men.[27]

Gradually, as concern has grown about these trends, a small, international cluster of researchers on issues of anal HPV infection and anal cancer in both men and women has taken shape. "We're a very tight-knit group," commented Palefsky in 2009, adding, however, that interest had expanded beyond the original three loci of research—San Francisco, New York, and Seattle—enough that it had become possible to conduct multisite studies.[28] Palefsky now runs the Anal Neoplasia Clinic at UCSF (the first such clinic in the world), where four practitioners see patients, among whom men now outnumber women. According to Palefsky, "no matter how many people we add to our staff, we still can't keep up with the demand."[29]

Despite such developments, it must be said that anal cancer remains an unfamiliar illness to most people in the United States. While this may reflect its relatively low incidence in the overall population, I suggest that it also speaks to the stigma that surrounds anything associated with the anus in general and anal sex in particular. It is also striking that some of the most authoritative sites to which people might turn to learn about the disease stick to the "straight and narrow"—for example, the National Cancer Institute's Web page on anal cancer says nothing about gay men, MSM, or people with HIV infection.[30] In the sections that follow, I consider how gay health advocates have sought to change this situation and how the advent of Gardasil has brought new attention to the issue. But I also discuss why the efforts of these advocates have been, at best, only partially successful.

The "Tush Pap": Fighting to Make the Anal Pap Smear the "Right Tool" for the Job

Like many mundane medical technologies, the Pap smear is taken for granted nowadays as a crucial yet ordinary women's health screening tool. However, as Monica Casper and Adele Clarke have explained, it took the simultaneous effort of "different collective actors, both inside and outside the cancer arena, over the past half-century, to make the Pap smear 'work' as a screening procedure."[31] The Pap smear was not inherently the "right tool" for cervical cancer prevention. Rather, the Pap smear came to be seen as suitable as a consequence of compromise and tinkering.[32] Now a subsequent chapter of this history is being written, as physicians and activists argue for the appropriateness of the anal Pap smear—or what Palefsky says he and his team "affectionately call . . . 'tush paps' "[33]—as a tool to secure gay men's health.[34]

Much as the discovery of anal cancer as a gay health threat intersected with the emergence of HIV/AIDS, so the attempt to adopt and adapt the Pap smear has been promoted by gay health activism that has been shaped by the history of that epidemic. In the late 1990s, as the advent of highly active antiretroviral therapy caused HIV/AIDS to recede as a public health emergency for many gay men in the United States, a new, grassroots gay men's health movement emerged.[35] These activists drew on the lessons of previous gay and AIDS-related activism: the importance of defending sexual freedom, the refusal to defer to credentialed expertise, and the willingness to embrace aggressive responses to serious health threats proactively, even before all the evidence was on the table.[36] At the same time, they sought a comprehensive approach to the health issues affecting gay men.[37] The point was not that the HIV/AIDS epidemic had ceased to be a concern. Rather, as a leading activist, Eric Rofes, insisted, the goal was "to expand dramatically beyond a narrow focus on HIV and address the many, many health issues faced by gay men of all colors and all generations."[38] To promote a new vision of gay men's health, Rofes and other organizers convened a Gay Men's Health Summit that brought together more than three hundred individuals in Boulder, Colorado, in the summer of 1999 to discuss such topics as "health promotion in cyberspace," "drug use at circuit parties," "domestic violence as a gay men's health issue," and "anal sex, anorectal disorders, and STDs."[39] By the following year, attendance at the summit was up to 458 participants from thirty-seven states plus Australia, Canada, and Switzerland; several regional meetings on gay men's health were also held.

Within the array of health threats that seemed to demand attention, anal cancer was a timely issue. Not only did anal cancer once again raise the specter of a link between gay sex practices, a sexually transmitted virus, and a deadly disease, but also this disease deserved notice because it seemed to have been rendered invisible by heteronormative medical assumptions. Indeed, with regard to the use of Pap smears as a preventive tool, activists complained that such assumptions seemed to be hardwired into the very infrastructure of mainstream medical practice.[40] In his 1999 book *Smearing the Queer,* activist Michael Scarce described persuading his physician to order an anal Pap smear for him, only to find that the clinic's computer billing system couldn't process the procedure. Only after the information technology staff changed Scarce's sex online to "female" could the Pap smear request be entered into the system without triggering an error message.[41]

Scarce's physician was open-minded and was willing to be persuaded to order an anal Pap smear. But other physicians were significantly less so, or they insisted that they could not give such an order in the absence of hard evidence from clinical trials about the efficacy of Pap smears in preventing anal cancer. Richard Loftus, a physician and a former ACT UP activist, recalled at a public forum in San Francisco:

> I remember there was a big debate when I was a Fellow at the HIV Clinic at [San Francisco] General [Hospital] about whether we should be doing anal Pap smears, which *technically* we still haven't necessarily proven that they are of benefit in identifying anal cancers before they turn into cancers. . . . One of the doctors said, "I don't check the anal Paps. They're always positive!" . . . I notice that the people who consistently make these kinds of comments are heterosexual men. To a one. Not a single woman provider in that clinic would ever suggest that we not do anal Paps for the men. And I felt like saying at one point: "Dude, if it was your booty, you would have the Pap, even if the evidence base hasn't caught up with it just yet."[42]

Around the same time as gay men's health activists began seeking to change such perceptions on the part of physicians, researchers also began addressing the issue squarely in the pages of medical journals. From the standpoint of evidence, the problem was the lack of solid longitudinal data showing that performing anal Pap smears on MSM would be efficacious and cost effective in preventing anal cancer down the line. Working in conjunction with Palefsky, Sue Goldie and coauthors sought to sidestep this problem by simulating a longitudinal cohort study using cross-sectional data. In articles published in 1999 and 2000, the authors concluded that "in homosexual and bisexual men, screening every 2 or 3 years for anal squamous intraepithelial lesions with anal cytology would provide life-expectancy benefits comparable with other accepted preventive health measures, and would be cost-effective."[43] In an editorial commenting on this work, noted AIDS researcher Paul Volberding called it "an important, and large, first step," but he also observed: "A model of disease screening based on a hypothetic cohort is an important tool with which to begin, but it does not yet prove the point."[44]

Volberding was surely correct: by the present-day standards of evidence-based medicine, anal Pap smears hovered in the expansive gray area between "promising" and "proven."[45] This situation made it difficult to advocate such screening as a standard measure, given the resulting public health expenditures, even if the procedure was noninvasive and brought no risk to the patient. Yet, as advocates of more aggressive screening measures for anal cancer were at pains to point out, neither had the cervical Pap smear been proved efficacious at the time of its broad-scale introduction; and if the rigid rules of evidence-based medicine had been in place earlier in the twentieth century, many lives would have been lost to cervical cancer while evidence of efficacy gradually accumulated.[46] Nor did a strict assessment of the quality and quantity of the evidence on behalf of anal Pap smears consider the thorny "agnotological" issue of why certain topics receive scholarly attention and funding while others linger, unaddressed and out of the spotlight. With relatively few researchers invested in a health issue that was burdened with the metaphorical baggage associated with anal intercourse between men, it was hard to amass data of a sort that might move the condition more toward center stage. In this regard, gay men confronted not only the stigma surrounding anal sex but also an obstacle that other medically underserved groups have likewise run up against: biomedical neglect of a population makes it difficult to generate the evidence of disparity that is typically demanded to justify a campaign to reverse such neglect.[47]

Palefsky underscored the issue of social priorities in comments he made to the *New York Times* in 2003: "We've decided as a society that it's important to spend billions of dollars to test for and treat cervical dysplasia before it turns into cancer. . . . And we should also be testing for and treating anal dysplasia in high-risk populations."[48] Yet in his own professional role as drafter of the HPV screening guidelines for HIV-positive patients issued by the Centers for Disease Control and Prevention (CDC) in 2004, Palefsky could go no further than a roundabout recommendation: "Although formal guidelines recommending anal Pap smear screening have not been adopted, certain specialists recommend anal cytologic screening for HIV-1-infected men and women. High-resolution anoscopy (HRA) . . . should be performed if a person has [evidence of lesions] on anal Pap smear. Visible lesions should be biopsied to determine the level of histologic changes and to rule out invasive cancer."[49] Palefsky later commented, "I've been allowed to put more and more information over time into these guidelines, but of course I have to be explicit about the level of evidence that supports them. And to the degree that we don't have the 'A' level of evidence . . . there's limitations on how much we can say or do. . . . I find it personally frustrating, but I understand the rationale for that."[50] While some in the gay community have accused the CDC of homophobia in refusing to endorse anal Pap smears for MSM,[51] Palefsky disagrees, arguing that governmental agencies are simply being strict about the level of evidence that they require. Of course, as I have already suggested, a failure to invest in the accrual of evidence can itself be a manifestation of social bias. Hence one prominent, community-based gay health organization, San Francisco's STOP AIDS Project, has endorsed the value of "evidence-based standards of care" while arguing that "until standards of care are implemented, something needs to be done"— namely, to "support and fund research on anal-cancer prevention."[52]

Meanwhile, in the absence of clear guidelines, much depends on the familiarity of individual physicians with the issue of anal screening and their inclination to act aggressively with patients at higher risk. From Palefsky's perspective, the orientation of a physician "changes very dramatically" once the practitioner actually encounters a patient with anal cancer: "For many [physicians], it's invisible and they don't feel the need to screen because they've never seen a case. But when they have one, we start to get a big rush of referrals."[53] However, when physicians do become motivated to use anal Pap smears more aggressively, one result is the increased demand for anoscopy (to follow up on abnormal test results), the training for which is lacking in most of the United States. Said Palefsky: "If you compare the number of anoscopists to colposcopists, it's a . . . small fraction. In most of the major cities where there are gay concentrations of men, there is at least one, but certainly not more than a handful in any of those cities."[54] Thus, as Casper and Clarke also noted in relation to the historical stabilization of the cervical Pap smear, the anal Pap smear can become the "right tool for the job" only when an infrastructure is in place to support its use.[55]

Enter Gardasil

How has the development, licensing, and marketing of Gardasil affected the social and medical dynamics concerning anal HPV infection and anal cancer? For those who were already seeking to bring visibility to the threat of anal cancer for MSM, all the attention devoted to Gardasil has increased the opportunities to talk publicly about the relation between HPV and cancer, while also raising the prospect of an effective preventive measure, particularly for future generations of gay men. At the same time, gay male health concerns have played a submerged role in a larger story: public consideration of whether Gardasil should be endorsed for all boys as well as all girls.

As early as 2001, AIDS-treatment information newsletters began to speak of the potential benefits of a (then still hypothetical) vaccine against HPV.[56] Among gay health advocates, serious interest in Gardasil appears to date to about 2005, as news began to filter out about Merck's promising vaccine. HIV/AIDS groups were among the first to take note, because of the complex causal pathways between HIV and HPV-related diseases. As the New York City–based Community HIV/AIDS Mobilization Project (CHAMP) explained in its June 2005 newsletter, on one hand, immune-compromised men are at greater risk of contracting HPV, and HIV infection may result in more rapid progression of diseases associated with HPV. But on the other hand, active HPV infection is believed to make people more vulnerable to contracting HIV. Hence CHAMP's important question in the headline to the article: "HPV Vaccine on the Way—But for Whom?"[57]

That same month, the UK version of the popular Web site Gay.com also speculated that Merck's new vaccine might protect gay men from anal cancer.[58] By 2007, such speculation had gone mainstream. Although the vast majority of mass-media coverage of HPV and Gardasil has not even mentioned any forms of cancer other than cervical,[59] articles did appear in the early part of that year in both the *New York Times* and the *Los Angeles Times* that discussed gay men's interest in Gardasil in relation to the question of whether men in general should receive the vaccine.[60] According to the *Los Angeles Times* article, "high-risk men, such as gay and bisexual men, [were] reportedly requesting and receiving Gardasil vaccination from their physicians."[61] Once the vaccine received U.S. Food and Drug Administration (FDA) approval for administration to young women, nothing prevented physicians from giving it "off label" to such men. To be sure, insurers generally do not pay for off-label uses, which meant that gay men who sought this expensive vaccine were obliged to pay the full price for the three doses.[62]

These articles presented somewhat contradictory assessments of the uncertainties associated with extrapolating findings from women's bodies to men's. Whereas the *Los Angeles Times* quoted an HPV expert's claim that "there is 'no guarantee' an HPV vaccine will work in men . . . because the skin cells infected by the virus differ greatly in men and women," the *New York Times* cited Joel Palefsky: "The cervix is similar biologically to the anus, so there's plenty of hope that it will work there [in the anus] also."[63] Such concerns about generalizing across sex categories at least had a basis in biological arguments. By contrast, much of the discussion in the mass media seemed to position the issue as being whether it made sense for men to be using a vaccine that was "meant for women"—as if there were some deep gendered intentionality embedded in the medical technology itself, or perhaps as if male homosexuality was so inherently "feminine" that such appropriations might be deemed sensible.[64] At least some of the media representations also seemed to portray the story as little more than a quirky curiosity and to trivialize the concerns of gay men who, it was claimed, were saying, "Gardasil isn't just for girls anymore."[65]

Gay health advocates were well aware that the efficacy of Gardasil had not yet been determined for men and that neither the efficacy nor the safety had been investigated for people who are HIV-positive. Still, many gay men began to take an interest in the issue. In part responding to community demand for information about anal cancer and Gardasil, and in part seeking to foment such demand, advocates stepped up their efforts. In San Francisco, on April 9, 2007, the STOP AIDS Project held a public forum entitled "Hole Health: A Forum for Gay Men on HPV, the Vaccine, and Anal Cancer"; the event was recorded and subsequently posted on the Internet in the form of a podcast.[66] "I'm really, really happy that people are starting to take notice of this problem," said Joel Palefsky, one of the speakers, adding that he had been working on the issue for fifteen years.

Public attention to Gardasil made it easier to focus gay men's attention on the general issue of anal cancer, even if the vaccine itself would likely not be of tremendous practical value to any except future generations. "In the long term, this is the real answer," Palefsky told the audience. Gardasil "is not going to help anybody who's already got the virus. So for any of us, who've already had sex, it's too late. But in the long term, over the next 10 or 20 or 30 years, hopefully we can get rid of anal cancer altogether, if everybody who needs this vaccine gets this vaccine."[67]

Palefsky was quick to remind the audience that the efficacy of Gardasil in preventing anal cancer was unproven. However, Richard Loftus, the physician and former ACT UP activist, was less circumspect when the time came for his presentation at the forum. Loftus, much like AIDS-treatment activists speaking almost two decades earlier about antiretroviral drugs, was prepared to accept uncertainty in the hope of saving lives. "I may be running out ahead of the data a little bit," he acknowledged: "Maybe not everybody needs Gardasil, and we certainly need to figure out what benefits are to be had from Gardasil, especially for gay men. But I think that the research, as Joel said, looks pretty promising—that there probably is going to be a benefit for some number of men who have sex with men—and I think we need to be supportive of the research efforts to prove that, but I also think that on an individual level a lot of people should be asking themselves, 'Huh, should I be getting Gardasil?'"[68]

As Loftus's comments indicate, one goal of the forum was to bring the community up to date about the status of ongoing clinical research on Gardasil in men. This was where the "undiscussable" story of anal HPV and anal cancer met up with, or was folded into, the much more publicly visible debate concerning the administration of HPV vaccines to boys. Public health experts had long understood that the fastest route to the eradication of HPV infection at the population level would be universal vaccination. By this logic, boys should be vaccinated along with girls, even if boys stand little to gain directly, because vaccinating boys will benefit girls through the creation of "herd immunity." However, the dilemma faced by Merck (the maker of Gardasil) and GlaxoSmithKline (the maker of Cervarix)—each of whom presumably would be happy, in principle, to double the size of their potential market—is that it would be a hard sell to convince parents that boys should be vaccinated with a "cervical cancer vaccine" when the boys were not themselves at risk of that disease. The companies could have pointed to the threat of penile cancer, which is also linked to HPV, but the incidence rates simply seemed too low to justify mass inoculation.[69] Nor was either company about to go anywhere near the loaded topic of anal cancer: not only is the condition too stigmatized, but even mentioning it would immediately sexualize (and homosexualize) the vaccines through the association with anal sex, whereas the key marketing strategy has been to forestall moral panic precisely by desexualizing the vaccines and marketing them purely as vaccines against cervical cancer (see chapter 12). In confronting this dilemma, Merck and GlaxoSmithKline took divergent paths.

Glaxo's solution (one reflected quite explicitly in the company's choice of a name for its vaccine, Cervarix) was to forgo the potential male market for the vaccine, at least initially. The company has not conducted any clinical trials in men (except to study the possibility of creating herd immunity that would benefit girls).[70] Indeed, Glaxo's stance prompted Richard Loftus, at the "Hole Health" forum, to half-seriously float the idea of "boycotting Glaxo for not supporting men's health by investigating their vaccine for our community."[71] By contrast, Merck sought a creative way to rope in young men. In addition to the two HPV viral types that are associated with cancer—targeted by both Cervarix and Gardasil—Merck built into Gardasil protection against two additional types that are associated with genital warts.[72] "We have a very clear benefit that we offer to men," Merck's executive director of clinical research told the *New York Times;* boys therefore don't need to perceive "an altruistic reason to get the vaccine."[73]

Merck's explicit attention to boys has double-edged implications for discussions of gay men's health. On one hand, there is the very real risk that the former will simply drown out the latter. This risk was well suggested by a 2008 *New York Times* article by Jan Hoffman that appeared (remarkably enough) on the front page of the Styles section in the *New York Times.* The article asked whether there would soon be a "Gardasil Boy" to accompany the "Gardasil Girl": "How cool are those Gardasil Girls? Riding horses, flinging softballs, bashing away on drum sets: on the television commercials, they are pugnacious and utterly winning. . . . But someone's missing from this grrlpower tableau. . . . Ah, that would be Gardasil Boy."[74] The article went on to consider the pros and cons of vaccinating boys, but without any discussion of whether the boys might be, or might grow up to be, gay. The issue of anal cancer was ignored in the article except for a brief aside in the twenty-seventh paragraph (out of thirty-four).

On the other hand, as other media coverage in the *Times* and elsewhere exemplifies,[75] interest in the question of whether boys should be vaccinated can have the effect of opening up space for the discussion of anal cancer as a substantial health risk for MSM. For example, the CDC has created a Web page titled "HPV and Men" and also distributes a fact sheet on the topic, dated December 2007. While these CDC resources primarily interpellate heterosexual men, they do nonetheless discuss the causal link between HPV and anal cancer, and they present the statistical claim that "gay and bisexual men are 17 times more likely to develop anal cancer than heterosexual men."[76]

This paradoxical pattern—in which attention to (presumptively heterosexual) boys may either obscure or shed light on the situation of MSM—is equally well represented in Merck's ongoing clinical research on Gardasil in males. With the goal of expanding its FDA labeling, Merck launched a large, randomized controlled trial of Gardasil in four thousand healthy men, ages 16 to 26. The primary endpoint in the trial is genital warts. However, very much below the radar screen, the company included within the larger study population a subset of six hundred young MSM, in order to assess the vaccine's efficacy in preventing AIN.[77] Palefsky credits Eliav Barr, the head of Merck's HPV vaccine program, for being forward-thinking on this issue: "He's a very smart fellow and also very sympathetic. And in early discussions I had with him, we decided together to try and convince Merck management to support an MSM study. And to their credit they agreed, because one can imagine all kinds of reasons why they might not want to."[78] In advertising that appeared online and in the gay press, anyone who was a "sexually active, healthy young man who has sex with men" and who was willing to make a three-year commitment to an HPV study was directed to a Web site, which provided six study locations in Boston, Chicago, New York, Philadelphia, San Francisco, and Washington, D.C.[79] These included prominent gay men's health centers, such as Fenway Community Health in Boston, the Howard Brown Health Center in Chicago, and the Whitman-Walker Clinic in Washington, D.C.

It was an open question whether investigators would meet the practical challenges involved in rounding up such a cohort. Given how ubiquitous HPV infection is among MSM, how does one locate and recruit sufficient numbers of 16–26-year-olds who understand themselves to be gay or MSM but who have not already had enough sex to thereby be infected with HPV? And even if one can find them, is it possible to overcome barriers to their participation in research studies? Loftus observed at the San Francisco community forum: "We really need to support the men's research study. And one of the big obstacles is, how the hell do you reach gay teenagers who may not even be 'out' to their family and friends. You come home [claps], 'Hi, Mom and Dad, hope you don't mind, I have a lot of anal sex and I need you to sign this permission slip so I can be in this clinical trial 'cause I'm under the age of 18.' It doesn't happen that much."[80] Despite these considerable obstacles, Merck eventually succeeded in reaching its recruitment target for the MSM subset, though to do so the company had to line up additional research sites outside the United States (including Germany and Brazil).[81]

However, little mention was made of the MSM subset when the news media reported on the first efficacy data from the men's study, released by Merck in

November 2008. The data indicated that the vaccine was 90 percent efficacious in preventing genital warts—an excellent result and sufficient for Merck to move forward the following month with its request that the FDA license the vaccine for use in males.[82] On October 16, 2009, the FDA acted on this request and approved the use of Gardasil for the prevention of genital warts due to HPV in boys and men, ages 9 through 26.[83] A few days later, the CDC's Advisory Committee on Immunization Practices (ACIP) declined to recommend the vaccine for standard use in boys as part of the approved vaccine schedule. Instead, ACIP left the decision about administration to boys to the discretion of physicians.[84] However, according to Palefsky, a lead investigator for the men's study, it will be some time before data are available from the MSM subset and anything definitive can be said about the vaccine's ability to prevent AIN.[85] Thus at least some gay men will now have greater access to the vaccine but without the certainty that, by obtaining it, they will significantly cut their risk of anal cancer.

Stigma, Shame, and the Politics of Gender Difference

To expose one's penis is a shameful act, but it is also a glorious one, inasmuch as it displays some connection with the Great Social Phallus. Every man possesses a phallus which guarantees him a social role; every man has an anus which is truly his own, in the most secret depths of his own person. The anus does not exist in a social relation . . . one does not shit in company.

GUY HOCQUENGHEM, *Homosexual Desire* ([1972] 1993), 97.

Despite the efforts of health advocates and researchers to inform gay men and MSM about the risk of anal cancer, despite a certain measure of media coverage of gay men's interest in Gardasil, and despite Merck's inclusion of an MSM subgroup in its study of Gardasil in men, it is noteworthy how invisible these issues are and how much general ignorance surrounds them. Anal cancer in general remains a topic that few people know much about, and even gay men who are closely integrated into gay communities are not well informed about the HPV–anal cancer link.[86] While certainly there are many topics that remain socially invisible, anal cancer seems to be one where lack of knowledge is compounded by the workings of stigma. Just as the anus is a deeply private and taboo region, associated with waste and filth, so anal cancer is a taboo disease about which one does not speak.[87] This association appears true for women, who suffer from anal cancer in higher numbers than do men: breast cancer prompts walkathons; anal cancer is shrouded in silence.[88] Reports of the actress Farrah Fawcett's struggle against anal cancer certainly raised the profile of the disease slightly, but neither Fawcett's eventual openness about her condition, nor her death in June 2009, prompted significant public discussion of the specific kind of cancer that claimed her life (and still less of HPV as a causative agent).[89] Stigma clings all the more to gay men with the disease—yet in discussion groups about HPV, women who engage in anal sex report that the "gay" connotations of the practice make them feel that much more "queered."[90]

Authorities and activists have sought to normalize anal cancer and to destigmatize its associated organ. At the aptly named "Hole Health" forum, UCSF researcher Palefsky presented his first slide and said, "This is the anus. You all have one."[91] His tone was matter-of-fact, but his point was clear: if we want to save lives, we need to bring the asshole into the light of day. Yet many gay men who suffer from anal cancer have described how isolated they feel with the condition and how much shame appears to cling to it.[92] One reader of an article about Gardasil that was reprinted on the popular Web site Gay.com added his own blog comments to describe his experiences with anal cancer: "This seem to be a very hidden secret in the gay community. At times I feel that I am alone with this. It's almost like having HIV again. The stigma is similar. I know that there are many gay men suffering with this and it shouldn't be kept a secret. We need to demand that this vaccine be available to the many men that could benefit from this."[93] Similarly, Richard Loftus commented at the "Hole Health" forum:

> There's this old-fashioned term for gay men called "Friends of Dorothy." I almost think we should have a term for gay men in San Francisco called "Friends of Naomi," because Naomi Jay is a member of the Dysplasia Center, and it seems like a large number of my social network actually know Naomi, although no one actually mentions that they know her, 'cause nobody likes to talk about the fact that they're going to a dysplasia clinic for their HPV. I can't tell you how many patients I see during the day who think they're the only gay man in the world who has HPV, and I have to tell them, "Oh no, honey, let me tell you, you know, American men have an 80 percent lifetime risk of getting an HPV infection sooner or later. . . ." And I have to tell them again and again, this is extremely common, there's nothing to be embarrassed about, there's nothing to be ashamed about, probably if you asked all your friends, you'd find out they are all also

friends of Naomi. [Laughter] Lots of people go to the Dysplasia Center for their HPV. They just don't talk about it at cocktail hour.[94]

While Loftus sought to end isolation and construct a new sense of community in relation to the health threat posed by HPV, many sufferers seem to find themselves more in the position of the Gay.com blogger, whose comments suggest how the existence of "hidden secrets" can create new divides within a community. The blogger's comparison of anal cancer at the present moment to HIV infection at an earlier moment in that epidemic's history is revealing. In both cases, it has been hard for many of those gay men who have become ill to escape a pervasive sense that there is something shameful about their having been stricken by an illness. This sense of shame well exceeds any that might cling to other sexually transmitted infections such as gonorrhea or syphilis.

Perhaps anal cancer has simply inherited the historical residue of AIDS stigma, which was propagated so effectively by antigay spokespersons in the 1980s. (At the time, some even sought the public shaming of the HIV-infected by making stigma visible on the body, as when William F. Buckley Jr. proposed, in a *New York Times* op-ed, that "everyone detected with AIDS should be tattooed in the upper forearm to protect common-needle users, and on the buttocks, to prevent the victimization of other homosexuals.")[95] Perhaps, with anal cancer as with AIDS before it, many gay men confront or fear an intolerable judgmentalism (or plain, old-fashioned homophobia) from the doctors and health professionals whom they consult[96]—a concern that Palefsky and his colleagues have tried to address, in part, by designing an anal Pap smear that can be self-administered.[97] Or perhaps, in regard to both HIV/AIDS and anal cancer, shame is accentuated by the association of disease risk not just with the anus and with gay sex but, even more specifically, with the position of the "bottom" in anal sex. (Never mind that "tops" may in fact also be at high risk of HPV infection.) Gay men are by no means immune to the effects of the homophobic and, at root, misogynistic cultural discourses that construe the insertive partner in anal sex as more manly and more socially acceptable and the receptive partner as more effeminate and aberrant.[98]

If it is true that when a gay man "comes out" as having anal cancer he accepts a sort of pejorative feminization, then this sheds new light on the cultural meanings that adhere to his adopting medical interventions, such as Pap smears and HPV vaccines, that have become coded as female. On one hand, simply adopting such technologies in the hope of preventing anal cancer is to further mark oneself as feminized. On the other hand, ironically, to the extent that gay men are understood as being "like women," it may be easier for them to argue that "women's health" interventions should be extended to them as well. In this regard, the discussions in the mass media about whether the cells of the cervix and the anus are sufficiently similar for Gardasil to prevent HPV infection in both places may have as much to do with the cultural meanings of gender and sexual identity as with cell biology.

Conclusion: The Outer Limits of Citizenship

Since 1982, when the link was first made between anal cancer and gay men's health, a confluence of scientific, political, and cultural developments have combined to make this health issue salient in certain circles.[99] Yet anal cancer remains relatively invisible in general, while the link to gay men remains undiscussable in the broader public sphere. The advent of Gardasil has sharpened this dilemma, both by raising the prospect of an effective preventive measure (particularly for future generations of gay men) and by vastly increasing the amount of public discourse about HPV infection. Much more is now known about anal HPV infection and anal cancer, thanks to the concerted efforts of a few, and public interest in the vaccine has made it somewhat easier to talk about matters such as anal Pap smears (which may be of more practical value than the vaccine to men who have already been sexually active with other men). Yet an agnotological analysis reveals the knowledge gaps that result from inadequate biomedical attention compounded by stigma, shame, and invisibility. As a consequence, in relation to HPV, gay men remain incomplete biomedical citizens—benefiting, to some degree, from publicly funded research and from services provided at public clinics, but achieving only partial success in their demands that their health issues be placed on a par with those of other groups confronting comparable health threats.[100]

Visibility is not a simple either-or. In relation to any health threat, there are likely to be complex hierarchies of visibility that may line up, however imperfectly, with hierarchies of honor and shame, as well as with degrees of incorporation of social groups into a health polity. Certainly, gay men in the United States may face serious obstacles in calling attention to their health needs. But anal cancer among that group is still relatively less undiscussable than, say, anal cancer in prison, or anal cancer in the global South.[101] Nor are "gay men" or "MSM" wholly undifferentiated categories of men who all confront obstacles to full biocitizenship on an equal footing. Indeed, as I mentioned, there is some evidence to suggest that rates of anal cancer are rising most rapidly among (and prognoses are worse for) black gay and bisexual men—yet these men are likely to be medically underserved, are less effectively targeted by gay advocacy projects, and appear to be underrepresented in the MSM subset of Merck's clinical trial of Gardasil in men.[102]

In considering such hierarchies, it is worth examining the particular plight of gay adolescents (or adolescent MSM) who are not yet (very) sexually active. These young men are positioned right at the boundaries of sexual and biopolitical citizenship.[103] And, among MSM, they are precisely the ones who stand to benefit the most from vaccination, assuming its efficacy in preventing anal cancer. As Richard Loftus put it, perhaps exaggerating somewhat: "For our generation, you know, the horse is out of the barn. The vaccine isn't going to be an option . . . to protect us."[104] Yet, as the earlier discussion of the difficulties involved in recruiting participants for the clinical trial suggested, attempts to reach out to young men raise a whole new set of problems. In the words of one researcher in the United Kingdom: "It is bad enough suggesting to people that their 12-year-old daughter might need a vaccine against a sexually transmitted infection. . . . I would be interested to see the response of suggesting to parents that they should vaccinate their boys at 12 in case they become gay."[105]

It seems safe to say that this is an unlikely scenario.[106] No one is about to advocate a targeted effort to vaccinate "pre-gay" boys. A more plausible scenario is that these young men will find themselves vaccinated along the way, as part of a broad-scale effort to vaccinate all boys. While neither the promotion of herd immunity nor the prevention of genital warts may prove sufficient motivators to provoke such an effort, the rationale for vaccinating boys took a significant leap forward in 2007, when the risk of oropharyngeal cancer linked to HPV transmission during oral sex (heterosexual or otherwise) became front-page news.[107] *Time* magazine quoted Johns Hopkins researcher Maura Gillison, who has been studying the HPV-oral-sex link for some time: "When you look at the cancers associated with HPV in men—including penile cancer, anal squamous cell carcinoma, oral cancers—it's very close to the number of cases of cervical cancer that occur in the U.S. in women every year. We need to adjust the public's perception . . . that only women are at risk."[108] Whether vaccinating all boys would actually prove to be an advisable and cost-effective public health intervention remains an open question on the basis of current evidence, and so far the CDC has been equivocal about endorsing such action.[109] Yet, *if* such a policy is adopted, and *if* parents agree that their boys should be vaccinated, and *if* the vaccine proves efficacious in preventing a range of cancers, then future generations of gay men may find their interests addressed in part, even while the anal sex–anal cancer link may remain a publicly undiscussable issue.

ACKNOWLEDGMENTS

I am grateful to Carlos Decena, April Huff, Héctor Carrillo, Keith Wailoo, Julie Livingston, Robert Aronowitz, Nayan Shah, David Serlin, S. Lochlain Jain, and the participants in the "Cancer Vaccines for Girls?" conferences held at Rutgers University for helpful suggestions on earlier drafts. I also want to thank audiences at the "Dialogues in Sexuality Studies" seminar (at the University of California, San Diego), the "Queer Studies Conference" (at the University of California, Los Angeles), Stanford University's Center for Advanced Study in the Behavioral Sciences, and the "Rethinking Sex" conference at the University of Pennsylvania for their comments. This work was supported by the Department of Sociology and the Academic Senate Committee on Research at the University of California, San Diego. A residency fellowship at Stanford University's Center for Advanced Study in the Behavioral Sciences permitted me to complete this work.

NOTES

Epigraph. *Gay and Bi-Sexual Men Taking Gardasil,* a video clip of a television news segment, CBS-5, San Francisco, ca. April 5, 2007, posted on YouTube by *vidwatch654,* www .youtube.com/watch?v=WSEfPYzOBd8 (accessed May 8, 2008).

1. As I explain later in this chapter, public discussions of the health issues I consider here have been systematically unclear about whether the referent is "gay men" (a category of collective identity) or "MSM" (a category based on behavior, independent of identity). This lack of clarity is by no means unusual; see Rebecca M. Young and Ilan H. Meyer, "The Trouble with 'MSM' and 'WSW': Erasure of the Sexual-Minority Person in Public Health Discourse," *American Journal of Public Health* 95, no. 7 (2005): 1144–1149. In this chapter I do not attempt to resolve such confusion. Instead, I simply employ the terminology used by actors whom I study.

2. Charlotte J. Haug, "Human Papillomavirus Vaccination—Reasons for Caution," *New England Journal of Medicine* 359, no. 8 (2008): 861–862; Elizabeth Rosenthal, "The Evidence Gap: Drug Makers' Push Leads to Cancer Vaccines' Rise," *New York Times,* Aug. 19, 2008; chapters 1 and 9, this volume.

3. Chapters 5 and 13; TAC Issues Global Call for Affordable Access to HPV Vaccines in the Developing World (Web page), Oct. 28, 2008, Treatment Action Campaign, www.tac.org.za/community/node/2428 (accessed Nov. 11, 2008).

4. I should be clear that my goal is not to juxtapose issues of access to Gardasil by *women* in the global South with those of access by *MSM* in the global North in a way that would occlude attention to the health needs of MSM in the global South. The South African HIV/AIDS treatment advocacy group Treatment Action Campaign connected these issues explicitly in its 2008 call for access to HPV vaccines—emphasizing the problem of cervical cancer but also calling for clinical trials in men, with a focus on MSM. See TAC Issues Global Call Web page.

5. In other words, I see this study as involving the politics of citizenship, broadly construed, and the politics of biopolitical and sexual citizenship in particular. I understand *citizenship* to refer to differentiated modes of incorporation of individuals or groups fully or partially into the national polity through the articulation of notions of rights and responsibilities. Stuart Hall and David Held, "Citizens and Citizenship," in *New Times: The Changing Face of Politics in the 1990s*, ed. Stuart Hall and Martin Jacques (London: Verso, 1990), 173–188; Gershon Shafir, ed., *The Citizenship Debates: A Reader* (Minneapolis: University of Minnesota Press, 1998). In a polity where health disparities are rampant, and where the biosciences have become influential in defining what is taken to be the essential natures of individuals and groups, biomedical authorities and technologies have come to play consequential roles in reproducing or challenging practices of social stratification and exclusion. That is, the struggles around what I have called *biopolitical citizenship* reflect not only that health and health care have been made into scarce resources that are inaccessible to many, but also that biomedical debates may be central to the processes by which lines of social cleavage are drawn. Steven Epstein, *Inclusion: The Politics of Difference in Medical Research* (Chicago: University of Chicago Press, 2007), 20–21, 277–302. In recent years, scholars also have become increasingly cognizant of the centrality of sexuality to the modern politics of belonging, and the concept of *sexual citizenship* has been put forward to describe both how sexual rights are pursued or denied and how distinct "sexual minorities" demand equal standing as political actors. For one of many examples, see Diane Richardson, "Sexuality and Citizenship," *Sociology* 32, no. 1 (1998): 83–100. On the specific dilemmas that confront gay health activists as they seek biopolitical and sexual citizenship, see also Steven Epstein, *Impure Science: AIDS, Activism, and the Politics of Knowledge* (Berkeley: University of California Press, 1996); Steven Epstein, "Sexualizing Governance and Medicalizing Identities: The Emergence of 'State-Centered' LGBT Health Politics in the United States," *Sexualities* 6, no. 2 (2003): 131–171.

6. David Hess, *Alternative Pathways in Science and Industry: Activism, Innovation, and the Environment in an Era of Globalization* (Cambridge, MA: MIT Press, 2007).

7. Robert N. Proctor, *Cancer Wars: How Politics Shapes What We Know and Don't Know about Cancer* (New York: Basic Books, 1995), 8.

8. Anal Cancer (Web page), 2008, National Cancer Institute, National Institutes of Health, www.cancer.gov/cancertopics/types/anal/ (accessed Sept. 15, 2008).

9. On biopolitical and sexual citizenship, see note 5.

10. David P. Ryan, Carolyn C. Compton, and Robert J. Mayer, "Carcinoma of the Anal Canal," *New England Journal of Medicine* 342, no. 11 (2000): 792.

11. J. R. Daling, N. S. Weiss, L. L. Klopfenstein, L. E. Cochran, W. H. Chow, and R. Daifuku, "Correlates of Homosexual Behavior and the Incidence of Anal Cancer," *Journal of the American Medical Association* 247, no. 14 (April 9, 1982): 1988–1989. The article began by noting "recent reports . . . that male homosexual behavior predisposes" to a variety of sexually transmitted infections and diseases (1988). On the emergence of biomedical concern with sexually transmitted infections among gay men in the late 1970s and early 1980s (before or coincident with the public emergence of the AIDS epidemic)—and how the sometimes stigmatizing forms that these medical claims took influenced initial conceptualizations of AIDS—see Epstein, *Impure Science*, 50–52.

12. Daling et al., "Correlates of Homosexual Behavior," 1990.

13. Young and Meyer, "Trouble with 'MSM' and 'WSW.'"

14. On the implications of the identity-practice distinction in gay men's health, see Epstein, "Sexualizing Governance," 158–160.

15. T. Croxson, A. B. Chabon, E. Rorat, and I. M. Barash, "Intraepithelial Carcinoma of the Anus in Homosexual Men," *Diseases of the Colon and Rectum* 27, no. 5 (1984): 325–330; I. H. Frazer, G. Medley, R. M. Crapper, T. C. Brown, and I. R. Mackay, "Association between Anorectal Dysplasia, Human Papillomavirus, and Human Immunodeficiency Virus Infection in Homosexual Men," *Lancet*, Sept. 20, 1986, 657–660.

16. Joel Palefsky, professor of medicine at UCSF, interview by the author, San Francisco, Dec. 10, 2008.

17. J. M. Palefsky, J. Gonzales, R. M. Greenblatt, D. K. Ahn, and H. Hollander, "Anal Intraepithelial Neoplasia and Anal Papillomavirus Infection among Homosexual Males with Group IV HIV Disease," *Journal of the American Medical Association* 263, no. 21 (June 6, 1990): 2911. The CD4 count is a measure of immune function that reflects the progression of HIV disease.

18. Daling et al., "Correlates of Homosexual Behavior," 1990.

19. J. M. Palefsky, E. A. Holly, M. L. Ralston, and N. Jay, "Anal Human Papillomavirus Infection in HIV Positive and HIV Negative Men (Conference News)," *AIDS Weekly Plus*, Dec. 8, 1997, 30. Palefsky indicated later that he was the one who popularized this statistical claim. Palefsky interview. Elsewhere Palefsky has noted that the incidence of cervical cancer in U.S. women before the widespread use of Pap smears was 40 to 50 per 100,000 and that the introduction of Pap smears has brought the figure down to 8 per 100,000. Joel Palefsky, *Human Papillomavirus Vaccines and HPV-Associated Anogenital Disease in HIV-Seropositive Men and Women: A New Era?* (Webcast), May 7, 2007, International AIDS Society Advanced CME course on Improving the Management of HIV Disease, http://69.9.172.142/Mediasite/Viewer/?cid=b6c27abe-ac90-494f-94da-bba6bbcd7367 (accessed Sept. 12, 2008).

20. On the "black-boxing" of statistical claims about gay populations, see also Tom Waidzunas, "Gay Youth Suicide and Dynamic Nominalism: The Troubled Project of Defining a Social Problem with Statistics," 2008, Department of Sociology, University of California, San Diego.

21. Some of the best incidence data come from long-running cohort studies of HIV-positive gay men. Palefsky interview.

22. Palefsky et al., "Anal Human Papillomavirus Infection," 30.

23. Alan McCord, "Risk for Invasive Anal Cancer High among HIV-Positive People" (Web news report), Feb. 11, 2009, Project Inform, www.projectinform.org/news/09_croi/021109.shtml (accessed March 13, 2009).

24. Palefsky, *Human Papillomavirus Vaccines*.

25. C. Piketty, T. M. Darragh, M. Da Costa, P. Bruneval, I. Heard, M. D. Kazatchkine, and J. M. Palefsky, "High Prevalence of Anal Human Papillomavirus Infection and Anal Cancer Precursors among HIV-Infected Persons in the Absence of Anal Intercourse," *Annals of Internal Medicine* 138, no. 6 (2003): 453–459.

26. Alan Nyitray, Carrie M. Nielson, Robin B. Harris, Roberto Flores, Martha Abrahamsen, Eileen F. Dunne, and Anna R. Giuliano, "Prevalence of and Risk Factors for Anal Human Papillomavirus Infection in Heterosexual Men," *Journal of Infectious Diseases* 197, no. 12 (2008): 1676–1684; Palefsky, *Human Papillomavirus Vaccines*; Joel M. Palefsky, "HPV Infection in Men," *Disease Markers* 23, no. 4 (2007): 261–272. See also the data provided on UCSF's Anal Neoplasia Clinic Web site, www.analcancerinfo.ucsf.edu/.

27. Joel Palefsky, Jen Hecht, Jason Riggs, and Michael Scarce, "Needed: Routine HPV Vaccines and Pap Smears for Gay and Bisexual Men," *San Francisco Chronicle*, April 24, 2007, B-7.

28. Palefsky interview.

29. Ibid.

30. See National Cancer Institute, www.cancer.gov/cancertopics/types/anal (accessed March 20, 2009).

31. Monica J. Casper and Adele E. Clarke, "Making the Pap Smear into the 'Right Tool' for the Job: Cervical Cancer Screening in the USA, circa 1940–95," *Social Studies of Science* 28, no. 2 (1998): 256.

32. Ibid., 257–258. On the introduction of cervical Pap smears, see also chapter 15.

33. Palefsky, *Human Papillomavirus Vaccines*.

34. The idea of performing a Pap smear of the anus may be seen as an example of how the users of technologies may often adopt and adapt them in ways that designers did not intend. See Nelly Oudshoorn and Trevor Pinch, eds., *How Users Matter: The Co-Construction of Users and Technology* (Cambridge, MA: MIT Press, 2003). The case also raises questions about the extension of a technology across gender categories. For another example, see Nelly Oudshoorn, *The Male Pill: A Biography of a Technology in the Making* (Durham, NC: Duke University Press, 2003).

35. Epstein, "Sexualizing Governance," 140.

36. Epstein, *Impure Science*.

37. Kenneth Mayer, "Beyond Boulder: A Glance Back and the Road Ahead," *Journal of the Gay and Lesbian Medical Association* 4, no. 1 (2000): 1–2; Eric Rofes, "Resuscitating the Body Politic: Creating a Gay Men's Health Movement," *Baltimore Alternative*, Aug. 8, 2000, 21; Eric Rofes, "Why Boulder? Why Gay Men's Health? Why Now?" opening plenary at Gay Men's Health Summit, Boulder, CO, July 29 to Aug. 1, 1999, Boulder County AIDS Project, www.bcap.org/html/pages/education/healthsummit/hs1999/keynote.html (accessed June 1, 2000); Michael Scarce, "The Second Wave of the Gay Men's Health Movement: Medicalization and Cooptation as Pitfalls of Progress," editorial, *Journal of the Gay and Lesbian Medical Association* 4, no. 1 (2000): 3–4; Michael Scarce, *Smearing the Queer: Medical Bias in the Health Care of Gay Men* (New York: Harrington Park, 1999).

38. Eric Rofes, "What Is a Healthy Gay Man?" opening plenary address, presented at the Gay Men's Health Summit, Boulder, CO, July 19–23, 2000.

39. The Boulder Summit—July 29-August 1, 1999: Launching a Multi-Issue, Multicultural Gay Men's Health Movement, www.managingdesire.org/boulder.html.

40. On the durability of infrastructures and the consequent, potentially harmful effects on the individuals who run up against them, see Geoffrey C. Bowker and Susan Leigh Star, *Sorting Things Out: Classification and Its Consequences* (Cambridge, MA: MIT Press, 1999).

41. Scarce, *Smearing the Queer*, 116.

42. *Hole Health: A Forum for Gay Men on HPV, the Vaccine, and Anal Cancer* (podcast), April 9, 2007, STOP AIDS Project, www.archive.org/details/STOP_AIDS_Project_HPV_podcast (accessed May 7, 2008).

43. Sue J. Goldie, Karen M. Kuntz, Milton C. Weinstein, Kenneth A. Freedberg, and Joel M. Palefsky, "Cost-Effectiveness of Screening for Anal Squamous Intraepithelial Lesions and Anal Cancer in Human Immunodeficiency Virus–Negative Homosexual and Bisexual Men," *American Journal of Medicine* 108, no. 8 (2000): 644. See also Sue J. Goldie, Karen M. Kuntz, Milton C. Weinstein, Kenneth A. Freedberg, Mark L. Welton, and Joel M. Palefsky, "The Clinical Effectiveness and Cost-Effectiveness of Screening for Anal Squamous Intraepithelial Lesions in Homosexual and Bisexual HIV-Positive Men," *Journal of the American Medical Association* 281, no. 19 (1999): 1822–1829.

44. Paul Volberding, "Looking Behind: Time for Anal Cancer Screening," *American Journal of Medicine* 108, no. 8 (2000): 674–675.

45. On the rise to prominence of evidence-based medicine as a medical movement, see Stefan Timmermans and Marc Berg, *The Gold Standard: The Challenge of Evidence-Based Medicine and Standardization in Health Care* (Philadelphia: Temple University Press, 2003).

46. Palefsky interview.

47. Epstein, *Inclusion*, 53–73.

48. David Tuller, "Some Urge Type of Pap Test to Find Cancer in Gay Men," *New York Times*, Feb. 18, 2003.

49. Constance A. Benson, Jonathan E. Kaplan, Henry Masur, Alice Pau, and King K. Holmes, "Treating Opportunistic Infections among HIV-Infected Adults and Adolescents," *Morbidity and Mortality Weekly Report* 53 (2004): 1–112, available at www.cdc.gov/mmwr/preview/mmwrhtml/rr5315a1.htm; Palefsky interview. By contrast, the New York State Department of Health AIDS Institute now recommends anal Pap smears annually for MSM, any patient with a history of anal warts, and women with abnormal cervical or vulvar histology. *Human Papillomavirus (HPV)* (Web document), Sept. 2007, Office of the Medical Director, New York State Department of Health AIDS Institute, http://std.about.com/gi/dynamic/offsite.htm?zi=1/XJ&sdn=std&cdn=health&tm=39&gps=29_732_1020_588&f=00&su=p284.9.336.ip_p736.8.336.ip_&tt=2&bt=1&bts=1&zu=http%3A//www.hivguidelines.org/GuideLine.aspx%3FguideLineID%3D102 (accessed May 11, 2008).

50. Palefsky interview.

51. Tim Horn, "Covering Our Asses: A Healthcare Provider Questions CDC HPV Information" (Web news), May 10, 2006, POZ, www.poz.com/articles/761_11149.html (accessed Sept. 16, 2008).

52. Palefsky et al., "Needed," B-7.

53. Palefsky interview.

54. Ibid.

55. Casper and Clarke, "Making the Pap Smear."

56. Nicholas Cheonis, "Anal Neoplasia: A Growing Concern," *BETA*, Winter 2001.

57. "HPV Vaccine on the Way—But for Whom?" *CHAMP HHS Watch*, June 2005, 1–3.

58. Ben Townley, "Study Links Anal Cancer with Gay Sex Habits" (Web news), June 16, 2005, Gay.com U.K., www.gay.com/news/article.html?2005/06/16/4 (accessed Sept. 14, 2008).

59. In an analysis of news articles published between 2003 and 2008, Lundy Braun and Ling Phoun (chapter 3) found that only 9% of the articles mentioned that HPV causes other types of cancer than cervical cancer.

60. David Tuller, "New Vaccine for Cervical Cancer Could Prove Useful in Men, Too," *New York Times*, Jan. 30, 2007; Shari Roan, "HPV: Men Can Get It Too," *Los Angeles Times*, March 19, 2007, F1.

61. Roan, "HPV."

62. A slightly different but functionally similar situation prevails in Australia, Mexico, and parts of Europe: Gardasil has actually been approved for use in boys, but (to the best of my knowledge) government health services will not pay for it. Tuller, "New Vaccine."

63. Roan, "HPV"; Tuller, "New Vaccine." On the recent salience of debates about the legitimacy of medical extrapolation across sex (or other) categories, see Epstein, *Inclusion*.

64. "They don't know if it will stop them from getting cancer, but gay and bisexual men in San Francisco are taking a vaccine meant to prevent cancer in females," was how the local San Francisco TV news put it, in a version posted on its Web site under the heading "Some Gay Men in S.F. Getting HPV Vaccine for Women," April 6, 2007, CBS, http://cbs3.com/health/HPV.Gardasil.gay.2.282840.html (accessed May 11, 2008). See also Michelle Roberts, "Gay Men Seek 'Female Cancer' Jab" (Web report), Feb. 23, 2007, BBC News, http://news.bbc.co.uk/go/pr/fr/-/2/hi/health/6342105.stm (accessed May 13, 2008). The coding of a vaccine as "female" or "male" is reminiscent of the gendering of hormones early in the twentieth century; see Nelly Oudshoorn, *Beyond the Natural Body: An Archeology of Sex Hormones* (London: Routledge, 1994) (thanks to Nancy Cott for reminding me of this connection).

65. Gay and Bi-Sexual Men Taking Gardasil.

66. *Hole Health.*

67. Ibid. From a public health standpoint, it may still be advisable to vaccinate those who have already been infected with HPV in order to protect them from viral types to which they remain unexposed. However, in one recent, Internet-based study of gay and bisexual men, interest declined when men were told that the vaccine's effectiveness declined after the onset of sexual activity. Paul L. Reiter, Noel T. Brewer, Annie-Laurie McRee, and Jennifer S. Smith, "Acceptability of HPV Vaccine among Gay and Bisexual Men," paper presented at the annual meeting of the American Public Health Association, Philadelphia, PA, Oct. 10, 2009.

68. *Hole Health.*

69. More recently, oral cancer in males has also emerged as an important concern linked to HPV infection. I return to this issue in the conclusion.

70. In 2008 the *New York Times* reported that Glaxo was conducting such a trial in Finland. Jan Hoffman, "Vaccinating Boys for Girls' Sake?" *New York Times*, Feb. 24, 2008, 1.

71. *Hole Health*.

72. On the economic logic that governs pharmaceutical companies' decisions when designing vaccines, see chapter 2.

73. Hoffman, "Vaccinating Boys for Girls' Sake?"

74. Ibid.

75. Tuller, "New Vaccine"; Roan, "HPV."

76. "CDC Fact Sheet: HPV and Men," 2007, U.S. Centers for Disease Control and Prevention, Atlanta, GA. See also Centers for Disease Control and Prevention, www.cdc.gov/std/hpv/stdfact-hpv-and-men.htm.

77. Tuller, "New Vaccine."

78. Palefsky interview.

79. The Web site, which is no longer active because the trial is no longer recruiting research participants, was www.hpvvaccinetrials.com. Archived versions dating from 2006 can be found at Internet Archive Wayback Machine, http://web.archive.org/web/*/www.hpvvaccinetrials.com.

80. *Hole Health*.

81. Palefsky interview.

82. "HPV Vaccine Can Prevent Genital Warts in Men" (Web news), Nov. 13, 2008, msnbc.com, www.msnbc.msn.com/id/27697369/ (accessed March 13, 2009); Rob Stein, "A Vaccine Debate Once Focused on Sex Shifts as Boys Join the Target Market," *Washington Post*, March 26, 2009.

83. "FDA Approves New Indication for Gardasil to Prevent Genital Warts in Men and Boys." U.S. Food and Drug Administration News Release, October 16, 2009, http://www.fda.gov/NewsEvents/Newsroom/PressAnnouncements/ucm187003.htm.

84. David Mitchell, "ACIP Supports 'Permissive Use,' but Not Routine Use, of Gardasil in Males." American Academy of Family Physicians, Oct. 27, 2009, www.aafp.org/online/en/home/publications/news/news-now/clinical-care-research/20091027acip-hpv-vacc.html.

85. Palefsky interview. On February 17, 2010, as this volume was in production, Merck reported that Gardasil was found to be 77.5 percent efficacious in preventing AIN in young MSM. The data were presented at a conference and had not yet been published.

86. In one Internet-based survey, 79% of gay and bisexual men (versus 61% of heterosexual men) had heard of HPV, and of them 33% of gay and bisexual men (versus 14% of heterosexual men) identified it as causing anal cancer. Terence W. Ng, Noel T. Brewer, Paul L. Reiter, Jennifer S. Smith, and Annie-Laurie McRee, "Men's Knowledge and Beliefs about HPV-Related Diseases," paper presented at the annual meeting of the American Public Health Association, Philadelphia, Nov. 9, 2009.

87. Historically colon cancer has shared some of this stigma, but in that case the stigma appears to have receded over time because of greater familiarity with the illness. Perhaps a turning point in visibility occurred in 1985, when President Ronald Reagan

had surgery to treat colon cancer. See Martin L. Brown and Arnold L. Potosky, "The Presidential Effect: The Public Health Response to Media Coverage about Ronald Reagan's Colon Cancer Episode," *Public Opinion Quarterly* 54 (1990).

88. I am indebted to Julie Livingston for this pithy observation.

89. An exception was an ABC news report, two days after Fawcett's death, about the stigma surrounding anal cancer, which mentioned the probable causal link with HPV infection. Dan Childs and Chitale Radha, "Farrah Fawcett's Anal Cancer: Fighting the Stigma," June 27, 2009, ABC News report, http://abcnews.go.com/Health/story?id= 7939402&page=1#.

90. These discussions groups were conducted by Giovanna Chesler as part of her "Tune in HPV" project promoting critical awareness and discussion of HPV (personal communication; see also chapter 8 and Tune in HPV, www.tuneinhpv.com/home/home .htm).

91. *Hole Health.*

92. My discussion in this section is informed by recent work in queer studies on the shame-pride dialectic. As David Halperin and Valerie Traub have observed, modern mainstream gay politics is founded on a notion of "gay pride" that both presupposes and never fully moves beyond a historical residue of shame. David M. Halperin and Valerie Traub, eds., *Gay Shame* (Chicago: University of Chicago Press, 2009). On the politics of sexual shame, see also Janice M. Irvine, "Shame Comes out of the Closet," *Sexuality Research and Social Policy* 6, no. 1 (2009): 70–79.

93. "New HPV Vaccine Hoped to Benefit Men" (Web article), Jan. 31, 2007, Gay. com, www.gay.com/news/article.html?2007/01/31/4 (accessed May 13, 2008).

94. *Hole Health.*

95. William F. Buckley Jr., "Steps in Combating the AIDS Epidemic" (op-ed), *New York Times*, March 18, 1986. On this point, see also Irvine, "Shame Comes out of the Closet," 75.

96. One gay man, identified as R.H.K., describes such an experience in a narrative posted on Tune in HPV, "a web television channel designed to raise awareness about the human papillomavirus through expression and entertainment" that was created by Giovanna Chesler (see chapter 8). In the narrative, R.H.K. expresses great frustration when a dermatologist who diagnoses his anal warts lectures him for having had unprotected anal sex and appears disinclined to believe his denial of such activity. R.H.K., "A Queer Story" (submitted narrative), Dec. 2008, Tune in HPV, www.tuneinhpv.com/ home/home.htm (accessed March 13, 2009).

97. Palefsky interview.

98. On the cultural meanings assigned to the anus and anal sex within the gender and sexual order, see Hocquenghem, *Homosexual Desire*, chap. 4; Leo Bersani, "Is the Rectum a Grave?" *October* 43 (1987): 197–222.

99. More recently, the National LGBT Cancer Network, founded in 2007, has sought to highlight the risk of a variety of cancers among LGBT individuals. See National LGBT Cancer Network, www.cancer-network.org/.

100. It is noteworthy that one of the most explicit efforts to conduct research on HPV that may specifically benefit gay men—the inclusion of a subset of MSM in Merck's randomized clinical trial of Gardasil in men—is a private initiative and not an effort funded by the state.

101. The South African HIV/AIDS treatment advocacy group Treatment Action Campaign has called for clinical trials of HPV vaccines in men, with a focus on MSM. TAC Issues Global Call Web page.

102. Joel Palefsky has suggested that white gay men may be overrepresented in the MSM subset. *Hole Health.* Of course, the underrepresentation of members of racial minority groups is by no means atypical in clinical research in the United States; see Epstein, *Inclusion.*

103. Matthew Waites, *The Age of Consent: Young People, Sexuality, and Citizenship* (Houndmills, Basingstoke, England: Palgrave Macmillan, 2005).

104. *Hole Health.* As noted previously, vaccination may be advisable for those who have already been infected with HPV in order to protect them from viral types to which they have not yet been exposed.

105. Anne Szarewski, quoted in Roberts, "Gay Men Seek 'Female Cancer' Jab."

106. On the cultural disavowal of the "pregay" child, see Eve Kosofsky Sedgwick, "How to Bring Your Kids Up Gay: The War on Effeminate Boys," *Social Text*, no. 29 (1991): 18–27.

107. Gypsyamber D'Souza, Aimee R. Kreimer, Raphael Viscidi, Michael Pawlita, Carole Fakhry, Wayne M. Koch, William H. Westra, and Maura L. Gillison, "Case-Control Study of Human Papillomavirus and Oropharyngeal Cancer," *New England Journal of Medicine* 356, no. 19 (2007): 1944–1956; Nicholas Bakalar, "Oral Cancer in Men Associated with HPV," *New York Times*, May 13, 2008; Stephanie Desmon; "HPV-Related Oral Cancers Rise among Younger Men," *Baltimore Sun*, April 14, 2008, A1; Jeremy Manier, "Researchers Blame HPV for Rise in Throat Cancer," *Chicago Tribune*, June 8, 2008, 1; Coco Masters, "Oral Sex Can Add to HPV Cancer Risk," *Time*, May 11, 2007. Somewhat in contrast to the optimistic tone of mass media reports, Joel Palefsky believes that "proving that the vaccine will reduce oral cancer will be even more challenging than anal or cervical cancer," because there is "no easily recognizable oral cancer precursor lesion." Palefsky interview.

108. Masters, "Oral Sex Can Add to HPV Cancer Risk."

109. Stein, "Vaccine Debate"; Rachel Zelkowitz, "HPV Casts a Wider Shadow," *Science* 323, no. 5914 (2009): 580–581.

CHAPTER FIVE

Cervical Cancer, HIV, and the HPV Vaccine in Botswana

Doreen Ramogola-Masire

It is late in the afternoon and the dry African heat is starting to wear off. It has been a long day in the clinic, and only one patient remains to be seen, a typical Motswana woman of traditional build, anywhere between 30 and 40 years old, with an open, pleasant face. Her present complaint is not unusual; she has been referred by her local clinic because she has not been able to conceive since the birth of her only child ten years ago. She is single, and since her sexual debut at age 18, she has had three or four partners. She reports having had two or three episodes of vaginal discharge in the past and remembers being given treatment and a little slip of paper asking her partner at the time to come to the local clinic for similar treatment.

Three years before that she had gotten very sick and had lost a lot of weight, and by the time she got to the nearest Infectious Disease Care Clinic (IDCC) providing HIV care, she could not walk without assistance. The doctor said she had HIV (*mogare*), and she was immediately put on treatment. At her recent checkup, the doctor was very happy with her progress; her CD4 count ('the soldiers of her body,' or *masole*) was very good, 285, and the *mogare* was undetectable. In other words, the viral load had been suppressed. She felt well, just like her old self.

For the past year, she has been in a steady relationship but is not sure of her boyfriend's immune status and has not disclosed her own status to him. Although the doctors and nurses at the IDCC tell her constantly to use condoms at all times to protect herself and her partner—as well as to prevent pregnancy—she has not used any for several months because she now wants to have a child. Besides, her boyfriend wants her to have "his" baby before he commits to marrying her. The rest of her history is unremarkable.

As I listen to her, I start to run through the likely diagnoses and mentally prepare the "infertility" work-up. The two most likely diagnoses are: (1) inadequate frequency of intercourse, and (2) tubal blockage from previous inadequately treated sexually transmitted infections. She has never had a Pap smear in her life, so I plan to do one for her.

I am happy with the examination until I get to the pelvis, which reveals the unmistakable cancer on her cervix. At a glance I can tell that it has already spread beyond the cervix. With dread certainty I know that she is most likely to be dead within five years, leaving behind devastation—a daughter still at school who will be left with an ailing grandmother, who herself is HIV-infected. The grandmother contracted the *mogare* while caring for her other daughter back in the day when most people did not disclose their HIV status because of stigma and when very little information was given to the carriers about infection control.

This is the typical profile of a patient with advanced cervical cancer in Botswana: she is fairly young, HIV-infected, and has been on highly active antiretroviral therapy for some time. Despite several contacts with health care workers, she has never been offered a cervical screening in her life. Her case testifies to the powerful and awful reality of cervical cancer in most developing nations, especially those of sub-Saharan Africa. More than 500,000 new cases of cervical cancer were diagnosed worldwide in 2007; more than 300,000 died of the disease (Garcia et al. 2007), and more than 80 percent of the deaths occurred in developing countries such as Botswana. Out of the fifteen to twenty oncogenic (cancer-causing) types of human papillomavirus that are essential for the development of cervical cancer, subtypes 16 and 18 have been associated with around 70 percent of all invasive cervical cancers in the Western world. Yet, unfortunately, too little is known about which subtypes are implicated in the parts of the world where most of the burden of invasive cervical cancer is felt, and preliminary studies suggest that the situation may be more complex in

Africa (see, e.g., Didelot-Rousseau et al. 2006; Thomas et al. 2004). For example, recent data from urban Zambia (Sahasrabuddhe et al. 2007) found that HPV 52, 58, and 53 were far more common than 16 or 18 in women with high-grade squamous intraepithelial lesions or squamous cell carcinoma, while in rural Gambia 16 was the most prevalent type (Wall et al. 2005). Given these gaps in knowledge—combined with an inadequate health care delivery infrastructure, shortages in human and financial resources, and other competing health needs (Denny and Sankaranarayanan 2006)—it is no surprise that cervical cancer is the leading cause of cancer deaths in sub-Saharan Africa.

The health crisis of sub-Saharan Africa contrasts starkly with the situation in most developed countries, where cervical cancer is a relatively rare disease due to routine screenings, appropriate treatment of precancerous cervical lesions, awareness among the general public and health care workers, strong advocacy from civil society, and prioritization of women's health issues. Of course, this is just part of the story, as even in the developed world, poorer women face similar problems related to limited access to health care. In other words, the situation of poor women—like the Motswana woman I described—highlights the inequity inherent in this deadly disease.

Issues of Vulnerability

When Botswana gained independence from Britain in 1966, it was rated as one of the twenty-five poorest nations in the world. Just over three decades later, Botswana was rated as one of the middle-income countries, due to a combination of prudent use of diamond-derived wealth, sound democratic processes, and proper governance structures, which resulted in rapid economic growth. However, even with its impressive human development, a significant number of Batswana (people of Botswana), most of them women, still live in poverty; and in recent years much of this poor population has been impacted by HIV. According to the *Botswana Human Development Report 2000*, "Poverty is an important factor in the transmission of HIV. Perhaps more than anything else, it informs many of the undesirable choices made by poor people, including behaviours that increase the risk of HIV infection—alcohol abuse, multiple sexual partners, and sex for money" (Government of Botswana/UNDP 2000). Until recently, Botswana had the highest infection rate for HIV. In 2001, the then President of Botswana, Festus Mogae, said in his State of the Nation Address, "HIV/AIDS is the most serious challenge facing our nation, and a threat to our continued existence as a people" (Government of Botswana 2001).

Six years later, while still the president, Mogae spoke on the occasion of World AIDS Day, whose theme that year was "Women, Girls and HIV/AIDS." Addressing the particular vulnerability of women and young girls in Botswana, he explained, "Our experience to date is that many women and girls are vulnerable to HIV and AIDS due to high-risk behaviours, actions or inactions, largely by others." In particular, Mogae pointed to how intergenerational sex influenced vulnerability among the youth, where HIV prevalence rates in girls in the 15–19 age group are twelve times as high as in boys. The disparity in material resources between older men and younger women, according to Mogae, "is clearly demonstrative of the economic power imbalances at play, which leaves the young vulnerable and unable to negotiate safer sex" (Government of Botswana 2007).

HIV, we have learned, greatly increases the risk of precancerous lesions in women coinfected with HPV. A growing body of evidence suggests that HIV-infected women develop the disease at least ten years earlier if they are also infected with HPV (Peter et al. 2003). Furthermore, the course of cervical cancer may be more aggressive in these women (Maiman et al. 1993). Currently, prevalence rates for HIV in Botswana are 28 percent in women in the 15–49 age group, and about 34 percent in pregnant women (Government of Botswana Ministry of Health, personal communication). By extrapolation, a large proportion of Batswana women can be assumed to be at increased risk of developing precancerous cervical lesions—making cervical cancer, next to tuberculosis, a new public health crisis in Botswana.

The complex issues underlying the spread of HIV hold the key to understanding how cervical cancer became the number one cancer killer among Batswana women. The *Botswana Human Development Report* published in 2000 (UNDP 2000) highlighted poverty, intergenerational sex, culturally based gender inequalities, high population mobility and urbanization, and stigma and denial as key factors contributing to the spread of HIV in Botswana. The first three factors directly affect women and girls and their risk of being coinfected with HPV. In other words, the story of HPV and cervical cancer in Botswana is deeply interlinked with that of HIV; just as the latter is at once "a social issue, a human rights issue, [and] an economic issue" (as described by the UN secretary general on World AIDS Day, 2007), so also the high incidence of HPV-induced cervical cancer is very much the result of existing socioeconomic and cultural disparities.

Challenges to Treatment and Prevention

The impressive initial gains made in postindependence Botswana's human development have been seriously undermined by the HIV/AIDS pandemic, as evidenced by the notable deterioration of major social indicators such as child and maternal mortality rates. This development has made the fight against HIV/AIDS Botswana's national development priority and has prompted the government's decision in 2001 to implement the national antiretroviral therapy (ART) program. Since its initiation, most of the development resources have been diverted toward this program. By 2007 the Government of Botswana (GOB) was funding nearly 80 percent of the ART program. Although the bulk of the funding comes from the GOB, the program's sustainability depends on the ongoing donor support. One of the most significant donors is the African Comprehensive HIV/AIDS Partnerships, a collaboration between GOB, the Bill and Melinda Gates Foundation, and the Merck Foundation that was established in July 2000 with the primary objective of supporting Botswana's HIV/AIDS response through the year 2009. The two GOB partners have each committed $50 million toward the project, and the Merck Foundation additionally donates two antiretroviral agents.

The GOB's decision to put most of its resources into HIV management coincided with the elevation of Botswana to the status of a middle-income nation. Since that status was reached, most traditional donors have either scaled down or pulled out their resources, dealing a double blow to a nation already reeling from unprecedented loss of productivity and skilled human resources, especially in the 24–35 age group. And yet, before 2002, the majority of those diagnosed with HIV/AIDS were sure to be dead within five to ten years. Now, with around 110,000 Batswana people currently eligible for treatment, 90 percent of them are already receiving it.

All in all, however, we are a nation of contrasts: just as Botswana features the sparkling Okavango Delta, with its lush vegetation, right in the middle of the desert, one of the most ambitious, extensive, and advanced national HIV treatment programs in the world stands in marked contrast to an overloaded cytology-based cervical cancer screening program. There are many challenges that impede the development and sustainability of a robust screening program—ranging from shortage of skilled manpower, inadequate laboratory systems and poor procurement processes for supplies, and uncoordinated cancer planning services. Most of the health dollars in cancer management currently are spent on the curative side of the spectrum, which is the most expensive and the least effective. It takes three to six months for a Pap smear to be read, and it can take even longer for the results to reach a patient. Access to gynecologic services is very limited; they are offered at the only two referral hospitals in the country. These services themselves suffer from extreme shortages of skilled manpower, equipment, and capacity to deal with the sheer numbers of patients who walk through the doors each day.

The national HIV guidelines in Botswana, released May 1, 2008, recommend annual pelvic examinations and Pap smears for all HIV-infected women, or about 85,000 smears per year. There are no guidelines for HIV-uninfected women, but if these are included in the screening service, there will be a need for many more smears annually. The current national laboratory infrastructure can barely support the current 20,000 smears a year. The intention is good, and the country's leadership should be applauded for taking this bold step, but the stark realities for frontline health care workers reveal another story altogether.

Furthermore, it is not clear whether immune reconstitution (or the resurrection of the immune system) associated with potent antiretrovirals is a factor in the regression of precancerous lesions; evidence on this point is conflicting. But if indeed that is the case, then extensive availability of ART in Botswana is likely, paradoxically, to permit progression to cancer in more women. All these factors taken together lead to the clear conclusion that cervical cancer is rapidly becoming a secondary epidemic in the wake of HIV in Botswana.

Vaccination

The two new HPV vaccines (the quadrivalent Gardasil from Merck, and the bivalent Cervarix from GlaxoSmithKline) have brought with them a lot of hope as part of a comprehensive cervical cancer prevention strategy. Both vaccines, when given prior to infection with HPV subtypes 16 and 18, have been found to have almost 100 percent efficacy in preventing cervical cancer. Many young women stand to benefit from this new technology; but how many of these beneficiaries are likely to come from regions with the highest needs, such as Botswana? The inequality inherent in cervical cancer is only likely to deepen unless we find innovative ways of getting the vaccine where it is most needed and where its impact is most likely to be dramatic. Until these inequalities can be adequately addressed, it is not going to be within the power of individual Botswana women to *choose* to be "One Less" (as in the Gardasil ad) victim of cervical cancer. Furthermore, acknowledging these inequalities means also recognizing the limited power and resources of the state in poor nations like Botswana. The public-private partnership model that Merck, Gates, the President's Emergency Plan for AIDS Relief, and others have promoted and participated in is in jeopardy, especially if the government of Botswana does not have the cash to continue to enroll new patients after 2016, when the partnership ends. So we must ask, Are the kinds of partnership that are reflected in this new model of global health care (through which we might see potential access for poor women to Gardasil or Cervarix) sustainable, tenable, and promising, especially given the current economic climate?

Is the HPV vaccine a substitute for a broad-reaching, well-funded health care system? Far from it! And yet, one can begin to see this vaccine as a powerful tool that might be employed to very positive effect within contexts like Botswana. A conversation with my oncology colleague Alexander Von Paleske is very sobering and a good reminder of the dire situation in this part of the world. As the sole oncologist for the country's central tertiary hospital, he (with the help of two junior doctors) saw more than 6,000 patients in 2006, whereas he saw 2,050 in 2002. During this period he was working a seven-day week in a twenty-bed unit. The dramatic increase in the number of cancer patients can be directly attributed to either cancers that are caused by HIV/AIDS (such as Kaposi sarcoma, cervical cancer, and so forth) or those that become more aggressive with HIV/AIDS. Von Paleske sees this as "a challenge of unforeseen proportions."

A walk through his ward displays untold misery. Very few of the patients are elderly and HIV-uninfected. The majority are between ages 20 and 40 and are HIV-infected. One HIV-infected woman lies with her legs open, with a large fungating mass growing from her genital area. At the age of 25 she has end-stage vulval cancer (almost certainly HPV-related), something almost unheard of in this age group in other parts of the world. It is easy to be overwhelmed by the hopelessness of the whole situation, but Von Paleske has learned to focus on one patient at a time. I ask him, "What is the single most valuable thing we could do as a nation to begin to deal with this issue?" And without hesitation he says, "Provide the HPV vaccine." This is a sentiment echoed by many health care workers who have some knowledge about cervical cancer prevention.

With an overstretched health system and limited disseminated information about cancer, most believe Botswana is headed for a second epidemic. As a national newspaper puts it, "Whilst the nation's attention [and resources] is focused on HIV/AIDS, cancer has made inroads—unnoticed and unmarked"

(Toteng 2007). The gynecologic and oncological services are overburdened, the cost of end-stage cancer is debilitating financially and emotionally, and to set up a properly functioning screening system will take forever. The HPV vaccine is here now and is seen by many as a welcome new technology, imperfect as it is. When you are sitting in a clinic in Botswana and seeing yet another young woman with end-stage cervical cancer, your realities are different from those you would face in a clinic on the other side of the Atlantic, where your most vexing question is whether a vaccine that potentially protects against a sexually transmitted virus will lead to promiscuity.

Nevertheless, it is also true that there are numerous uncertainties surrounding the HPV vaccine in Botswana. Since no knowledge exists of the prevalent cervical-cancer-causing HPV subtypes in our country, it is hard to confidently argue for or against the current generation of HPV vaccines. The effectiveness of such a vaccine in an HIV-infected individual also remains unclear, as we know of no studies that have looked at the effects of the HPV vaccine on this specific population. Moreover, no information exists on the durability of the vaccine protection beyond five years. Setting up an infrastructure for the delivery of the new vaccine will be difficult. Even with the political will to make such an endeavor a priority, the biggest challenge will be funding it, because there are so many competing needs for Botswana's health dollars. The Global Alliance for Vaccines and Immunization, which was set up in 1999 to support immunization globally, assists seventy-two countries with funding mechanisms for new vaccine implementation. However, Botswana's recently acquired middle-income status will disqualify the nation for such funding, unless there is a change in the funding criteria.

As a nation that was clearly heading for disaster as a result of the HIV epidemic, our view of the pharmaceutical giant Merck is different from that held by many in the West. Well aware of Merck as a profit-making entity, the people of Botswana have nonetheless greatly benefited from its significant philanthropic effort; and identifying it as one of our key stakeholders, we see an opportunity for collaboration in the implementation of the HPV vaccine.

Conclusion

The argument for an HPV vaccine in Botswana is obviously more complicated than I portray it. But regardless of the numerous barriers to HPV vaccination thus far, the only hope of making any impact in cervical cancer prevention in the long run lies in a nationwide HPV vaccination of its adolescent population coupled with modification of the cytology-based screening. There was a time not so long ago when the concept of ARTs being available to persons in developing countries, along with an infrastructure to deliver them, was unfathomable. As we know, Botswana and many other developing nations have defied the odds. It is with the knowledge of our past that we can look forward to HPV vaccination and cervical cancer prevention with great expectation and hope.

Cervical cancer is one of the few cancers that can be prevented through vaccination and effective screening. Focusing on prevention rather than cure of invasive cervical cancer will not only save lives but will also free up valuable resources to deal with other health issues. But unless drastic changes occur in how quickly new vaccination technologies become truly accessible and affordable to resource-low communities, many women are still going to die from a theoretically preventable disease. Major strides can be made worldwide in curbing this silent epidemic only if global citizens across many disciplines pool their intellectual resources and work toward this goal.

REFERENCES

Denny, L., M. Quinn, and R. Sankaranarayanan. 2006. "Screening for Cervical Cancer in Developing Countries." *Vaccine* 24S3:S3/71–77.
Didelot-Rousseau, M. N., N. Nagot, V. Costes-Martineau, X. Valles, A. Ouedraogo, I. Konate, H. A. Weiss, et al. 2006. "Human Papillomavirus Genotype Distribution and Cervical Squamous Intraepithelial Lesions among High-Risk Women with and without HIV-1 Infection in Burkina Faso." *British Journal of Cancer* 95:355–362.
Garcia, M., A. Jemal, E. M. Ward, M. M. Center, Y. Hao, R. L. Siegel, and M. J. Thun. 2007. *Global Cancer Facts and Figures 2007.* Atlanta, GA: American Cancer Society.
Gichangi, P. B., J. Bwayo, B. Estambale, H. De Vuyst, S. Ojwang, K. Rogo, H. Abwao, and M. Temmerman. 2003. "Impact of HIV Infection on Invasive Cervical Cancer in Kenyan Women." *AIDS* 17:1963–1968.
Government of Botswana. 2001. State of the Nation Address, delivered at the first meeting of the third session of the eighth Parliament of Botswana on Oct. 29, 2001, by President Festus Mogae.
———. 2007. Message on World AIDS Day, Dec. 1, delivered by President Festus Mogae, Ghanzi, Botswana.
Government of Botswana/UNDP. 2000. *Botswana Human Development Report 2000: Towards AIDS-Free Generation.* Gaborone, Botswana: Government of Botswana/UNDP.
Maiman, M., R. G. Fruchter, Guy L. Levis, S. Cuthill, P. Levine, and E. Serur. 1993. "Human Immunodeficiency Virus Infection and Invasive Cervical Carcinoma." *Cancer* 71:402–406.

Sahasrabuddhe, V. V., M. H. Mwanahamuntu, S. H. Vermund, W. K. Huh, M. D. Lyon, J. S. A. Stringer, and G. P. Parham. 2007. "Prevalence and Distribution of HPV Genotypes among HIV-Infected Women in Zambia." *British Journal of Cancer* 96:1480–1483.

Thomas, J. O., R. Herrero, A. A. Omigbodun, K. Ojemakinde, I. O. Ajayi, A. Fawole, O. Olapedo, et al. 2004. "Prevalence of Papillomavirus Infection in Women in Ibadan, Nigeria: A Population-Based Study." *British Journal of Cancer* 90:638–645.

Toteng, T. 2007. "A Silent Killer on the Loose." *Sunday Standard*, April 22.

UN Secretary General. 2007. Message on World Aids Day, Dec. 1. http://data.unaids .org/pub/PressStatement/2007/071126_sg_wad_statement_en.pdf (accessed Jan. 15, 2010).

Wall, S. R., C. F. Scherf, L. Morison, K. W. Hart, B. West, G. Ekpo, A. N. Fiander, et al. 2005. "Cervical Human Papillomavirus Infection and Squamous Intraepithelial Lesions in Rural Gambia, West Africa: Viral Sequence Analysis and Epidemiology." *British Journal of Cancer* 93:1068–1076.

Part II / Girls at the Center of the Storm

Marketing and Managing Gendered Risk

CHAPTER SIX

Safeguarding Girls

Morality, Risk, and Activism

Heather Munro Prescott

In a March 2007 article entitled "Who's Afraid of Gardasil?" published in the progressive journal *Nation*, columnist Karen Houppert condemned the widespread skepticism about attempts to mandate vaccination of all adolescent girls. These "strange bedfellows," Houppert said, included Christian conservatives, critics of Big Pharma, the antivaccine movement, parental-rights libertarians, and feminist health activists whom Houppert had expected would be "chomping at the bit" to approve this new innovation to protect women's lives.[1] Houppert's diatribe not only trivializes the long history of feminist health activism but overlooks other key historical issues regarding child health, parental autonomy, and female sexuality as well.

Laura Mamo and colleagues (chapter 7) offer a visual and narrative content analysis of the "One Less" advertising campaign produced by Merck, the manufacturer of Gardasil. According to them, these ads convey both an image of "risky girlhood" and a message of health empowerment. Girls declare, "I choose" the vaccine "because my dreams don't include cervical cancer." Mothers too are included in this "call to action," to show that wise mothers choose to have their daughters vaccinated. As opposed to Gardasil's campaigns, numerous parents and religious groups have objected to mandatory vaccination against the human papillomavirus (see chapter 9). Some have argued that this public health initiative violates parental rights and fosters sexual activity by protecting girls from the consequences of their sexual behavior.[2]

Examining the similarities between the HPV vaccine campaign and earlier efforts to protect child and adolescent health, this chapter explains that the public debate about the HPV vaccine is not the first time cultural anxieties about adolescent female sexuality have led to critiques of a public health initiative. Similar issues arose when some gynecologists proposed routine pelvic examinations for adolescent girls during the mid-twentieth century. Since that time, the pelvic exam has become a routine procedure for adolescent girls, as well as an opportunity to inform the young patient about sexually transmitted diseases and contraception. In her outright rejection of some of the skeptics of the HPV vaccine, Houppert ignores how the reservations of contemporary feminist health activists grow out of earlier feminist critiques of the medical profession's treatment of women. Since the 1960s, feminist health organizations such as the National Women's Health Network have exposed abuses of female patients in the development and testing of new drugs and medical devices. Feminist health activists have also demonstrated sexism in public health efforts. Since the early twentieth century, public health campaigns have focused on the behavior of individuals, or what Allan Brandt calls the "moral valence of individual risk."[3] This focus has been especially evident in efforts to control sexually transmitted infections, which efforts have tended to place most of the blame on women's sexual behavior. The emphasis on female sexuality has often ignored larger public health issues such as the long-term efficacy or inadequacy of vaccination treatments, the need to continue to insist on routine health screenings even with vaccination, and the persisting realities of unequal access to regular gynecologic health care.

Scientific Mothers and Daughters

Beginning in the early twentieth century, medical experts blamed high rates of infant and child mortality on women's lack of accurate scientific knowledge about how to prevent and manage childhood diseases. The solution to this problem was to make motherhood more "scientific" by instructing women to rely on the advice of pediatricians and other scientific experts. These principles of "scientific motherhood" were promoted in high school and college courses in home economics, advice manuals, government pamphlets issued by the U.S. Children's Bureau, and popular advice columns in women's magazines. By the 1920s, women had become accustomed to seeking child care advice from medical and scientific experts rather than from neighbors, friends, and relatives.[4]

Mothers were not the only target of medical advice: physicians and other health experts also enlisted children and youth in these public health campaigns. Primary and secondary schools, organizations such as the YWCA and the Girl Scouts, settlement houses, and other organizations and agencies serving young people in the early twentieth century all became venues for disseminating basic information on disease prevention, nutrition, physical fitness, and personal hygiene. Nonprofit organizations such as the Child Health Organization and the National Tuberculosis Association developed games and school curricula to make classroom instruction fun and interesting to young students. Children received "health report cards" that gave them points for acquiring health habits such as regular hand washing and tooth brushing. These child health campaigns were closely tied to efforts to "Americanize" massive waves of immigrants who arrived in the United States during the late nineteenth and early twentieth centuries. Public health experts believed that "Old World" practices were to blame for many child health problems. They hoped that when children and adolescents were taught how to live more "American" (i.e., hygienic) lives, they would serve as "health ambassadors" to their families, who would in turn adopt these health practices themselves.[5]

Adolescent girls were central to these preventive efforts, partly because of their future roles as mothers, but also because in many immigrant families, girls were in fact "little mothers," responsible for caring for younger siblings while their mothers were at work. Many public schools and settlement houses created after-school "little mothers clubs" to ensure that girls were "intelligently trained" in the healthful care of infants and young children. Public health experts hoped that the "little mothers" would not only become faithful health consumers when they had children of their own but would also serve as "health emissaries" to take American methods of child-rearing to other mothers in their homes and neighborhoods.[6]

This connection between health and responsible American citizenry appears in various campaigns for childhood vaccinations. For example, according to Leslie Reagan, after widespread media reports of babies with birth defects who were born to mothers who contracted German measles during pregnancy, public health experts launched a nationwide effort to vaccinate all children against the disease. These experts urged boys and girls to be good health citizens and accept vaccination as a way to protect future babies from disabilities.[7] For example, a poster issued by the Oregon chapter of the March of Dimes warned girls, "Someday . . . You might have a Baby! Don't risk Birth Defects from Rubella!"[8]

As James Colgrove observes (chapter 1), these vaccination campaigns did not go unchallenged. Like the modern-day parents described in chapter 9, there were parents who rebelled against requirements that their children be vaccinated in order to attend public school. Yet, in light of recent objections to mandatory HPV, a remarkably large majority of parents and children willingly engaged in these vaccination programs against major childhood diseases. Nowhere was this eagerness more apparent than in the field trials of the Salk polio vaccine in 1954, which included 650,000 schoolchildren in what one newspaper called the "biggest public health experiment ever."[9]

Social Hygiene and Sex Education

In contrast to their success with promoting the polio vaccine, public health experts encountered major obstacles when they tried to prevent more controversial diseases—namely those spread through sexual contact. Indeed, use of the term *venereal* signaled the sinful origins of these afflictions. Although sexually transmitted infections (STIs) had vexed medical experts for centuries, infection rates as high as 25 percent during World War I gave new urgency to campaigns to eliminate these deadly scourges. At the same time, it was becoming clear that young people's sexual practices were radically changing. Max J. Exner, public health officer for the YMCA, reported that more than half of his sample of 948 college men surveyed during the second decade of the twentieth century had engaged in sexual practices of some kind. Studies of the college sex behavior of female college graduates conducted by Clelia Duel Mosher at Stanford and by New York City Correction Commissioner Katherine Bement Davis between 1910 and 1929 indicated that similar changes were occurring on a smaller scale among female undergraduates.[10]

At historically black colleges, particularly those sponsored by religious denominations, undergraduates were also rebelling against the rules aimed at preserving their health and respectability. Students demanded an end to compulsory chapel attendance, dress codes, mandatory physical education, and strict regulation of social interaction between male and female students.[11] Black physicians and educators found these trends especially disturbing because their students encountered the same racial stereotypes about black health and sexuality as did lower-class African Americans. To combat white prejudice, black reformers made an explicit link between racial health and middle-class standards of sexual respectability. The noted black activist and club woman Mary Church Terrell, for example, told students at Howard University that "blacks could not afford to abandon traditional morality even if many young whites were doing so."[12]

Reformers had long been concerned about "declining morals" among working-class youth. The fact that "respectable" young people were also engaging in such "radical" sexual practices was disturbing to many parents, health professionals, and educators in public schools. To address these concerns, public health experts created new initiatives in what they called "social hygiene," a term that represented both a euphemism for the control of "venereal diseases" and a new approach to the prevention of these afflictions. Social hygiene differed from earlier "purity crusades," which shielded young people from moral and physical dangers via a "conspiracy of silence" about sexual matters. Social hygienists believed education rather than scare tactics was the best way to protect the health of the nation's youth. Social hygienists believed sex education was especially important for female students because so many young women became teachers and mothers. At the same time, social hygienists challenged the sexual double standard, advocating total abstinence before marriage for both men and women.[13]

The social hygiene movement was strongly influenced by the ideas of Sigmund Freud and G. Stanley Hall, who emphasized that sexual impulses were normal parts of both male and female adolescent development. According to Hall, psychoanalysis demonstrated how the sex instinct pervaded "every sphere of life, even those in which it was not suspected." The challenge for educators, Hall concluded, was not to suppress sexuality but "to short-circuit, transmute it, and turn it on to develop the higher powers of man."[14] Hall claimed that students "should have the medical truth about sex diseases (not perversions) as plainly, directly and compactly" as possible, but "too much relevance should not be placed on this." "Warning methods" should be played down in favor of attempts to divert sexual energy through physical activity, "wholesome religious interest," "literary fervor," and other innocuous activities.[15]

Hall's work had a significant impact on theories about adolescent female sexuality and the psychological development of adolescent girls. Before World War I, most writers on adolescent female psychology argued that healthy female development involved protecting the young girl from premature awakening of sexual instincts and longings. By the 1920s, most experts in adolescent mental hygiene declared that sexual curiosity and a limited degree of sexual experimentation was a normal part of healthy female development. In fact, mental hygiene experts were more concerned about girls who did not develop an interest in the opposite sex by the middle years of adolescence, since these girls were more likely to become lesbians later in life.[16]

Social hygienists' efforts to control male sexual behavior, especially within in the military, proved largely futile. Sexual prowess continued to be the hallmark of masculine identity. Instead of controlling male sexuality, disease control efforts among servicemen focused on more pragmatic strategies such as condom distribution, chemical treatment after exposure, and crackdowns on houses of prostitution near military bases and camps. Public health experts also described a new "girl problem"—young females who were not prostitutes but who were so infatuated with men in uniform that they willingly engaged in sex to boost military men's morale. During World War II, U.S. Public Health Service physician Otis Anderson coined the term *patriotute* to describe these young women. He and other public experts warned that while these women looked "clean," they were just as likely as prostitutes to spread disease among the troops. Within this framework, all women and girls who lived or traveled alone—especially those who were nonwhite—became possible "reservoirs of infection."[17] Public health experts found that incarcerating these women failed to halt the spread of venereal disease among U.S. soldiers. Instead, women who tested positive for syphilis or gonorrhea were given the option of prison or confinement to a quarantine hospital, where they received accelerated treatment for their diseases. This strategy perpetuated the sexual double standard by scrutinizing and controlling the sexual behavior of "promiscuous" women and girls.[18]

Risky Girlhoods

These anxieties about the dangers of uncontrolled female sexuality continued in the period following World War II. Alfred Kinsey's study *Sexual Behavior in the Human Female* (1953) disclosed that more than 50 percent of the women in his sample had engaged in premarital sex.[19] Kinsey's findings were accompanied by the somewhat reassuring fact that the percentage of married teenage girls had increased markedly. By 1959, 47 percent of all brides had married before the age of nineteen, and the percentage of girls married between four-teen and seventeen had increased by one-third since 1940.[20]

These findings led to significant changes in adolescent gynecology. According to Monica J. Casper and Adele E. Clarke, the cytological cancer-screening vaginal smear, developed by George Papanicolaou in the 1920s, became a fundamental part of the move toward routine annual gynecologic checkups following World War II.[21] Yet, my work in the history of adolescent gynecology indicates that routine pelvic examinations for teenage patients, especially those who were white and middle-class, were highly controversial.[22] Until the 1960s, most gynecology textbooks recommended that vaginal examinations of girls should be avoided, and many even advised against performing pelvic examinations on all unmarried women. These views were based on Victorian views about young women, which stated that girls should remain chaste and pure until marriage. Many physicians believed that a vaginal exam would not only damage the hymen, the physical proof of a young woman's virginity, but could also arouse sexual passions in the patient by drawing unnecessary attention to this area of the anatomy.[23]

Nevertheless, during the 1940s and 1950s, a small number of gynecologists began to suggest that regular gynecologic exams and Pap smears should be made part of standard medical care for adolescent girls so that diseases and abnormalities could be detected early. One of the leaders in pediatric and adolescent gynecology was Goodrich C. Schauffler of the University of Oregon Medical School, who published the first textbook on the subject in 1942 and who promoted the benefits of regular gynecologic examinations for girls and women in his advice column in the *Ladies' Home Journal*. Schauffler argued that the taboo against examining girls resulted in a lack of basic knowledge not only of disease but also of normal adolescent anatomy, physiology, and development.[24]

Edward Allen, professor of obstetrics and gynecology at the University of Illinois, likewise observed in 1943 that too often "morals came before scientific medicine" in office practice, thereby inhibiting study and treatment of gynecologic problems in adolescent girls. Allen placed some of the blame on mothers, who were accustomed to visiting the gynecologist for a premarital exam and for obstetrical care but had not been educated as to the importance of regular pelvic examinations. "Mothers can hardly be expected to request something for their daughters which they dread so much that they will not accept it for themselves," he observed. Yet Allen also blamed doctors, who he believed need to reeducate themselves "to a sane, unbiased attitude toward these problems connected with the sex organs of their own daughters as well as the daughters of their patients. The father-daughter, brother-sister attitude toward these problems has no place in scientific medicine" Allen concluded.[25]

As I have argued elsewhere, pediatricians began to expand their expertise into the adolescent years starting in the early 1950s.[26] Yet, an interest in adolescent gynecologic issues was seldom included. Indeed, some experts in adolescent medicine believed gynecologic examinations could actually thwart healthy female development. For example, J. Roswell Gallagher, director of the Adolescent Unit at Boston Children's Hospital, warned that "these young girls are going through a period in their emotional development which involves their attempting to discard their tomboyishness and to accept a feminine way of life; at this particular time a maneuver such as a vaginal examination may represent a sexual attack and have a regrettable emotional effect."[27]

Allen commended pediatricians for recognizing the need to cover the ages of 12 to 18 years, but he added that "they have not asked for, or perhaps received, the needed cooperation of the gynecologist" in treating the adolescent girl. Allen observed that "mothers of these growing girls need and are asking for our best help in educating and protecting their daughters." He argued that "the triad of an educated mother, the pediatrician, and the gynecologist" should spearhead efforts to improve gynecologic care for this age group. Adolescent girls also had to be persuaded to visit the gynecologist. "Further advances in the detection of early pelvic cancer will probably not occur until we educate our young women as to the ease and necessity of routine pelvic examination before the sex inhibitions become so fixed."[28]

Allen's allusion to "sex inhibitions" indicates that more was at stake than accurate diagnosis of gynecologic disease: he and other gynecologists were worried that many of the adolescent girls they saw in their practices had an "unhealthy" attitude toward their genitals, expressed in excessive modesty or anxiety in regard to this area of their bodies. Whereas nineteenth- and early-twentieth-century physicians had viewed modesty as one of the cornerstones of adolescent female identity, by the 1940s physicians began to claim that a girl who found gynecologic examinations distasteful had a "morbid" attitude toward this area, which indicated an underlying ambivalence toward the feminine role. In elevating their own competence in the treatment of adolescents, gynecologists blamed the mother for a girl's failure to adjust to normal femininity. They suggested that mothers who protected their daughters from sexual knowledge and refused to educate them properly about female sexual development did their daughters no favor but actually prevented their daughters from developing healthy attitudes toward their bodies. According to Schauffler, "the fact that so many women do not understand these rudiments of sex is responsible for much unnecessary personal and parental trouble." The solution, said Schauffler and others, was for mothers to reject the "many old wives' tales about the impropriety, impracticality, and even indecency of doctors' examinations in the genital area" and allow physicians to provide their daughters with a thorough understanding of the female pelvic anatomy.[29]

Proponents of adolescent gynecology claimed that the pelvic exam helped foster normal sexual adjustment within marriage. A woman who entered marriage with "sexual inhibitions" would encounter a host of sexual problems, including frigidity, dyspareunia (painful intercourse), sterility, and hostility toward her spouse. Needless to say, such a woman would be unwilling or even incapable of bearing healthy, well-adjusted children. To prevent such problems, a "gentle, unprejudiced physician" could help young women to accept their femininity by eliminating the fears and anxieties they had about their genitals. In fact, some even suggested beginning routine pelvic exams during infancy so that girls would become as accustomed to this procedure as examination of their teeth and toes. The result of this "scientific" attitude, they claimed, would be the elimination of a host of marital and social problems that arose out of a young woman's "pathological" reaction to her developing body.[30]

Support for gynecologic examinations for adolescents did not entail support for teenage sexuality. In fact, these gynecologists claimed that extramarital sexual activity was the result of ignorance, not premature sexual awakening caused by a pelvic examination. Schauffler even suggested that girls who adopt a "conscious-virginal or adult attitude of self-protection" during a pelvic examination were probably also engaging in premarital sex. In contrast, girls who became accustomed to the gynecologist's touch before puberty would gain a "sensible" attitude toward these body parts and would not be tempted to engage in "aberrant" sexual behaviors.[31]

To further reassure concerned parents, proponents of gynecologic examinations claimed that the examination would not damage the hymen, and they even designed special instruments for adolescents and girls that could be inserted through an intact hymeneal membrane. The inventor of one of these devices, John W. Huffman of Northwestern University Medical School, was especially adamant that his device would allow examination of adolescent and even prepubertal girls without threatening their anatomical virginity.[32] Other proponents argued that using the intact hymen as a criterion of virginity was outdated. For example, Allen urged physicians to convince parents that "virginity is a state of mind and not an anatomic condition." Therefore, he concluded, pelvic examinations did not pose a threat to a young girl's sexual virtue.[33]

These gynecologists did not ignore the consequences of premarital sexual activity: like other Americans who were interested in addressing issues of poverty, they were concerned about high unmarried pregnancy rates among girls from ethnic and racial minorities as well as white adolescents from lower socioeconomic groups. Despite the declining age of marriage among women, the rate of "illegitimate" births tripled between 1940 and 1957. Views about how to address this problem were strongly shaped by racial biases. Physicians attributed unwed pregnancy among black teenagers either to the innate "moral incapacities" of black women or to an allegedly greater tolerance of illegitimacy among African Americans. White girls who became pregnant out of wedlock, in contrast, were depicted as neurotic or maladjusted, particularly if they insisted on keeping their babies without marrying the father. The "cure" for a white unwed mother's maladjustment, therefore, was to persuade her either to legitimate her pregnancy by marrying the child's father or to give up the baby for adoption and prepare for a "normal" path to marriage and motherhood.[34] Some gynecologists at this time suggested that helping African American girls adopt white, middle-class values would alleviate the problem of teenage pregnancy in this population.[35]

These anxieties about the dangers of adolescent sexuality also appeared in epidemiological studies of cervical cancer. Starting in the 1950s, scientists explored the link between adolescent sexual activity and the development of cervical cancer later in life. Several epidemiological studies published in the 1950s and early 1960s indicated that women who married before age 20 appeared to be at higher risk for cervical cancer. Some speculated that women who had multiple "broken marriages" were especially susceptible.[36] Isadore D. Rotkin of the Cancer Research Project at the Kaiser Foundation in California elaborated on this research and found that age at first coitus was the most significant variable distinguishing cervical cancer patients from controls. In a study of more than four hundred patients, 85 percent of whom were Caucasian, twice as many cancer patients as controls began coitus at ages 15 to 17, comparatively few began after age 21, and almost none began as late as age 27. Rotkin hypothesized that some kind of infectious agent transmitted by male partners was a contributing factor and that the adolescent cervix was especially vulnerable to cell abnormalities caused by exposure to such an agent. Although Rotkin acknowledged that measures aimed at improving male sexual hygiene could reduce the incidence of the disease, he argued that postponement of marriage and limitation of the number of sexual partners for young women were the most effective means of prevention. In addition, given that a disproportionate number of patients were nonwhite, non-Jewish women of low socioeconomic status, Rotkin recommended that routine Pap smears were especially important for nonvirgins from underprivileged groups.[37]

Rotkin's work echoed many of the paternalistic attitudes toward America's "underclass" during the 1960s. Like the political liberals who spearheaded President Lyndon Johnson's "Great Society" programs, Rotkin realized that poverty was a contributing factor in cervical cancer risk and suggested that ideally the way to prevent disease "might be the transference of underprivileged populations into higher socioeconomic levels."[38] At the same time, he tended to reinforce prevailing stereotypes about the links between disease risk, race, and class.

Sexual Revolutions and Feminist Health Activism

These stereotypes started to unravel as it became apparent that the sexual behavior of "nice" girls from white, middle-class backgrounds was also changing.[39] As early as the late 1950s, obstetricians and gynecologists found that growing numbers of girls from white middle-class backgrounds were becoming sexually active—and pregnant—at increasingly younger ages, much to the dismay and embarrassment of their status-conscious parents. Jerome S. Menaker, an obstetrician from Wichita, Kansas, observed in 1958 a "marked shift of early unmarried pregnancies into the more privileged classes," a phenomenon he attributed to "the lessening of parental supervision plus the increased availability of privacy provided by the automobile." Rather than viewing teenage pregnancy as a problem limited to impoverished ghetto youth, physicians like Menaker found alarming incidence of "promiscuity in the youngsters of even the most privileged and well-to-do."[40]

Within this context, objections to gynecologic examinations for adolescents seemed outdated and even dangerous. Many physicians believed it was ridiculous to withhold gynecologic examination from girls who were sophisticated in both sexual knowledge and experience. According to Anna L. Southam of Columbia University College of Physicians and Surgeons in 1966, most adolescents were quite knowledgeable about reproductive anatomy and function, and a young woman who consulted a gynecologist would "wonder how you can know if there's anything wrong with her uterus or her ovaries if you don't examine that system as well as her heart and lungs and her breasts."[41]

These changes in adolescent gynecology were furthered by the growth of the women's health movement. The popular book *Our Bodies, Our Selves* (1971) and the later publication for adolescents, *Changing Bodies, Changing Lives* (1981), promoted self-determination in women's health care and challenged traditional ideas about femininity.[42] Gynecologists trained in the 1970s, many of whom were women, were exposed to this new feminist message about women's bodies. Consequently, many of them began to challenge traditional views about adolescent sexuality, particularly the notion that adolescent gynecologic problems were rooted in conflicts over feminine role identity.[43]

The women's health movement also investigated and critiqued sexism in the medical treatment of women and girls. The National Women's Health Network, for example, arose out of the controversy surrounding the use of diethylstilbestrol (DES) as a morning-after pill. In 1972, Belita Cowan, a graduate student at the University of Michigan, told the Health Research Group of Ralph Nader's consumer watchdog organization Public Citizen that DES was being given as a morning-after pill to students and other women in Ann Arbor. Cowan was also horrified that doctors continued to administer DES as postcoital contraception even after a study published in 1970 reported a high incidence of vaginal cancer in the daughters of women who had consumed DES while pregnant. Although there was no link between the use of DES as a morning-after pill and vaginal cancer in women who took it, Cowan and her peers were sufficiently alarmed to begin investigating the matter. In 1971, Cowan joined with women who had been patients at the University of Michigan Health Service to form Advocates for Medical Information (AMI), the forerunner of the National Women's Health Network. Hoping to inform women at both the University of Michigan and other colleges, AMI first warned of the dangers of postcoital use of DES in its feminist newspaper, *her-self.* Members also promoted the need for routine gynecologic screening for girls and women who had been exposed to DES in utero.[44]

The women's health movement also created new vocabulary to describe teenage sexual behavior. Until at least the mid-1960s, most physicians used morally charged terms such as *promiscuity* and *sexual delinquency* to describe sexual activity among teenagers. By the early 1970s, physicians were increasingly using the

term *sexually active* to describe girls who engaged in extramarital sexual intercourse. As Joan Jacobs Brumberg argues in her history of adolescent girls and
their bodies, the term "sexually active" was "an important semantic innovation
because it described a social state without reference to morality." Brumberg also
demonstrates that this semantic shift represented physicians' recognition that
the "family claim" to a daughter's virginity was no longer viable. Although parents continued to ask gynecologists to "see if my daughter is a virgin," most
gynecologists by the late twentieth century considered this question completely
inappropriate.[45]

Conclusion

As we all know, not all parents have given up their right to protect their daughters' sexual virtue. This fact is all too apparent in claims that mandatory vaccination "promotes promiscuity."[46] These concerns about morality seem to have
prompted a "semantic shift" in press coverage and advertising of Gardasil and
Cervarix; that is, the drugs are described as preventing *cancer*. This emphasis
diverts attention from the STI that causes this form of cancer.

There is much in the HPV vaccine controversy that replays older public debates about disease, prevention, and social morality. In particular, the focus on
girls in the HPV vaccination campaigns tends to perpetuate the earlier myths
about girls and women as "reservoirs" of sexually transmitted disease. Although
it makes sense to vaccinate girls if one wants to prevent cervical cancer, herd
immunity requires vaccinating 85 percent of the total population. Furthermore,
as Epstein describes in chapter 4, there appear to be clear links between HPV
infection and various cancers in men, including penile, anal, oral, and possibly
prostate cancers. Nevertheless, a recent article in the *New York Times* suggested
that the best way to get "Gardasil Boys" on board was to appeal to male chivalry
and persuade boys to protect their future female partners against disease.[47] This
strategy, too, is reminiscent of earlier vaccination campaigns urging children
to be "responsible health citizens."

We need to take prior history into account in analyzing the controversy
stirred by the HPV vaccine—the myths it perpetuates and the realities it obscures. For instance, one of the aspects that are obscured in the controversy is
the impact that the vaccine will have on routine gynecologic examinations.
The official position of the Society for Adolescent Medicine is to endorse the

recommendation of the Centers for Disease Control and Prevention (CDC) that HPV vaccine "be administered routinely to all females 11 to 12 years of age as well as 13–26 year olds who have not previously received the vaccine." Yet, the society also supports the CDC's advice to continue routine Pap screening for their patients. Their rationale is that not all cancer-causing HPV strains are contained in the vaccines, recipients may not complete the full series of vaccinations or may not receive them in a timely fashion, they may have been infected before vaccination, and it is possible that immunity may not persist over their lifespan.[48] I would add that gynecologic visits are significant ways to inform adolescent girls about sexuality and reproduction. If public health policy emphasizes vaccination *instead of* routine health screening, will these educational opportunities be lost?

Returning to the *Nation* magazine article with which I began this chapter, I am concerned with Karen Houppert's use of the term "strange bedfellows" to describe the various politically opposed constituencies that are skeptical of the HPV vaccine. Her remark glosses over some legitimate issues raised by the National Women's Health Network's representative, Amy Allina, during the approval hearings for Gardasil. Allina supported approval but also urged "follow-up research with the study population and additional post-approval research . . . so that we can learn more about the safety and efficacy in the general population and real world use, as well as longer term efficacy." She also urged Merck "to support programs that will ensure access to the vaccine for those women." Finally, she asked that the U.S. Food and Drug Administration "mandate some kind of labeling or other mechanism for communicating to health care providers and patients the necessity of continued regular screening for cervical cancer."[49]

In ignoring the legitimate concerns of Allina's position toward the vaccination campaign, Houppert's diatribe trivializes the efforts of the women's health movement past and present. Rather than see this kind of vigilance on behalf of female consumers and human subjects as a hindrance, we need to take these criticisms of new pharmaceuticals and reproductive technologies seriously and to focus on ways we can foster rapprochement between health care professionals, representatives from the pharmaceutical industry, and members of feminist women's health advocacy groups. In particular, it is important to seriously consider issues of access and health care disparities. My literature survey indicates that to date, implementation of the CDC policy by adolescent health care professionals has focused on vaccine acceptance by parents, practitioners, and to a lesser extent, adolescent and young adult women.[50] I suggest that by focusing on individual attitudes and choice, this policy overlooks larger public health issues such as socioeconomic status and access to health care services. It is mainly the women's health activist Web sites, such as the fabulous *Our Bodies, Our Blog*, www.ourbodiesourselves.org/, that are going beyond this individualist model. Until there is a greater social commitment to meeting the health needs of uninsured and underinsured women, a disproportionate number of whom are from racial minorities, these women will lack the routine preventive care that more privileged women take for granted.

NOTES

1. Karen Houppert, "Who's Afraid of Gardasil?" *Nation*, March 26, 2007, www .thenation.com/doc/20070326/houppert, accessed March 18, 2008.

2. R. Alta Charo, "Politics, Parents, and Prophylaxis—Mandating HPV Vaccination in the United States," *New England Journal of Medicine* 356 (May 10, 2007): 1905–1908.

3. Allan M. Brandt, "Behavior, Disease, and Health in Twentieth-Century America: The Moral Valence of Individual Risk," in *Morality and Health*, ed. Allan M. Brandt and Paul Rozin (New York: Routledge, 1997).

4. Rima Apple, *Perfect Motherhood: Science and Childrearing in America* (New Brunswick, NJ: Rutgers University Press, 2006).

5. Elizabeth Toon, "Teaching Children about Health," in *Children and Youth in Sickness and in Health: A Historical Handbook and Guide*, ed. Janet Golden, Richard Meckel, and Heather Munro Prescott (Westport, CT: Greenwood Press, 2004), 93–95.

6. Apple, *Perfect Motherhood*, 45.

7. Leslie Reagan, "An Epidemic of Anxiety: Maternal Rubella, 1963–65," paper presented at the American Association for the History of Medicine conference, May 2003.

8. "Someday . . . You Might Have a Baby! Don't Risk Birth Defects from Rubella! (German Measles)." Image ID no. 208455, National Library of Medicine, Images from the History of Medicine Digital Collection. Bethesda, MD.

9. David Oshinsky, *Polio: An American Story* (New York: Oxford University Press, 2006), 188–189.

10. Clelia Duel Mosher, *The Mosher Survey: Sexual Attitudes of Forty-five Victorian Women* (reprint, New York: Arno Press, 1980). Mosher described similar changes among Stanford coeds in her diary, Clelia Duel Mosher Papers, SC 11, box 1, folder 10, Stanford University Archives; Katherine Bement Davis, *Factors in the Sex Life of Twenty-two Hundred Women* (New York: Harper, 1929).

11. Martin Summers, Manliness and Its Discontents The Black Middle Class and the Transformation of Masculinity, 1900–1930 (Chapel Hill: University of North Carolina Press, 2004), 272, 244, 276.

12. Christina Simmons, "African Americans and Sexual Victorianism in the Social Hygiene Movement, 1910–1940," *Journal of the History of Sexuality* 4 (1993): 51–75.

13. Jeffrey Moran, *Teaching Sex: The Shaping of Adolescence in the 20th Century* (Cambridge, MA: Harvard University Press, 2000), 23–32.

14. G. Stanley Hall, "Education and the Social Hygiene Movement," *Journal of Social Hygiene* 1, no. 1 (1914): 30.

15. G. Stanley Hall, "The Teaching of Sex in Schools and Colleges," *Social Diseases* 2, no. 4 (1911): 19.

16. Ruth M. Alexander, *The "Girl Problem,": Female Sexual Delinquency in New York, 1900–1930* (Ithaca, NY: Cornell University Press, 1995), chaps. 2 and 4.

17. The quotation is from Leslie Reagan, "An Epidemic of Anxiety: Maternal Rubella, 1963–65," paper presented at the American Association for the History of Medicine conference, May 2003; Marilyn E. Hegarty, Victory Girls, Khaki-Wackies, and Patriotutes: The Regulation of Female Sexuality during World War II (New York: New York University Press, 2008).

18. John Parascandola, "Presidential Address: Quarantining Women: Venereal Disease Rapid Treatment Centers in World War II America," *Bulletin of the History of Medicine* 83, no. 3 (Fall 2009): 431–459.

19. Alfred Kinsey, *Sexual Behavior in the Human Female* (Philadelphia: Saunders, 1953).

20. Beth L. Bailey, *From Front Porch to Back Seat: Courtship in Twentieth-Century America* (Baltimore: Johns Hopkins University Press, 1988), 43.

21. Monica J. Casper and Adele E. Clarke, "Making the Pap Smear into the 'Right Tool' for the Job: Cervical Cancer Screening n the USA, circa 1940–95," *Social Studies of Science* 28, no. 2. (April 1998): 255–290.

22. Heather Munro Prescott, "Guides to Womanhood: Gynecology and Adolescent Sexuality in the Post World War II Era," in *Women, Health, and Nation: Canada and the United States since 1945*, ed. Georgina Feldberg, Molly Ladd-Taylor, Alison Li, and Kathryn McPherson (Montreal: McGill-Queen's University Press, 2003).

23. This desire to protect young women from premature sexual awakenings extended well beyond the profession of gynecology. See Joan Jacobs Brumberg, *The Body Project: An Intimate History of American Girls* (New York: Random House, 1997); and Constance Nathanson, *Dangerous Passage: The Social Control of Sexuality in Women's Adolescence* (Philadelphia: Temple University Press, 1991).

24. Goodrich C. Schauffler, *Pediatric Gynecology* (Chicago: Year Book, 1942), 7.

25. Edward D. Allen, "Gynecological Problems of the Adolescent Girl," *Medical Clinics of North America* 27 (1943): 17.

26. Heather Munro Prescott, *"A Doctor of Their Own": The History of Adolescent Medicine* (Cambridge, MA: Harvard University Press, 1998).

27. J. Roswell Gallagher, *Medical Care of the Adolescent* (New York: Appleton-Century Crofts, 1960), 196–197.

28. Edward D. Allen, "Pelvic Examination of the Preadolescent and Adolescent Girl," *Transactions of the American Gynecological Society* 77 (1954): 109–110.

29. Goodrich C. Schauffler, *Guiding Your Daughter to Confident Womanhood* (Englewood Cliffs, NJ: Prentice-Hall, 1964), 65, 96. For similar views, see Allen, "Gynecological Problems of the Adolescent Girl," 17–25; Edward D. Allen, "Examination of the

Genital Organs in the Prepubescent and in the Adolescent Girl," *Pediatric Clinics of North America* (Feb. 1958): 19–34; and Laman Gray, "Gynecology in Adolescence," *Pediatric Clinics of North America* (Feb. 1960): 43–63.

30. Schauffler, Pediatric Gynecology, 15.

31. Ibid., 17–18.

32. John W. Huffman, "Gynecologic Examination of the Premenarchal Child," *Postgraduate Medicine* 25 (1959): 169.

33. Allen, "Examination of the Genital Organs," 21–22.

34. Nathanson, Dangerous Passage; Rickie Solinger, Wake Up Little Susie: Single Pregnancy and Race before Roe v. Wade (New York: Routledge, 1992).

35. For more on this point, see Solinger, *Wake Up Little Susie*, 214–215. I describe the impact of these attitudes on pregnancy-prevention programs in further detail in "*A Doctor of Their Own*," 149–150.

36. These studies are summarized in I. D. Rotkin, "A Comparison Review of Key Epidemiological Studies in Cervical Cancer Related to Current Searches for Transmissible Agents," *Cancer Research* 33 (June 1973): 1353–1367.

37. I. D. Rotkin, "Adolescent Coitus and Cervical Cancer: Associations of Related Events with Increased Risk," *Cancer Research* 27, no. 4 (April 1967): 603–617. See also I. D. Rotkin, "Relation of Adolescent Coitus to Cervical Cancer Risk," *Journal of the American Medical Association* 179 (Feb. 17, 1962): 486–91.

38. I. D. Rotkin, "Epidemiology of Cancer of the Cervix. 3. Sexual Characteristics of a Cervical Cancer Population," *American Journal of Public Health and the Nations Health* 57, no. 5 (May 1967): 827.

39. Beth Bailey, "Sexual Revolution(s)," in *The Sixties: From Memory to History*, ed. David Farber (Chapel Hill: University of North Carolina Press, 1994), 235–262.

40. Jerome S. Menaker, "Teenage Obstetrics," *Pediatrics Clinics of North America* (Feb. 1958): 141–142.

41. Anna L. Southam, "Metropathia Hemorrhagia and Nonpsychogenic Amenorrhea," in *Adolescent Gynecology*, ed. Felix P. Heald (Baltimore, MD: Williams and Wilkins, 1961), 51.

42. Boston Women's Health Book Collective, *Our Bodies, Our Selves: A Book By and For Women* (New York: Simon and Schuster, 1971). For more on this publication, see Sara Elaine Hayden, "Re-Claiming Bodies of Knowledge: An Exploration of the Relationship between Feminist Theorizing and Feminine Style in the Rhetoric of the Boston Women's Health Book Collective," *Western Journal of Communication* 61 (1997): 127.

43. S. Jean Emans and Donald Peter Goldstein, *Pediatric and Adolescent Gynecology* (Boston: Little, Brown, 1982), 10.

44. Belita H. Cowan, "Special Report: The 'Morning-After' Pill," *her-self,* Sept.–Oct. 1974, 6–7.

45. Brumberg, *The Body Project*, 171–192, quotes on 173.

46. Charo, "Politics, Parents, and Prophylaxis," 1905–1908.

47. Jan Hoffman, "Vaccinating Boys for Girls Sake?" *New York Times*, Feb. 24, 2008, www.nytimes.com/2008/02/24/fashion/24virus.html?pagewanted=1&_r=1.

48. Society for Adolescent Medicine, "Position Statement: Human Papillomavirus (HPV) Vaccine," www.adolescenthealth.org/PositionStatemsent_HPV_Vaccine.pdf, accessed Jan. 17, 2008.

49. Transcript of May 18, 2006, meeting, Vaccines and Related Biological Products Advisory Committee, www.fda.gov/ohrms/dockets/ac/06/transcripts/2006-4222t1.pdf.

50. "Teen Girls Voice Mixed Thoughts, Info over HPV Vaccine," *New York Amsterdam News* 98, no. 17 (April 19, 2007): 18; Donald R. Hoover, Beth Carfioli, and Elizabeth A. Moench, "Attitudes of Adolescent/Young Adult Women toward Human Papillomavirus Vaccination and Clinical Trials," *Health Care for Women International* 21 (2000): 375–391.

CHAPTER SEVEN

Producing and Protecting Risky Girlhoods

Laura Mamo, Amber Nelson, and Aleia Clark

Medical innovations and practices have long been viewed as suspect sites for women's health and bodies (Reismann 1985), and medicine has been seen as a site of social control (Zola 1972). As medicine gradually extended beyond mandates to cure the ill, a medicalization of daily living emerged; an increasing number of classifications and treatments were applied to those deemed "healthy," and as a result medicine has come to play an expanded role in our human existence. Putting drugs into healthy bodies requires the production and acceptance of the notion that our bodies are always and already "at risk." To ensure success in the adoption and thus in sales of Gardasil, Merck needed to produce girls and their bodies as being at risk of acquiring an illness.

This chapter analyzes the ways Merck sought to garner public acceptance for the Gardasil vaccine through a promotion of girls' bodies as at risk for cervical cancer and girls and women's subjectivities as being ideal consumers of vaccine technology. In analyzing the branded and nonbranded direct-to-consumer (DTC) advertising campaigns for Gardasil alongside other text advertising created by Merck to produce "awareness" about the human papillomavirus and cervical cancer, we ask these questions:

- What are the sociocultural mechanisms through which Merck promoted its medical innovation as the "right tool" for U.S. cancer prevention?
- In what ways did Merck produce U.S. girls as the ideal users of the Gardasil vaccine?
- What are the cultural, social, and political implications of this emergent technology and its associated productions of gender and sexuality?

Biomedicalization theory and girlhood studies frame our analysis. Biomedicalization allows a lens into the shifting mechanisms of health care discourse and practices and their co-constitutive effects on subjectivities and bodies, and girlhood studies pay particular attention to the dynamics of gender and age along with race, class, and sexuality embedded in and through discourses that construct meanings of girls and girlhood.

We utilized a feminist discursive analytical method to examine the print and online media for Gardasil trademarked from 2005 through 2007. We did not employ a content or media analysis, as is often done by sociologists. Instead, we analyzed the ways language produced, not described, meanings of gender, sexuality, and health.[1] Drawing on literary-cultural studies rather than a linguistic-sociolinguistic tradition (Mills 2004, 9), we understand discourse as producing certain, albeit contradictory, ways of thinking and speaking about gender and power relations in a particular moment and location. Our reading, therefore, examined the discursive frameworks producing the ideas about gender and health employed in the Gardasil advertisements.

We look at the cultural work employed to garner support for the HPV vaccine found in Merck's "One Less" advertising campaign and its nonbranded *awareness* campaigns variously named "Tell Someone," "Make the Connection," and "Make the Commitment."[2] The awareness campaign began before U.S. Food and Drug Administration (FDA) approval was granted to Merck for the release of Gardasil. The DTC media promotion was launched on September 30, 2005, as a nonbranded campaign entitled "Make the Connection." This campaign focused on raising public awareness about the etiological role of HPV in cervical cancer. The American "awareness" slogan "Tell Someone" encouraged girls to tell someone about the virus that causes cancer. Merck also partnered with several health-oriented nonprofit organizations and funded nationwide events featuring celebrities who made and wore beaded "Make the Connection" bracelets to promote cervical cancer awareness (Siers-Poisson 2007). In 2006, Merck received FDA approval for Gardasil and then unleashed a high-gloss DTC marketing campaign titled "One Less." We analyze this campaign alongside the education and awareness marketing that occurred before FDA approval.

Through a discourse analysis, we show the multiple ways Gardasil marketing produced a gendering of health message that, we argue, relied upon the production of cervical cancer as a disease of innocence, replacing discourses of sexually transmitted infection (STI) risk and their associated construction of "at risk" sexual actors. Instead of categories of "at-risk" girls, a universal girlhood is produced. Girls are produced not as at sexual risk but as passing through a normal life stage from childhood to adulthood. Girls, a generic unmarked category, possess a moral obligation to be healthy as they move toward their presumed future heterosexual lives. This construction, we argue, is relationally produced along with the invisible others: past and present deviant girls (i.e., sexually promiscuous and queer) and queer boys. Ideal users of Gardasil, heterosexual yet not quite heterosexually active girls, are empowered: not as agents of their sexual desires, behaviors, and choices, but as agential health consumers who, by making the "right choices," can realize their imagined disease-free adult bodies.

Biomedicalization and Health

Contemporary sets of health care processes and ways of knowing frame the release and adoption of the human papillomavirus vaccination. Today, innovations and hybridizations in molecular biology and informatics technologies shift the target of medical intervention from pathology (i.e., illness) to risk assessment and risk reduction (i.e., health). Dynamics involved in the production, marketing, and adoption of pharmacotherapies include convincing populations to adopt a new technology without the immediate risk (or desperation) that once accompanied drug recommendations. Further, in the context of U.S. for-profit, consumer-based medical care, drug innovations are no longer driven by greatest need but by greatest market potential.

In this context, health, not illness, has emerged as the object of medical intervention and the active goal of consumers and their consumption practices. Individuals have a moral responsibility not only to be healthy but also to improve themselves, their bodies, and their lives through consumption of available biotechnological interventions. In a consumption rubric, medical products are ways to achieve a desired appearance, identity, self, and degree of physical control. This is what Deborah Lupton terms an efficient performance of the self (Lupton 1995).

Within this context, we increasingly come to understand our selves through the language of biomedicine and experience our lives through biology. Beyond individual subjectivities, these same biological risk identifications become a basis for the formation of group and collective social bonds and actions (Rose [2001] 2006). Groups and their shared collective identifications are key actors in biomedical discourse and practice, resisting certain interventions and identifications and selectively choosing others. As a result, "Risk-based, genomics-based, epidemiology-based, and other technoscience-based identities" are produced and mobilized through medical classifications and interventions (Clarke et al. 2003, 187). Subjects are brought forward, identities are shaped, and bodies are inscribed with meaning as biomedical knowledges, technologies, and practices are brought into the hands and minds of their users (Balsamo 1996; Haraway 1991; Martin 1994).

Vaccination, including vaccination with the Gardasil vaccine, becomes another do-it-yourself (DIY) biological project enabling girls to transform their identities, bodies, and futures. Shifting from the normalization of individuals to the customization of bodies, potential future selves are brought into the present through biotechnologies such as Gardasil.[3] Significantly, a consumer imperative accompanies this customization: individual girls and girlhood collectivities are brought forward to consume in the name of health, adhering neatly to neoliberal ethics of individual responsibility and normative constructions of gender and sexuality. New biomedical interventions, in turn, are "technologies of optimization," designed to improve life for the healthy rather than cure the sick (Rose [2001] 2006). And it is the moral imperative of consumers to heed the call. This imperative is especially evident as the role of public health is increasingly privatized and delivered by the pharmaceutical "education" campaigns preceding and accompanying product advertising campaigns. With this increasing privatization of public health along with health care knowledge and medical care services, economic production has been relocated at the genetic, microbial, and cellular level: life has been drawn into the circuits of value creation (Cooper 2008). That Gardasil emerged on the U.S. market as a very expensive (and profitable) risk treatment is an important part of this politico-economic constitution. Further, not only are biomedical innovations such as Gardasil constitutive of capital; they are sites of politics. These are not neutral entities but places of cultural, social, and political contestation.[4]

The contemporary U.S. health care context is one in which biomedical innovations are not only material entities; they are culturally symbolic, increasingly shaping people's conceptions of their health, identities, and bodies. The way in which technologies are discursively constructed "matters" for the acceptance and resistance of any technology. Putting drugs into (healthy) bodies requires the preproduction of the body itself as in need of intervention and the production of subjects ready and able (and even desiring) to consume the products offered. Once bodies and their associated subjectivities are produced, they are made meaningful (and potentially profitable) within a medical discourse.

Before the medical innovation of an HPV vaccine such as Gardasil could be realized, corporeal markets had to be envisioned and public acceptance created. In the United States, girls are the first market for Gardasil and the early adopters of this vaccine. While the market could have been established otherwise (see especially chapter 4), Merck needed to produce all (not some) girls as the ideal users of Gardasil. We ask, In what ways did Merck produce girls, girls' bodies, and girlhoods as symbolically meaningful? And how did Merck produce girls as the ideal objects and agential users of this vaccine intervention?

Do-It-Yourself Feminism: Girlhood Power and Crisis

The production of empowered girls necessitates these girls' relational others: girls in crisis. We argue that invisible yet implicated others, including deviant, sexually promiscuous girls; "at-risk" girls; and queer boys and girls of both the past and present, are part of the Gardasil advertising campaign. Girlhood studies offer an important angle of vision here. The concept "girlhood" has replaced "youth" as a symbolically meaningful category bringing feminine specificity forward (Gonick 2004; McRobbie 2001, 201). Feminist girlhood scholars challenge the limited discursive representations of girl embodiments, degrees of masculinity and femininity, and options for feminine adolescent subjectivity, including sexual agency and sexual health (Halberstam 1995; Tolman 1994). These contrast with representations of earlier generations of girls; those girls were primarily depicted as newly passive and experiencing a "loss of voice" from their preadolescent selves (Gilligan 1982) or entering a crisis of self-esteem (Pipher 1994). Rejecting such treatments, a discourse of girls as subjects via "Girl Power" emerged and coincided with the rise of DIY feminism (see Driscoll 2002). "The idea of girl power encapsulates the narrative of the successful new young woman who is self-inventing, ambitious, and confident" (Harris 2004, 17). Contrasting the "can-do" girl who possesses girl power to the "at-risk" girl in need of protection, Harris explains that the former conception represents ideal or hegemonic (white, middle-class) girlhood. Resilient and self-reliant, the "can-do" girl creates her present self and her future possibilities as she strives for success defined in terms of normative adulthood. With girl power, "can-do" girls embrace a DIY feminism of self-empowerment.

As Marnina Gonick (2006) argues, the girl-in-crisis subject is fragile, vulnerable, passive, and in need of protection. Both "Girl Power" and what is termed "Reviving Ophelia" discourses are psychological knowledge forms that produce certain kinds of feminine subjectivities (18) characterized by Elizabeth Marshall (2006) as "wounded girls." Marshall shows how psychological and popular discourses are never neutral and often position girls as compelled toward psychically injured subjectivities. Yet, such discourses of injured girls are both produced and resisted through cultural texts. Working in tandem, these discourses produce a dual focus that normalizes individual choice and personal failure and resonates with requirements for a neoliberal subject. Gonick (2006) asserts that neoliberal discourse demands that citizens remake themselves through consumption.

In popular media, girlhood discourse often serves as a device for understanding, monitoring, and regulating girls, their bodies, and their presumed feminine subjectivities (Gonick 2006, 18). S. Lamb and L. M. Brown (2006) argue that too often when the term *girl power* is used, it refers to the power to negotiate "choices" regarding product consumption. As a result, what started as a powerful movement to enhance girls' power slowly eroded as corporate entities coopted empowerment messages as "warped versions of girl power" and fantasized identity stories (1–2).

Beyond monitoring and regulating girlhood and feminine adolescence, these discourses produce legibility, as well as lived subjectivities and embodiments.[5] Absent are real girls situated in material social locations as well as agential adolescent sexual subjects.[6] It is our argument below that girlhood is produced by Gardasil as embodying a group-based precancer risk identification and one that can be overcome. Individual girls, in turn, are produced as biomedical citizens and consumer subjects pivotal to the production of their health.

Gendering the "Will to Health" in Advertising
The Rightness of Vaccine Technology: Making Gardasil the Right Tool for the Job

In what ways was Gardasil promoted as the "right tool" for cancer prevention, when condoms and Pap smear diagnostic technology had already successfully reduced U.S. cervical cancer rates by 70 percent?[7] We found that the construction of girlhood as an at-risk category moved from sexual risk to cancer risk and thereby produced Gardasil as the right tool for cancer prevention. This construction followed, yet sidestepped, a common gendered narrative and one that is often simultaneously raced and classed of girlhood sexual risk. Girls' sexual desire and sexual activity prior to adulthood have long been produced as problematic entities in need of protection and regulation. Some girls *are* problems (those who engage in so-called early sexual activity and risk or become pregnant teens), while others are in need of protection from boys' and men's assumed unsolicited sexual desires. We argue that relying on, yet sidestepping, these sexual narratives was a first step in the cultural work ensuring the promotion of Gardasil as the right tool for cancer prevention in girls. How so? First, Merck produced cervical cancer not only as the "object" of intervention but as a major public health concern for *all* girls. Second, Merck advertisements captured and reproduced the existent discourse of cervical cancer as a "silent killer" of U.S. women along with the generalized media attention and fear of reproductive cancer for all women. With slightly fewer than four thousand U.S. deaths per year, cervical cancer was transformed from a relatively rare cancer with an effective health care infrastructure for prevention and early detection into a death sentence, a "disease of innocence," *and* a major public health concern. In all, Merck produced girls and their bodies as being at risk *not* for HPV, but for the specter of cervical cancer, through its marketing of Gardasil as the presumed "right tool for the job."

While cervical cancer *is* a normal health risk for all girls, generalized risk and universal girlhoods do not exist. The advertising campaigns had some cultural work on their hands. Some girls *are* at increased risk by virtue of a lack of access to routine health care services; unequal gender, race, and class power relations; their social networks; their exposure to misinformation or a lack of sexual health education; and the structural and cultural obstacles to condom use. Visually, the campaign utilizes a multiethnic cast to simultaneously and symbolically blur and distinguish innocence from risk, calling upon latent ideas of deviant girls and at-risk teens. Racialization and class imagery made this tactic effective. What results is a Gardasil education and advertising campaign that successfully converts the STI risk symbolically associated with deviant sexually active girls and boys into normal cancer risk for *all* preadolescent girls. In the process, Gardasil becomes an advanced biomolecular therapy in the war against (women's) cancer. This HPV vaccine does not manage sexual risk or risk groups; it manages a girl's life and even "life itself." Such a move displaces any need for sex education and contraception access as necessary prevention strategies for STIs and simultaneously minimizes the importance of Pap smear technology in early detection for not only precancerous HPV but all forms of HPV disease (see Nack 2008 for an interview-based study of women living with chronic and incurable HPV disease).

Yet, as Gardasil provides viral protection and erases the need for sex education and sexual health literacy (see chapter 8), it compels individuals to know their risk and modify their bodies. Mothers are compelled not to think about sexual risk but to consider the risks of cancer. Daughters, in turn, are compelled to inform themselves and become health care consumers willing and able to engage in biomedical practices of risk reduction. Thus, it is individuals and their fears and risks that are called forward as objects and actors in contemporary consumer-based medicine. These individuals are not asked to think about their sexual selves and desires, their population-level risk, or their social network, topics usually considered in STI research; instead they are asked to think about (and become) individual health care consumers engaging in biocitizenship (see chapter 4 for a definition of biocitizenship). Girls are part of a collectivity of normal girls at risk for cancer.

Averting a Sex Panic: Cancer of Innocence as a Major U.S. Public Health Concern

In the months before FDA approval of Gardasil, controversy erupted over the cervical cancer vaccine as some perceived that this vaccine would encourage teenage sex (Stein 2005). This reaction was surprising for vaccine researchers and, we believe, led Merck to emphasize Gardasil as the "right tool" for cancer prevention over HPV prevention. It was the HPV–cervical cancer link discovered by Harald zur Hausen, who was later awarded the Nobel Prize in Medicine, that drove the development of Gardasil. Yet, despite the proven link, medical researchers in search of cancer prevention found themselves at odds with social conservatives who interpreted HPV immunization as the encouragement of adolescent (even preadolescent) sexual activity. An HPV vaccine brought to mind sexual agency in girls and, we argue, in the promiscuous girls of previous decades. Abstinence-only education faded away. As Merck entered the early FDA approval process, it seemed to have a "sex panic" on its hands and a panic that conjured notions of girls as problems. Yet, this vaccine did not follow this path (see chapter 1). By the time Merck secured FDA approval, the controversy had abated, and in its wake emerged not only widespread support for the HPV vaccine but calls to mandate its use for girls. And within a few years, these mandates were mostly abated as a new risky girl emerged: the girl at risk of Gardasil side effects. However, mandates for girls entering the sixth grade were passed in Washington, D.C., and the state of Virginia (with opt-out provisions) and for girls and women seeking immigration to the United States.

The controversy, nonetheless, has much to tell us about the sociocultural intersections of gender, sexuality, and drug innovations. Although short-lived, the controversy was not straightforward: many of the same social-conservative positions that called for the rejection of Gardasil were at the forefront of efforts to mandate vaccination for all school-aged girls. In Texas, Governor Rick Perry issued an executive order mandating Gardasil for sixth-grade girls and requiring free vaccination to uninsured and underinsured girls of ages 9 to 18. The decision was overturned in May 2007, but this legal battle demonstrates a conservative shift from containing girls' sexual agency to promoting girls as the ideal users of HPV vaccination. These policy initiatives capture some girls through public mandates and produce a consumption imperative for others, raising questions about the (uneven) role of governments, public health, and private enterprise in the health of the nation.

The early unfolding of these events signaled to Merck a need to promote Gardasil as the "right tool" for cancer prevention even before FDA approval. Through "awareness and education" campaigns, Merck positioned Gardasil as an intervention into cervical cancer rather than protection from the consequences of unprotected sexual behaviors. Despite the message that all girls are at risk of cancer, it seemed that some girls (and their parents) required a mandate to receive vaccination. Government-provided public health was a default health care service for certain girls. The diffusion of both the original controversy around encouraging sex and the subsequent controversy around the mandate reinforced the production of girls' bodies in need of intervention not because of risky sexual behaviors but because of the risks inherent in girls' bodies. A corporeal market was created when risky sexual behaviors no longer in need of prevention were replaced by the innocent yet somehow "risky" bodies of 9-to-26-year-olds, a risk group imagined and created by Merck. This age range was used in the hope of capturing pre-sexual girls, or intervening before sexual activity began, yet it is also culturally symbolic of what sociologists term "the lifecourse" and what queer cultural studies scholars' term "heterosexual time": a period of girl time en route to normative adulthood (see Elder 1998; Halberstam 2005). As Chesler and Kessler argue in chapter 8, this girl-like period occurs prior to monogamy and "settling down." We argue here that displacing sexual risk with cancer risk was a necessary precondition to averting these and any other sex panics that might have accompanied an HPV vaccine. Usual associations with the time and space between innocent childhood and normal adulthood (i.e., raging hormones, sexual promiscuity, and girls at risk) need not be brought into the minds of potential users. The effacement of sexual risk required the production of universal girls empowered through the moral obligation to knowledge, health, and consumption.

A Will to Knowledge, A Will to Health: "Know the Link"

The "Make the Connection" and "Tell Someone" campaigns reveal cultural mechanisms that paved the way for Gardasil to become the "right tool" for cancer prevention. The media blitz called forward a "will to knowledge" that cervical cancer is a risk for all girls and women. The phrase "know the link," of HPV and cancer, begins this process. A peer-to-peer communication visual is used as a strategy for conveying knowledge; the personalized phrasing "Gardasil and You" engages "you" directly as a recipient of HPV knowledge. Public health is thus personalized: girls' bodies are produced as at risk and as a collective, and personalized girls' agency emerges as available to generate protection. Girls are compelled by the ads to know their bodies as at risk of HPV and subsequently of cervical cancer. Before the approval of Gardasil and the more prominent "One Less" campaign, girls were encouraged to know the facts and know the risks. The risks, however, were constructed as not getting vaccinated, and they were explicitly not constructed as the sexual health risks of engaging in unprotected sexual activity.

Significantly, boys and men were not included in this educational campaign. Their absence is not surprising given the longstanding medical and cultural pathologization of women's bodies and its concomitant production of their bodies as ripe for medical interventions, including pills for "healthy" bodies, as with the birth control pill. Young women were and continue to be a "blockbuster" market for many medical interventions. Gardasil merely extends this responsibility to girls.

Instead, as a vaccine protecting girls from cervical cancer, Gardasil is shaped by and through biomedicalization's emphasis on consumption and risk reduction. The rhetorical strategies that produce cervical cancer as a major health concern and a disease of innocence position all girls in a cancer-risk group, while the imperative or "will to health" insists that individuals take charge of risk reduction. Gardasil becomes the right tool by fulfilling this risk-reduction need and sidestepping sexuality. Girls become ideal users because they possess innocent (yet monitored and regulated) bodies not yet affected by the risks of HPV and complex sexual desires and lives. Further, they are the daughters of women who are actually at risk for developing cancer (increased age is the most common risk factor for cancer). The specter of vulnerable adolescent girls (and predatory men and boys) diminishes for this new generation, replaced with practices to assess and reduce health risk. Risky girls do not harbor the same risks that their mothers did (unwanted sexual advances and encounters by predatory men as well as risks of HIV infection that accompanied most sexual encounters); instead, they harbor the risk of cancer.

Such a health-risk discourse is concomitant with biomedicine's economic imperative and turn to "health." The HPV vaccine not only utilizes molecular technologies; it does so to transform "risk" and therefore bodies. The scientific and cultural understandings of cervical cancer risk *before* the FDA approval of the HPV vaccine targeted sexually active women and their bodies. Today, that risk is transported onto female bodies before sexual activity occurs and as they are imagining their future adult selves and bodies: the so-called problems associated with teenagers are absent. As Giovanna Chesler and Bree Kessler (chapter 8) demonstrate, girls are presented with a limited binary: to be vaccinated or not to be vaccinated. Such a choice is false, as it ignores other possibilities, such as protecting oneself from more than four strains of HPV. Chesler and Kessler are concerned that such a false choice negates or reduces the need for safe-sex practices and for expanded sexual health education. We agree, and we also see another false binary: Health or Cancer. Reducing the alternatives to either vaccination or cancer risk effaces the full range of HPV illnesses to which all sexually active or previously sexually active girls, women, boys, and men are at risk of exposure.

It is girls who are called forward with a moral responsibility to enhance their well-being through reconfigurations of the self and its bioattributes. On the way to becoming normal adult women, girls must become empowered consumer citizens rationally reducing perceived risks associated with tumultuous adolescence, young adulthood, and normal womanhood. In this production, preadolescent girls are singled out as the subjects and bodies to reduce, if not eliminate, this life-threatening disease and the larger public health crisis now associated with it.

The focus on pre–sexually active girls allows one to blink one's eye and skip over the specters of risky sex, predatory men, and complex issues of sexuality altogether. Issues such as age of consent, sexual agency, diverse sexual desires and practices, homophobia, the specter of HPV disease, genital warts, and other sexually transmitted infections, including HIV, can and do disappear with the cultural work of Gardasil.

This cultural work is accomplished in the advertising campaign titled "Tell Someone." As a new public health educator, Merck encourages girls to become educated and act in particular ways. Girls learn that they are bound together by their risky female embodiment and their moral obligation to educate themselves and others. The discourse solicits girls to will themselves to health and take on future (disease-free) identities of mothers sharing a burden of health care for their families and the nation.

The Merck pamphlet and other media sources under the "Tell Someone" campaign rubric feature images of girls standing, most often smiling at the camera, either alone or accompanied by other girls. The only adult pictured is a woman doctor shown advising a girl patient. The girls are mostly engaged in physical contact with one another, arms entwined or holding hands. This "girl-like" representation is an important move to efface the specters discussed above. Further, the idea of telling someone evokes girlhood culture as a period of intense girl-talk wherein girls engage in frequent communication, sharing their lives and learning from one another. With Gardasil the girls are also connected by their shared risk of cancer status; they become a social group with a collective bio-identity. Once girls are produced as a corporeal market for Gardasil, girlhood allows the necessary conditions for knowledge dissemination and subsequent consumption.

Beyond the Medicalization of Deviance:
The Biomedicalization of Health

Cindy Patton (1995) documented the ways media coverage around HIV produced deviant and normal adolescents moving through the passage from innocent child to sexually responsible adult. Media produced three cultural groups: (1) "adolescence": the dominant white, innocent youth moving appropriately through a stage of "raging hormones" on their way to "civilized adulthood" (read: married heterosexual procreation); (2) "gay youth": moving from innocence to a subcultural (male homosexual) world that foreclosed "civilized adulthood"; and (3) "at-risk teens": young people of color never associated with innocent childhood or civilized adulthood, but instead positioned outside the concerns of protecting innocence and of individuals at the brink of responsible adulthood. Gender distinctions, she argued, were subordinated to racial and class hierarchies, with gender most pronounced for poor youth of color and with nonwhite boys associated with HIV/AIDS and potential violence (Patton 1995).

While these discourses continue, serving as "past in present" ideas about feminine and masculine sexuality (Collins 2004, 20), the Gardasil campaigns produce preadolescent girls able to elide "truths" about their present and pending sexualities. The culture work allows other groups demarcated by age, race, sex, and class to occupy positions of deviance in contrast to innocent, generic, and unmarked girlhood. Racialized girls, black, Latina, and white, are interspersed throughout the advertisements not only as a multiethnic marketing technique but also as a means to symbolically demarcate the deviant from the innocent, pushing viewers into a "colorblind racism" in which they see but do not see the ways the advertisements depict "at risk" teens and "adolescents," the former black, brown, and poor and the latter white and affluent. As girlhood scholars assert, all girls occupy the generic (read: white, middle-class, and empowered) girl of neoliberalism.

Until recently there were no popular accounts of girls' risk of contracting cervical cancer.[8] Instead, risk identifications emerged in sex panics associated with (1) "deviant girls," youth at risk of unwanted pregnancy, and STIs as a result of promiscuity; (2) innocent girls at risk of sexual predators; and more recently, (3) sexually deviant and gender-deviant boys and girls expressing same-sex desires or engaging in gender nonconforming practices. Responsible girls emerge relationally through the contrasted irresponsible "at-risk teens" and "gay youth" of the 1980s and the more recent queer (and not-yet-queer) boys and girls.[9]

The deviant girl "at risk" for HIV in Patton's study represents the "past-in present" idea of girlhood sexuality that penetrates and is mobilized by the discursive apparatus through which Gardasil becomes the right tool for the prevention of cervical cancer. The "Tell Someone" campaign began the work of effacing the sex in a sexually transmitted disease by the seemingly contradictory rhetorical move of redefining HPV as a normal result of (hetero)sexual inter-course that is common and ostensibly unavoidable. A Merck pamphlet advises, "HPV is easily transmitted, so any exposure puts you at risk. It's estimated that many people get HPV within their first two to three years of becoming sexually active" (2006, 4). All (heterosexual) sex is thus redefined as risky for contracting HPV and therefore as cancer risk. Furthering the rationale for Gardasil, the pamphlet cites the Centers for Disease Control and Prevention: "The only way to protect yourself against HPV is to *avoid* any sexual activity that involves geni-tal contact" (ibid., original emphasis).

Producing Ideal Vaccine Users: Gendered Call to Action

We found that Gardasil brings forward the "can-do" girl ready and able to undergo vaccination to rid her body of the potential risk of cervical cancer. This "can-do" girl is not one of sexual empowerment but of consumer empow-erment. Strong girls are depicted intermittently with mothers at their sides as they negotiate health care knowledge and select appropriate body practices. Gardasil is produced by Merck as the right tool by enlisting girls and their mothers as active health care consumers reducing later risks of developing cer-vical cancer (not as risky sexual actors reducing STI risk). In doing so, Merck captured what is a well-known gender disparity in health care consumerism—girls and women are primed to receive health care products and messaging in ways that men and boys are not.

This enlisting required several discursive moves: (1) sidestepping girls' sex-ual agency in favor of empowering them as healthy and engaged consumers; (2) constructing mothers as playing a vital role in not only their daughters' health but the health of the nation: mothers are valorized as essential actors, women's health advocates, feminist activists, and moral pioneers protecting girls from harms of adulthood; and (3) inviting mothers and daughters into a collaborative effort to optimize health and manage life. Each of these moves produces a feminized collectivity responsible for protecting the nation's health through body practices.

Crafting One's Future: Becoming One (1) Less Life

Once Gardasil was approved by the FDA, the "One Less" DTC campaign promoted its use. The television version of the campaign depicts racially diverse girls speaking directly into the camera while engaged in activity. Scene one is a girl skateboarding and doodling "one less" on her sneaker as she tells the "facts" of cervical cancer. The commercial quickly cuts to two girls playing basketball. Then it switches to a girl engaging the viewer and then to a skilled drummer sounding out, "Now there is Gardasil, the only vaccine that may help protect you from the four types of human papillomavirus that may cause 70 percent of cervical cancer." A white adult woman (presumably a mother) appears, stating the side effects of Gardasil. As she speaks, we see images of a white girl, presumably her daughter, juggling a soccer ball. Then we see a black girl saying to the camera, "Become One Less," her words spliced with images of a group of four black girls dancing.

Merck's "One-Less" slogan for selling Gardasil (re)produces constructions of girls as in crisis (at risk of getting cervical cancer), yet also as able to imagine and produce their healthy future bodies. These girls are winners. They not only can do it all; they are doing it all: designing clothes, winning races, and crafting health education messages. They possess a girl power accessible through biomedical interventions into their presumably risky bodies.

Confidently holding "one less" signs above their heads—a Number 1 train subway sign in New York City and a Route 1 road sign in a rural area—girls stand and shout, "I want to be one less, one less!" The voices chant in harmony, and we see young black girls jumping rope as they sing, "I want to be one less, one less!" In all, what we find are messages of individualized yet collective engagements: individual girls, active, engaged, strong, and independent, with healthy risk-free bodies now and in the future. Sports capture a range of girlhood experiences including social classes, regional places, and interests. Creative, entrepreneurial girls design clothes, win races, ride horses, and assert their voices.

While the "One Less" slogan moves seamlessly among both crisis and empowerment, it ultimately declares "the girl" as already able to make the "right choice" offered through a rubric of consumption. Girls are brought forward by an imperative to implement biomedical health and a consumer responsibility to be(come) healthy adult women nurturing themselves and their families. The "can do" girl is willing and able to transform herself at the level of the body to step into this American ideal of family and ensure her own future health. The do-it-yourself girls shown in these commercials make (feminist) mothers proud. These girls of the "One Less" campaign also resonate with contemporary American advertising that focuses on the individual as the locus of moral, political, and public health leverage.

In this advertising, popular cultural narratives of girlhood as a life stage in which individual girls are vulnerable yet empowered are co-opted by Merck to sell Gardasil. The rapid cuts of the commercial are characteristic of the "MTV aesthetic," an editing style well known for capturing the attention of young viewers (Messaris 1997, 88). The sports and music are specific to girlhood empowerment cultures. Shown within "play" contexts (i.e., basketball and rope-jumping), these social peer-group settings conjure meanings of childhood girlfriends. There are no boys or men and not many adult women present. Gardasil is a gendered vaccine and a socially sanctioned space for girls to perform ideal girlhood and for mothers to perform ideal womanhood.[10] This is a generation poised to take Gardasil. As Chesler and Kessler (chapter 8) point out, it is these girls who are primed as users of hormonal birth control pills, not for pregnancy prevention, but for bloating, acne, depression, and premenstrual disorders. Girls are thus primed for taking pills that our mothers understood as part of sexual liberation but which are lifestyle medications for this generation.

Sidestepping Sexuality on the Route to Normal Adulthood

Girls' sexuality has traditionally been produced as innocence in need of protection and regulation. Teen girls lack a strong voice that enables them to say "no" to sex or, alternatively, to articulate and direct their sexual desires (Tolman 1994). Such lack of agency signals girls' weak feminine positions in gendered and sexist contexts (see also Holland et al. 1994; Tolman 1999). Scripts of girlhood often include narratives of girls as "objects" of boys' "raging hormones" and thus sexually at risk and in need of protection. These are heterosexual constructions as well.

Gardasil relies on this discourse of girl innocence in need of protection yet simultaneously sidesteps it: sexuality is not an attribute of a "strong girl"; it is a presumed future. One pamphlet defines the age group for Gardasil by specifying that HPV is contracted within two to three years of sexual debut. The back of the pamphlet (Merck 2006, 22) reads, "Cervical cancer is caused by a virus many people get in their teens and 20s." The debut, as social scientists like to call it, refers to a girl's first penetrative heterosexual sex. While the advertisements depict young teens and even preteens, the risky actions will take place down the road. As Merck states, "Of the approximately 6 million new cases of genital HPV in the United States, every year, it is estimated that 74 percent of them occur in 15–24 year olds" (Merck 2007). These assertions produce adolescent, not *pre*adolescent, girls as sexually active. The result is an ideal user: girls located before sexual agency and categorically different from older cohorts; thereby the advertising averts any focus on sexual behavior and reifies cervical cancer prevention (and HPV prevention) as separate from sex. Yet, the visual imagery draws the viewers into feminine embodiments. Girls' bodies are produced and lived in ways that are at once risky and able to ward off risk. They are not sexual but empowered: flexible bodies that can be transformed through the use of biomedical interventions, however permanent these may be. This construct contrasts with oppositional media projects, such as Chesler's Tune In HPV or the Down There Collective's *HPV Zine*, which reinsert discussions of sex and sexuality as well as of disease into the awareness and education available. The people on Tune in HPV are also empowered, but consumers they may or may not be, and their messages and actions move within and beyond those scripted by the Gardasil advertisements.

A Mother's Wisdom: Garnering Legitimacy for Putting Drugs into Healthy Bodies

While girls are called forward by Gardasil, their mothers are also implicated as actors. Mothers play a significant role in protecting not only their daughters but all girls and, by extension, the health of the nation. This production is of course not new, and the historical continuity of mothers' responsibility as both health care consumers and health educators cannot have been lost on Merck. As knowledgeable caregivers weighing the risks and benefits of vaccination, mothers are invited by Merck, through old and new feminist activist slogans, to participate in this women's-health campaign. Currently in their late thirties or forties, this generation of mothers came of age with hormonal birth control, the AIDS crisis, and safe-sex campaigns. These mothers are familiar with feminist activism and are well aware not only of cancer but of their own risk of cervical cancer. Familiar calls to action join this looming fear to implicitly target mothers, urging them to get their daughters vaccinated. The mothers of Gardasil can only hope that safe-sex practices and regular Pap smear and gynecologic exams will protect them, but their daughters have the opportunity to transform their bodies and their potential health risk via Gardasil. What mother wouldn't want this for her daughter?

When mothers appear in the advertisements, they are positioned as "experts" and "experienced"; thus legitimacy for vaccination is garnered by calling forward familiar health care consumers. For example, in a "One Less" commercial, a mother lists the potential side effects of Gardasil while standing next to her daughter. The side effects (pain, swelling, itching, and redness at the injection site; fever, dizziness, and nausea) are communicated through the protective role of motherhood. The moral imperative, a "will to health," constructs the side effects of Gardasil as acceptable risks in the pursuit of future health. Mothers expose their daughters to risks only if they believe the risks to be in their daughters' best interest. This presentation not only produces adolescent girls as in need of mother's protection, but it produces appropriate mothering practices and asserts the importance of maternal involvement and consent. Further, as Robert Aronowitz argues in chapter 2, the adoption of Gardasil for mothers' daughters also promises individual control over fear, offering relief from worry and perhaps from the mothers' own shame and stigma from HPV disease.

Targeting black mothers, an ad in the March 2007 issue of *Ebony* reads: "The Power to Help Prevent Cervical Cancer Is in Your Hands and on Your Daughter's Arm." This discourse relies on two ideologies: feminist empowerment with its invitation to participate in a social movement and moral motherhood (in contrast to inadequate or selfish mothers) (Blum 2007). A mother's power to protect conjures its silenced other: mother blame. Mother blame serves as a metaphor for a range of social and political fears, erasing race and class differences in the social and cultural capital that is needed to cultivate the knowledge and skills to acquire valued health care for one's daughter (Blum 2007; Lareau 2003). To prevent cervical cancer, a responsible mother must not only offer up her daughter's body to medical intervention but also encourage consumer practices to achieve desired states. The mother becomes responsible for her daughter's risk of getting cervical cancer and is valorized for the potential to prevent cancer for the nation. At the same time, the black daughter or girl in this discourse is disembodied—portrayed as a selfless object, a part of the process of preventing cervical cancer for society. Another advertisement in the same issue reads, "Get Your Daughter Vaccinated as a Girl. Help Prevent Her from Getting Cervical Cancer as a Woman." The mother-valor narrative is invoked by appealing to women accustomed to feminist health advocacy. This advertisement places the future health of daughters in the hands of mothers. Mothers are responsible for daughters' risky health, further erasing the subjectivity of girls and moving them into adulthood through a certain form of biomedical consumption.

"Roll Up Your Sleeves. It's Your Turn to Help Guard against Cervical Cancer," admonishes another advertisement, in the April and June 2007 issues of *Glamour*. Here, young, glamorous, entrepreneurial women who came of age with 1990s "girl power" are called forward to join with a new technology in the fight against cancer. Mothers, daughters, and high-style independent young women become coparticipants in cervical cancer prevention. The slogan "Roll Up Your

Sleeves" evokes a mental image of the iconic Rosie the Riveter / We Can Do It image for the generation of mothers whose daughters are the targets of this vaccine. Through "Make the Commitment" pledges, adult women are encouraged to talk with their doctor about their risk of cervical cancer; by this means, Merck simultaneously produces strong glamorous "girls" and "women" ready and able to make the "right choice" and ensure the rightness of Gardasil for the job of cervical cancer prevention. Gardasil advertising relies on a gendered collective and individualized "will to health," attainable via offering up your daughter's arm, finding out more about Gardasil and its viral protection, or rolling up your sleeve and taking your turn in protecting a social body (our nation) from the silent killer.[11]

Discussion: Producing and Protecting Risky Girlhoods

We posed three questions of the marketing campaigns for Gardasil: What are the sociocultural mechanisms through which Merck promoted its medical innovation as the "right tool" for U.S. cancer prevention? In what ways did Merck produce U.S. girls as the ideal users of the Gardasil vaccine? What are the social and political implications of this emergent technology and its associated productions of gendered sexual bodies?

We found that this vaccine was represented by Merck as the right tool for the job through its production of cervical cancer as a major health issue and a cancer of innocence for which all girls are at risk. Further, vaccination became another "do-it-yourself" feminine project enabling women and girls to transform their bodies, identities, and lives. Enterprising bodies through vaccinations are a route not only to health but also to achieving normal adulthood—to realizing one's imagined sense of self, one's goals and hopes for the future. Through vaccination—the reengineering of the at-risk body—girls become women ready and able to function in the life course as "normal" adults and as mothers taking their place in a long line of health care consumers. As gendered girls and soon-to-be women, these actors are at once produced as asexual and as simultaneously highly heterosexual. Subjects are made and identifications are produced through strategies of self-governance that include reconfiguring the body itself. On the way to becoming normal adult women, girls must become empowered consumer citizens rationally reducing their risk.

Merck's DTC advertising relies on dual constructions of girls as empowered and vulnerable and mothers as valorized or blamed for their daughter's future health or illness. Girls' own agency with regard to sex remains invisible and their own sexual identity negated, and instead girls are offered empowerment by choosing to get vaccinated with Gardasil. Gardasil produces a can-do girl compatible with neoliberalism, including rhetorics of self-responsibility, color-blind racism, and gender and sexual normativities. Further, equating cervical cancer risk with individual choice effaces the social and economic context that surrounds the women in the United States who have the highest rates of cervical cancer. In other words, the women who lack access to Pap smear screening and other sexual health care, who are, in fact, at higher risk for cervical cancer, are made invisible and replaced with a one-size model of cancer risk that relies on a false category of girlhood and a universal girl. The invisible young women are also those least likely to have access to the vaccine. Perhaps most troubling, girls who are most at risk for cervical cancer (those from lower-income families with less education; ethnic or racial minorities; and foreign-born girls) will continue to be at risk, not only because they cannot afford Gardasil but because the social-economic issues of unequal access to Pap smear screening continue to go unaddressed.[12]

Ignoring issues of sexual health literacy and reproductive health care access perpetuates girls-in-crisis and mother-blame metaphors that efface social, political, and economic barriers that must be negotiated. Silencing or erasing sexual subjectivities and enforcing a heterosexual assumption disempowers girls and women to enact their own sexual choices. Yet, alongside this sexual disempowerment is the self-empowerment offered to girls through consumption of biomedical technologies such as Gardasil.

Gardasil produces girlhood as a risk-identity group and simultaneously creates new ways to perform responsible and empowered girlhood. Through "Tell Someone" and "Make the Connection" rhetoric, a moral imperative is instilled in individuals to protect not only themselves, but also a friend. The imperatives to health and knowledge are bound up with consumption: reduced risk and a reengineered body are available for purchase. The Gardasil campaigns harness these cultural mechanisms and represent vaccines as an appropriate tool for both managing risk and optimizing bodies. Ideal girls are willing and able to make the right choices, to win the prize, to go for the gold. Their mothers are informed in the *Ebony* Gardasil advertisement, "The Power to Help Prevent Cervical Cancer Is in Your Hands and on Your Daughter's Arm."

Gardasil produces and maintains the simultaneous rhetoric of girls-in-crisis and girl-power subjectivities. Its rhetoric calls consumers forward and appeals to their free will. Merck's ideal consumers are girls at (future) risk, unaware of the tumultuous passage upon them but ready and able (with the wisdom of Mom) to rationally respond to the risks of adulthood and the demands of the free market. Both the narratives of girls in crisis and of girl power are brought forward and sustained by the advertisement campaigns.

Multiple subjectivities and embodiments for girls emerge: notably, a (pre) adolescent girlhood, a fragile (pre-sexual) innocence in need of protection as girls move through the transition from children to agential adults, *and* a girl power ready and willing to make the right health choices. Together, preventing women's cancer and producing preadolescent girlhood as being at future, yet preventable risk allowed the success of Gardasil. This success occurred in ways that messages of preventing HPV in sexually active youth could not. Safe sex, sexual literacy, and sexual health were not the goal of this public health message. Instead, Merck produced responsible, innocent girls able to avert potential illness through medical consumption. Yet, we argue, the responsible generic girls that emerge are contrasted with their absent references: irresponsible heterosexually active teenagers and all queer boys and girls. Relying on both a gendered ideology and its concomitant heteronormativity of women/mothers as nurturing caretakers, Gardasil reproduces this script yet brings in girls as also responsible for the nation's risky health.

While it is not far-fetched to ask why Gardasil is produced and marketed for girls and not boys, this analysis reveals the gendered patterns and productions that, we think, promoted Gardasil's public acceptance. Although Gardasil for boys is an obvious next step and will most likely arrive soon (clinical trials are under way in boys),[13] girls and their mothers are already ready for Gardasil. They are recipients of drugs, objects of intervention, and poised health care consumers. As was evident in the early advertisements for Gardasil, the production of girlhood and of women's cancer as depoliticized sites positioned girls and women well as early adopters of this vaccine. Further, a sex panic was averted; but so too was a homophobic backlash. We wonder what an advertisement campaign for boys might look like. Would it urge boys to "know the HPV–anal cancer link?" Would fathers be enrolled as health care consumers and protectors of their sons? Boys and men are clearly absent in the Gardasil campaigns to date, but they are implicated actors lurking in the background as either predatory men or risky queers. Gender is always relationally produced. Damaging gendered relations are produced through these cultural scripts as invisible yet implicated boys and men are inscribed with risky sexual subjectivities and as female sexual subjectivity is explicitly absent. The result is a full erasure of health advocates who demand gay men's attention to the HPV–anal cancer link. Of course, anal cancer may be the "great undiscussable," as Steven Epstein argues in chapter 4, but anal sex is not a health issue facing only gay men. Anal sex and anal cancer are sexual health risks for girls, boys, and all men and women who engage in anal sex. Effacing this issue not only harms girls through the silencing or erasure of multiple sexual practices, but also once again it stigmatizes and thereby threatens the health of gay men and prequeer boys.

The U.S. public health infrastructure needs to be wrestled away from capital (pharmaceutical and biotechnology companies) and returned to a multidisciplinary set of social thinkers and actors who weigh global and local STI risks, preventions, and treatments, including but not limited to the HPV-cancer link, in the interests of public good. Further, sexual education needs to be extended beyond a deviance and risk-reduction perspective to one of sexual health literacy including positive sexual health, practices, and desires; access to reproductive and sexual health care; and gender and power education (see also Nack 2008, chap. 8). Doing so will not only minimize health risks associated with misinformation and unprotected sex but will allow young people to emerge as responsible for their bodies and pleasures, not through biomedical consumption but by leading their sexual lives with the accurate information they need.

NOTES

1. For an excellent textual analysis of the Gardasil advertisement, see "On Drugs—What—and Who—Is behind the Marketing of Gardasil?" by Bree Kessler and Summer Wood, in *Bitch Magazine* 36 (2007).

2. Preapproval marketing of Gardasil was secured through "Make the Connection" and "Make the Commitment" events produced by the Cancer Research and Prevention Foundation (see http://preventcancer.org) and the celebrity charity Step Up Women's Network (see www.suwn.org; www.prwatch.org). "Tell Someone" commercials did not mention Gardasil but urged girls and women to "Tell Someone" that cancer is caused by a virus called HPV. After FDA approval, on Nov. 13, 2006, Merck launched "One Less" DTC commercials (see www.gardasil.com/).

3. The concept of "possible selves" was coined by Markus and Nurius (1986).

4. See Casper and Carpenter (2008), who eloquently argue that HPV vaccines—Gardasil and Cervarix—are gendered pharmaceutical technologies. Vaccines, they argue, are a distinctive kind of pharmaceutical, invoking notions of contagion and containment, and politics shape every aspect of the pharmaceutical life course. Their

argument about contagion and containment supports our finding that a generalized risky girlhood is produced concomitantly with an empowered-consumer girlhood.

5. For more on how discourses produce legible or illegible subjectivities, see the work of Judith Butler. For her theorization of political agency via the concept iterability, see Butler [1990] 1999. In the same work, Butler discusses how genders become intelligible .

6. Filling this gap, Adina Nack (2008) provides a thorough study of the lives of women living with sexually transmitted disease.

7. "Rightness" is produced through (inter)actions and negotiations and is not a special property of a particular tool or job (Clarke and Fujimura 1992).

8. A 2005 survey shows that 61 percent of American women had never heard of HPV (National Cancer Institute 2007).

9. We thank Steven Epstein for this point.

10. There are certain nonnegotiable markers, such as attractiveness and heterosexuality, and increasingly, girl power (see Adams and Bettis 2003).

11. In contrast, the Gardasil advertisement in *Better Homes and Gardens* of Nov. 2007 reads: "Calling Gardasil a Cervical Cancer Vaccine Is Only the Beginning of the Story." This message is much more indirectly calling on an adult (most likely a middle- or upper-class strong woman or mother) to learn more about the Gardasil vaccine and its potential for the future (the rest of the story). All of the initial Merck-sponsored educational campaigns and advertising call forward individualized, independent, and collective risk groups in a hierarchy. Ideal users are preteen, presexual innocent girls, but young, possibly strong women and mothers are called forward through a sisterly bond to cervical cancer disease. Once the can-do girls take their place in performing their potential cancer-free futures (the beginning of the story), the risky boys and gay men stand by to serve the rest of the story.

12. Several socioeconomic and racial disparities exist within and across the United States and globally, based upon poverty, access to care, and other social factors. Racial and ethnic minorities in the United States also have higher cancer incidence and mortality rates, in part because of increased likelihood of diagnosis during later stages of the disease rather than via precancerous lesions detected from regular Pap smear screening (see U.S. Cancer Statistics Working Group 2006; Shavers and Brown 2002; and Singh et al. 2004).

13. Two HPV strains targeted by Gardasil are linked to genital warts in both men and women.

REFERENCES

Adams, N., and P. Bettis. 2003. "Commanding the Room in Short Skirts: Cheerleading as the Embodiment of Ideal Girlhood." *Gender and Society* 17 (1): 73–91.
Balsamo, Anne. 1996. *Technologies of the Gendered Body: Reading Cyborg Women*. Durham, NC: Duke University Press.
Blum, Linda M. 2007. "Mother-Blame in the Prozac Nation: Raising Kids with Invisible Disabilities." *Gender and Society* 21 (2): 202–226.
Butler, Judith. [1990] 1999. *Gender Trouble*. New York: Routledge.

Casper, M., and L. Carpenter. 2008. "Sex, Drugs, and Politics: The HPV Vaccine for Cervical Cancer." *Sociology of Health and Illness* 30 (6): 886–899.

Clarke, A. E., and J. H. Fujimura. 1992. "What Tools? Which Jobs? Why Right?" In *The Right Tools for the Job at Work in Twentieth-Century Life Sciences*, ed. A. E. Clarke and J. H. Fujimura, 3–46. Princeton, NJ: Princeton University Press.

Clarke, A. E., J. Shim, L. Mamo, J. R. Fosket, and J. R. Fishman. 2003. "Biomedicalization: Theorizing Technoscientific Transformations of Health, Illness, and U.S. Biomedicine." *American Sociological Review* 68 (2): 161–194.

Collins, Patricia Hill. 2004. *Black Sexual Politics: African Americans, Gender, and the New Racism*. New York: Routledge.

Cooper, Melinda. 2008. *Life as Surplus: Biotechnology and Capitalism in the Neoliberal Era*. Seattle: University of Washington Press.

Driscoll, Catherine. 2002. *Girls: Feminine Adolescence in Popular Culture and Cultural Theory*. New York: Columbia University Press.

Elder, Glen H., Jr. 1998. "The Life Course as Developmental Theory." *Child Development* 69 (1): 1–12.

Gilligan, Carol. 1982. *In a Different Voice: Psychological Theory and Women's Development*. Cambridge, MA: Harvard University Press.

Gonick, Marnina. 2004. "The 'Mean Girl' Crisis: Problematizing Representations of Girls' Friendships." *Feminism and Psychology* 14 (3): 395–400.

———. 2006. "Between 'Girl Power' and 'Reviving Ophelia': Constituting the Neoliberal Girl Subject." *National Women's Studies Association* 18 (2) (Summer): 1–23.

Halberstam, Judith. 1995. *Female Masculinity*. Durham, NC: Duke University Press.

———. 2005. *In a Queer Time and Place*. New York: New York University Press.

Haraway, Donna. 1991. *Simians, Cyborgs, and Women: The Reinvention of Nature*. New York: Routledge.

Harris, Anita. 2004. *Future Girl: Young Women in the Twenty-first Century*. Routledge: New York.

Holland, J., C. Ramazanoglu, S. Sharpe, and R. Thompson. 1994. "Power and Desire: The Embodiment of Female Sexuality." *Feminist Review* 46:21–38.

Lamb, S., and L. M. Brown. 2006. *Packaging Girlhood: Rescuing Our Daughters from Marketers' Schemes*. New York: St. Martin's Press.

Lareau, Annette. 2003. *Unequal Childhoods: Class, Race, and Family Life*. Berkeley: University of California Press.

Lupton, Deborah. 1995. *The Imperative of Health: Public Health and the Regulated Body*. London: Sage.

Markus, H., and P. Nurius. 1986. "Possible Selves." *American Psychologist* 41 (9): 954–969.

Marshall, Elizabeth. 2006. "Borderline Girlhoods: Mental Illness, Adolescence, and Femininity in Girl, Interrupted." *Lion and the Unicorn* 30:117–133.

Martin, Emily. 1994. *Flexible Bodies: Tracking Immunity in American Culture from the Days of Polio to the Age of AIDS*. Boston: Beacon Press.

McRobbie, Angela. 2001. "Good Girls Bad Girls, Female Success and the New Meritocracy." Keynote address at "A New Girl Order: Young Women and the Future of Feminist Enquiry," Kings College, London, Nov. 14–16.

Merck and Company. 2006. *You'd Tell Her She Has Lipstick on Her Teeth. So Why Wouldn't You Tell Her about a Virus That Can Cause Cancer?* Marketing pamphlet.

———. 2007. Merck brochure.

Messaris, Paul. 1997. *Visual Persuasion: The Role of Images in Advertising*. London: Sage.

Mills, Sara. 2004. *Discourse: The New Critical Idiom*. 2nd ed. New York: Routledge.

Nack, Adina. 2008. *Damaged Goods: Women Living with Incurable Sexually Transmitted Diseases*. Philadelphia: Temple University Press.

National Cancer Institute, Health Information National Trends Survey. 2007. "In 2005, 61 Percent of American Women Had Never Heard of HPV." *HINTS Briefs* 5 (March), http://hints.cancer.gov/briefs.jsp.

Patton, Cindy. 1995. "Performativity and Spatial Distinction: The End of AIDS Epidemiology." In *Performativity and Performance*, ed. A. Parker and E. K. Sedgwick, 173–196. New York: Routledge.

Pipher, Mary. 1994. *Reviving Ophelia: Saving the Selves of Adolescent Girls*. New York: Ballantine Books.

Reissman, Catherine K. 1985. "Women and Medicalization: A New Perspective." *Social Policy* 14:3–18.

Rose, Nikolas. [2001] 2006. "The Politics of Life Itself." *Theory, Culture, and Society* 18 (6): 1–30.

Shavers, Vickie L., and Martin L. Brown. 2002. "Racial and Ethnic Disparities in the Receipt of Cancer Treatment." *Journal of the National Cancer Institute* 94 (5): 334–357.

Siers-Poisson, Judith. 2007. "Research, Develop, and Sell, Sell, Sell: Part Two in a Series on the Politics of PR of Cervical Cancer." Center for Media and Democracy Web site, PR Watch.org. Available at www.prwatch.org/node/6298.

Singh, Gopal K., Barry A. Miller, Benjamin F. Hankey, and Brenda K. Edwards. 2004. "Persistent Area Socioeconomic Disparities in U.S. Incidence of Cervical Cancer, Mortality, Stage, and Survival, 1975–2000." *Cancer* 101 (5): 1051–1057.

Stein, Rob. 2005. "Cervical Cancer Vaccine Gets Injected with a Social Issue: Some Fear a Shot for Teens Could Encourage Sex." *Washington Post*, Oct. 31, www.washingtonpost.com/wp-dyn/content/article/2005/10/30/AR2005103000747.html.

Tolman, Deborah L. 1994. "Doing Desire: Adolescent Girls' Struggles for/with Sexuality." *Gender and Society* 8:324–342.

———. 1999. "Female Adolescent Sexuality in Relational Context: Beyond Sexual Decision Making." In *Beyond Appearance: A New Look at Adolescent Girls*, ed. N. G. Johnson, M. C. Roberts, and J. Worell, 227–246. Washington, DC: APA.

U.S. Cancer Statistics Working Group. 2006. *United States Cancer Statistics: 2003 Incidence and Mortality*. Atlanta, GA: U.S. Department of Health and Human Services, Centers for Disease Control and Prevention, and National Cancer Institute.

Zola, Irving K. 1972. "Medicine as an Institution of Social Control." *Sociological Review* 20:487–504.

CHAPTER EIGHT

Re-Presenting Choice

Tune in HPV

Giovanna Chesler and Bree Kessler

Merck's 2008 Gardasil TV spots add two words to their "One Less" advertising mantra: "I Chose." In these spots, mothers repeat the line "I chose to get my daughter vaccinated" and young women boldly proclaim, "I chose to get vaccinated." The spots conclude with a tagline suitable for any feminist enterprise: "You have the power to choose."[1] In Merck's "I Chose" advertisements for Gardasil, the choice as presented seems a limited binary: "I receive the vaccine" or "I do not receive the vaccine." Yet as a product for consumption, this choice, further encouraged by interactive Web tools and lobbying groups, involves choosing to agree to a limited understanding of human papillomavirus. Choosing to purchase and receive the Gardasil vaccine also includes choosing to protect oneself from only some of the strains of HPV that may lead to cervical cancer and genital warts (see chapter 3). Further, by choosing to situate Gardasil in any conversation about HPV and in allowing Gardasil the power to construct the public discourse surrounding the virus, one limits (if not negates) a consideration of sex practices involved in contracting the human papillomavirus.

The Tune in HPV project (www.tuneinhpv.com) was developed as an alternative space for public discussion of HPV, one that might reinsert sex practice and sexual behavior into any consideration of the virus and one that would diversify representations of people who contract and spread HPV beyond the limited set of young girls featured in Gardasil advertisements. In its initial iteration, Tune in HPV began as a series of videos titled *HPV Boredom*, made and performed by Giovanna Chesler in October 2006 and posted to YouTube.[2] In these tapes, Chesler recovers from an HPV-related surgery, immobilized in her bed. In *HPV Boredom 2*, she receives a call from her "doctor," a robotic voice that tells her about HPV contraction and the Gardasil vaccine: "The vaccine will only cover you if you are between the ages of 9 and 26. It is our hope that if you are over the age of 26 that you are married, settled down, and monogamous." The videos have inspired multiple readings: "Is this the future of the public service announcement . . . in the form of a video made and edited on one's home computer and uploaded directly to the Internet? Or is this an ironic spoof of the public service announcement as a genre?" (Serlin 2010).

As an artist, an activist, and a scholar in gender and media, Chesler sought to expand the 2006 series to include considerations of HPV beyond those of her performed character and in a format that would build from the stories of anonymous authors who wanted to share their HPV experiences. Sexual health educators, nurses, and physicians affiliated with the Washington, D.C., Department of Health HIV/AIDS Administration and the Sexually Transmitted Disease Control Program served as preliminary advisers to develop a Web site design that could bridge educational content and personal expression.

Tune in HPV was launched in February 2008 in collaboration with Web architect Zulma Aguiar and students in Chesler and Aguiar's course "Communication and Social Change," taught at American University. The site they developed, www.tuneinhpv.com, is a participatory Web channel dedicated to providing information and entertainment related to the human papillomavirus. The site contains two streams of content: user-generated stories about HPV that are submitted anonymously through a story-submission tool and videos made by the producers of the site. The stories submitted by users provide examples of HPV transmission and treatment, while the video content enacts elements of the stories and conveys information on HPV contraction. As a project built from stories of people commonly affected by HPV, Tune in HPV experimentally identifies a public and provides a potential service platform to that public around a sexual health issue. Bree Kessler, a trained public health practitioner, analyzed users' stories to better understand the emerging themes related to sexual health. The site launch was supported in part by the Center for

Social Media at American University, whose ongoing projects and studies consider the future of public media as media that contribute to "helping people understand ongoing and complicated issues, both with content and through practices" (Clark and Aufderheide 2009, 18). In their study "Public Media 2.0: Dynamic, Engaged Publics," Clark and Aufderheide identify the trend in public media as media that is participatory, interactive, accessible, and egalitarian. They underscore the importance of identifying and mobilizing publics as central to any public media effort.

Advocates of sexually transmitted infection (STI) prevention point to the Internet's increasing accessibility and the usefulness of pooling STI prevention resources online (Rietmeijer and McFarlane 2008; McFarlane, Ross, and Elford 2004). Studies by the Pew Internet and American Life Project show that 80 percent of Internet users in the United States have looked for health information online (2008) and 22 percent of e-health surfers between the ages of 18 and 29 have looked for sexual health information specifically (2006). As authors in this volume illustrate, in the case of HPV, at the moment of the development of the vaccine, a public was produced through a target age range and gender category designed by the pharmaceutical product makers. The terms and limitations of this public initially encompassed teen girls, through a highly successful print and television advertising campaign (see chapter 7) reinforced online in the form of Merck's Gardasil Web site, www.gardasil.com.

The Gardasil site has at least twenty-four unique pages, two of the Gardasil TV spots, one behind-the-scenes video covering the "making of" the TV spots, and four videos produced by people who are survivors of cervical cancer. These videos are called *Our Stories* and stand as somber, homemade, do-it-yourself video testimonials of (mostly) young women's experiences with cervical cancer. Housed within this frame, they seek to provide visible, testimonial evidence of why the choice to receive the Gardasil vaccine is necessary. The Gardasil site, as relaunched in the summer of 2008, contains numerous interactive devices ranging from links to Gardasil-themed e-mails that site visitors can send to friends to downloads like "I Chose" wallpaper for personal computers. These tools assume a level of comfort between young female consumers and pharmaceutical brands, which is understandable in that young women have been routinely targeted as consumers of pharmaceuticals for many off-label purposes. Consider that young women in America have been marketed to take Yaz, a hormonal birth control that has been "proven to treat moderate acne" as well as premenstrual dysphoric disorder. Young women actors appear in Seasonale's early ad campaigns (campaigns that were pulled after U.S. Food and Drug Administration [FDA] disapproval) sporting white dresses free of pink spots (i.e., period "free"). Through advertising slogans and key words in promotional press releases, this form of pharmaceutical use has been successfully categorized as feminist activism; taking drugs for off-label benefits has become an empowered, consumptive choice.

The connection between brand and consumer is strengthened by Gardasil cyber friends, actors made familiar in the print and television ads who appear on the Web site and seek to interact with the viewer. When you click a thought bubble above one of their heads that reads "Hear what we have to say," the actors "come to life" and speak to you, encouraging you to learn more about the vaccine, as they did. These Gardasil girls linger at the site's edges, texting on cell phones, reading books, and chatting with each other while they wait for you, the visitor, to interact. All of this takes place within a warmly lit room, decorated with leather armchairs and modern beige sofas, which conveys a homey, friendly vibe (fig. 8.1). The goal is to seamlessly connect their home with your home, joining the lived real space where your Internet search occurs with the potential, consumptive space of the Gardasil vaccine.

Although Tune in HPV was not designed to directly respond to Merck's Internet marketing efforts (indeed, the Tune in HPV Web site launch preceded the Gardasil Web site's "living room" redesign by a few months), there is a similarity in the style of the two sites. Student designers Genna Duberstein and Maggie McGrath created an interface for Tune in HPV housed within an imagined living-room space. Their friendly but sagging couch and animated yet dusty analog television suggest intimacy and comfort and intend to draw viewers into sharing and exploring personal dialogues around this STI (fig. 8.2). The designers worked with hyperbright color and kitsch, incorporating a visual

Fig. 8.1. Gardasil girls, www.gardasil.com, September 2008

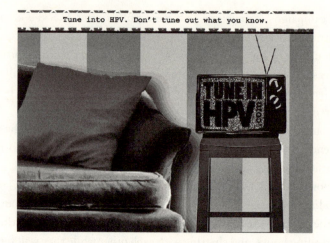

Fig. 8.2. Tune in HPV, splash page, www.tuneinhpv.com, May 2008

boldness that seeks to translate to boldness in discussing sexuality. Their analog television, which figures prominently in the splash page, the navigation area, and the logo for the project, echoes the site's slogan of encouragement: to drive viewers toward the act of "tuning in" to the site's video content, but further, to the idea of taking action while "dialing up" and "dialing into" HPV awareness.

In addition to highlighting the video content on the site, the television and the frame it provides for the stories and videos (fig. 8.3) was selected to reframe the act of consumption. Instead of tuning in to a TV for mindless amusement or reception of advertising campaigns, here the consumptive act of watching television is reappropriated and reimagined as an engaged form of "tuning in to" sexual health choices.

The Tune in HPV and Gardasil sites present mirrors of each other—heterotopias, Foucault's term for: "sites . . . that have the curious property of being in relation with all the other sites, but in such a way as to suspect, neutralize, or invert the set of relations that they happen to designate, mirror, or reflect." There are utopias and "real places—places that do exist . . . which are something like counter-sites, a kind of effectively enacted utopia in which the real sites, all the other real sites that can be found within the culture, are simultaneously represented, contested, and inverted" (Foucault 1998 [1967], 178). Merck's site is the feminist utopia—a world without men where women are sav-

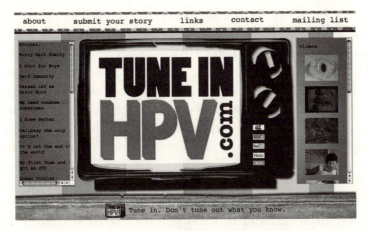

Fig. 8.3. Tune in HPV, navigation page, April 2008

ing each other (see Kessler and Wood 2008), while Tune in HPV represents the "counter-site" where oppositional knowledge is produced and reproduced and is distinct from that presented by the official narrative of HPV on the Gardasil site. Users' stories, although from a self-selecting sample of Internet users, portray the ways in which individuals, given the space, are challenging and recreating the discourses about HPV in the media and in their everyday lives.

Working within an experimental feminist film and video practice exemplified by Su Friedrich and Barbara Hammer, Chesler follows a tradition of artists who hone the personal narrative to intersect or intercept medical knowledge. The first story published on Tune in HPV, which invited users to submit their own, was an expression of Chesler's personal experience:

> After an intensive surgery related to an HPV strain that gave me pre-cancerous warts on the outside of my body, I was stuck in bed for six weeks. Thanks to my new computer, I was saved! Because I started making these tiny tapes from my recovery place—my bed. Then I posted them on You Tube. People who saw them told me their stories after they watched mine. I wanted to hear more.
>
> Human papillomavirus is the most popular sexually transmitted infection and there is a lot of misinformation and confusion surrounding it. I want visitors to this site to submit their own stories. Through our videos, made from stories, we share knowledge, experience and humor. (*g6, Washington, DC,* "HPV Boredom Begins in Bed")

As the viewer clicks on Chesler's story "HPV Boredom Begins in Bed" (fig. 8.4), a video launches below the text in the browser window. This video performance was part of the initial HPV Boredom series. In *HPV Boredom 2* (fig. 8.5), Chesler lies in bed and receives a phone call from her doctor. A robot voice (the doctor) tells her that there is a vaccine to protect her from HPV or, in fact, to protect her from only four of the more than one hundred strains of the virus. The "doctor" further reveals that her insurance will cover the vaccine only if she is between the ages of 9 and 26 because "it is our hope that when you are over the age of 26 that you are married, settled down, and monogamous." The video and the story are explicit about the links between sexual behavior and HPV as they depict a woman situated in the potential space of contracting STI, her bedroom, speaking about her genital warts, and suggesting nonmonogamous status.

After this initial posting, users began submitting their narratives of HPV contraction and diagnosis: "When I was younger I messed around a lot, and by younger I mean around 15. Probably throughout that year between being 15 and 16 I had around ten to eleven sexual partners, I can't remember anymore. One of those had HPV. I found out I had it during a regular pap, which came back abnormal. I had never heard of HPV before, no one had ever taught it to me in school, no one ever told me that most women get this disease at some point in their lives, no one ever told me most men don't even know they have it. Well, now I had it" (*Rachel, 18,* "I Messed Around a Lot").

The site is visited by several hundred people each month, 90 percent of whom come to the site directly. Around 6 percent come through Web searches via Google and Yahoo, and 4 percent come to the site from the videos posted on YouTube. Of the thousands of unique visitors who have visited the site, twenty-four had contributed stories between February 2008 and July 2009. Contributors ranged in age from 18 and 60, and their relationship status varied equally between married, single, nonmonogamous, and monogamous. Nearly all contributors had been diagnosed with HPV or believed that they may have had it. Of the contributors, most identified as women; 20 percent identified as male, and "trans" is also a category that can be selected. Half of the contributors were between the ages of 20 and 24 when they submitted their stories. Regardless of continuity in gender or age, the stories, which drive the community, have the potential to provide complex and individualized examples of an HPV public. While these stories are arguably therapeutic for the writer and perhaps even the reader, some provide a critical analysis of interpersonal and structural issues at

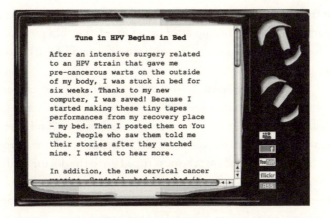

Fig. 8.4. Tune in HPV, May 2008

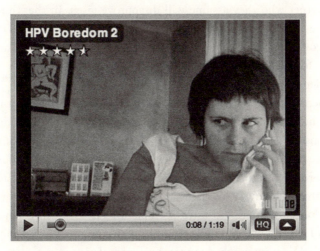

Fig. 8.5. Tune in HPV, October 2006

play in HPV contraction, diagnosis, and treatment that underscores and augments studies of the virus and the HPV vaccine controversies.

In this story, published as "My First Time and I Got an STD," a young woman recounts the moment when she believes she contracted HPV: "I was a virgin when I met this guy. We fell in love but took it slowly. After a while we moved in together and had sex on a regular basis mostly using condoms. 'Mostly.' I blame myself for being naive and giving in to him without any protection but birth control. I was diagnosed with high-risk HPV when I went to receive my Pap Smear results. We broke up after being in relationship for 2 years. I hate the fact that the only guy I ever slept with gave me an STD" (*A woman in Alabama, age 20–24*).

Several contributors echoed the difficulty of using protection during a sexual encounter, even when they knew it was the best choice to make. One writer from Washington, D.C., in her twenties, notes how her "brain freaked out" while she was having unprotected sex with "this older guy" she "barely knew." She adds, "I am determined that the next time I go to the doctor I will ask about getting an HPV test. I want to be tested, but I never dreamed it would be this hard." These themes agree with a study by Brown and colleagues (2008) of teen use of condoms; their research revealed that across region, gender, race, and ethnicity, teens have two prominent reasons for not using condoms: decreased pleasure during sexual encounters and discomfort in communicating with a sexual partner who they assume will disapprove (607). In the Tune in HPV frame, the woman in Alabama and the woman in Washington, D.C., reveal how miscommunication between sexual partners may manifest. This specificity allows for public health interventions around condom use and is addressed in the video content of this site.

The people who have participated in creating content through their stories are straight, gay, and bisexual. *R.H.K.* is a man who has anal condyloma and whose consideration of HPV, "A Queer Story," warrants several installments on the site. As evidenced by Steven Epstein (chapter 4), the call for an HPV vaccine for men resounds for men who have sex with men. *R.H.K.* adds to this chorus through his story by asking why Gardasil was released as a vaccine for women alone: "Men are just as responsible for HPV transmission as anyone else. And I can almost guarantee you this: if anal and/or penile cancer rates in heterosexual men approached the rates of cervical cancer in women, we would have had a vaccine and/or a cure at least fifteen years ago."[3] *R.H.K.* and another author, Ann (in her fifties), ask, "What about a shot for boys?" In addition, several male authors who believe they have passed HPV to their female sex partners participate in the discussion and understand their place in it, whether they have symptoms or not. A man in his forties wrote, "I didn't know I had it as a guy if you don't have big ass warts on some part of your body how would you know. Never even knew about it until my long term girlfriend got it from me." These contributors, straight, queer, male, and female, collectively problematize Gardasil's choice campaign by pointing to the lack of choice for men.

Other contributors to Tune in HPV see the vaccine as a panacea. In the posting "HPV Leads to a Breakup" by 20-year-old *Stephanie*, she suggests that by not choosing the vaccine, HPV contraction was inevitable and, perhaps, deserved: "I have now learned that I have dysplasia or abnormal cells because of HPV. I just told my boyfriend and we decided to call it quits because we don't want to risk infecting him for the rest of his life. It was the hardest decision I have ever had to make. And to think that I could have been vaccinated." As seen in Stephanie's story, misinformation represents the most disruptive, yet potentially fruitful, aspect of the Web site. Yet, stories with misinformation are published, and other users are able to react and correct their community. Another example is this story, "It's Not the End of the World": "My doctor said it's not just transmitted the typical heterosexual way, lesbians get it, and some think it could get under fingernails or the like. Condoms are great and people should use them, but I think of HPV more like the common cold—you just live with the fact its out there. I think people need to take this in stride. Get a PAP every year and look after yourself. Stop living in fear. There are far worse things in life" (*A woman in San Francisco, CA, 30–35*). In other stories, users explain how misinformation led to their infection. For example, in "I Knew Better," a young woman says, "We always used condoms. He said it would be fine, that the infected area was covered. I believed him. I trusted Him. We broke up on Tuesday, and I got diagnosed on the Thursday."

Video content, made by the producers of the site, seeks to augment these conversations and address miscommunication between sexual partners. In Jennifer DeRosa's *HPV Dykes* video (fig. 8.6), she wryly presents an example of how this conversation can take place. As two women are about to engage in oral sex, a television clicks on, interrupting their make-out session. On screen is one woman's embodied subconscious, telling herself that she should use a dental dam. "People can have and transmit HPV without any visible symptoms," her lips say, "and doctors don't test for HPV unless a woman had an abnormal pap." *HPV Dykes*, like the other Tune in HPV videos, links to the site from its post on YouTube. That forum allows for viewer comments, and this video drew both praise and ridicule from some of its thousands of viewers. Viewer comments range from "ugh disgusting" by *datoneportagee* to recitations of lines from the tape including "scissors!!!" by *tropdars* and "I wish I was a savvy political dyke," by *GymTeachersOnion*. These comments suggest a discomfort (or relief) with the insertion of lesbian sexuality into the public conversation around HPV.

Fig. 8.6. Tune in HPV, April 2008

Vanessa Bradchullis's video series presents numerous scenarios in which men and women of varying ages engage in obviously self-destructive activities (for example, putting a key in an electrical socket, running with scissors, or putting a hand on a hot burner (fig. 8.7). Her videos conclude with the tag line "Didn't see that coming, did you? Of course you did. Tune in. Don't tune out what you know." The videos speak directly to the story content, which reveals that a disconnect exists between individuals' awareness of "risky behaviour" and actual bodily practices, resonant in contributions that attest to the disconnect.

Video content seeks to pull together themes beyond prevention that emerge through the stories. Several writers describe the negative effect of HPV on their bodies. In "Just Some Cramping," *Amortentsia,* a woman in her early twenties, writes:

> I was diagnosed with HPV in the Fall of 2007. I went for my routine pap smear and when my results came back abnormal, the Student Health Center checked me automatically for HPV, and found that I am infected with the "high risk" strains. The strains that cause some forms of cervical cancer. So, the SHC referred me to a real gynecologist's office for "further testing." The doctor I was referred to, while nice, lacked the basic understanding of what reaction the tests and procedures would elicit from my body . . . being a man. Nothing against men, but when you tell me that I may experience "just some cramping" isn't really equivalent to what happened. After my colposcopy (you know, the one where

Fig. 8.7. Tune in HPV, July 2008

they cut off little bits of your cervix for testing for those pesky cancerous cells), the vaginal cramping was so severe that I couldn't reinsert my NuvaRing. Nevermind the pain. That's not "just some cramping." But it got better! The results showed that I did have some dysplasia (precancerous cells), so my doctor recommended cryosurgery. Basically, this boils down to sending super-cooled CO_2 at my cervix. For three minutes. Three minutes of the coldest thing I've ever felt at one of the spots in my body that is normally somewhere around 98.6 degrees Fahrenheit. Definitely the worst experience of my entire life. And my poor male doctor was laboring under the impression that I'd just have some cramps.

Stories including "Cervical (Not Breast) Cancer," "A Shot for Boys," and "I Messed Around a Lot" echo this writer's relationship to her cervix as both a site for testing and a site for treatment. In these stories containing descriptions of the battered cervixes, or in other stories in which writers describe their genitals with warts, or their HPV status in and of itself, their words reveal a type of self-flagellation. Some users are "so ashamed and broken down" ("Fears Are Coming True," by a man in his late twenties) by their inability to make informed choices, as is *A Woman in California*, in her early twenties, who writes: "I have warts on my genitals, and I knew better."

HPV Donut (fig. 8.8) was created in response to these stories whose writers speak of their cervixes under attack. *HPV Donut* begins with an image of a perfectly glazed strawberry donut, but titles indicate "This is not a strawberry

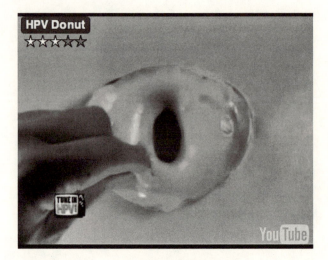

Fig. 8.8. Tune in HPV, April 2008

glazed donut" and "This is your cervix." After the donut has been torn apart by fingers and a knife, the text explains, "This is your cervix on HPV." As of January 2010, this video has been the most viewed in the Tune in HPV collection, with more than forty thousand views. As videos posted on YouTube and connected to the Tune in HPV site through their YouTube Universal Resource Locators (URLs), these media pieces have the potential to reach audiences well beyond the confines of the Tune in HPV Web site. By spring 2009, the Tune in HPV videos had been viewed tens of thousands of times and had received numerous comments. As with the stories, text comments added to the videos in the YouTube interface seek to inform the audience while correcting misinformation. In response to the *HPV Donut* video, *Tinktink264* asks, "Will you get hpv and all that warts and stuff only if your sexually active???" *DemmiXxX* writes: "Erm not all hpv cause cancer!!! type 6 and 11 are genital wart viruses. If warts are treated and people keep their immune system strong, their body can fight the virus and destroy it. this might take 2–5 years. And not only sexually active people get it. Oh and for the person who talked about mouth warts: Its Very Rare, but it does exist." *firstladyshine* adds, "Don't forget that the Gardasil (the hpv vaccine) is killing a lot of young girls. It's best to try to treat hpv naturally."

As evidenced above, YouTube comments like this, like the material submitted to Tune in HPV, has the potential to spread misinformation. Yet whether challenging the vaccine's efficacy or debating over modes of HPV contraction, these stories and comments as a whole provide a knowledge base that need not be unified to be effective. Through the act of discussion and expression, the users of Tune in HPV create a counterpublic to the Gardasil girls and articulate their sexuality and sexual behavior. They challenge each other (user to user through story submission) to become more informed, practical, and aware. Choice as manifested by this public becomes choosing to consider one's sexual health, choosing to have a Pap smear, and choosing to speak to a sexual partner about barrier methods during a sexual encounter.

"The discussion of HPV in the mainstream media is largely dominated by information from large pharmaceutical companies who currently focus their marketing on cervical cancer; we think this focus on making money instead of providing good health care for everyone is fucked up" (Down There Health Collective 2007). Another of the few alternative media sites for considering the HPV vaccine controversies. *HPV Zine*, created and self published by the Down There Health Collective of Washington, D.C., builds similarly from the impulse to intercept a dominant discourse.[4] This collaboratively authored fifty-one-page pamphlet, publicized on the group's MySpace page and available for free downloading by e-mailing the organization, is loaded with information about protection and treatment for genital human papillomavirus. The authors begin by placing their work within the context of the Gardasil media storm: "When the HPV vaccine came out in mid-2006, there was a flood of information. A lot of our questions about HPV infection, warts, and cervical cancer were answered but new ones arose. With healthy scepticism, we've tried to sort through the hype and drug company propaganda" (i). In directly challenging and addressing Gardasil, the collective makes explicit the higher rates of cervical cancer according to race and class. They note that "within the US, immigrants, people of color, and low-income people experience the highest rates of cervical cancer. African American women are more than twice as likely as white women to die from cervical cancer. Latinas are almost twice as likely as non-Latina white women to be diagnosed with cervical cancer. Vietnamese women have the highest rates of cervical cancer of any group in the US." Further: "We take issue with the common classification of race, ethnicity and economic status as 'risk factors' for cervical cancer. Is being African American an inherent risk factor? No, but living in a country founded on racist principles that continue to undermine access to health care *is* one. Poverty doesn't genetically predispose someone to developing cervical cancer, but high insurance costs can make it damn hard to get regular screenings" (23).

The Down There Health Collective acknowledges fluidity in sexuality and a variety of sexual health practices throughout the zine. In particular, they queer their discussion of HPV as they describe the visibility (and invisibility) of the virus across gender. Writing for an audience of readers who identify as male, female, intersex or transgender, Down There points to sites of infection without imagining a specific sexed body attached to the genitals, vagina, cervix, anus and anal canal wherein HPV symptoms may be visible. Options for treatment are traditional and nontraditional, and in keeping with their mission to "take control" of their own health, they present nutrition and lifestyle suggestions. For example, they encourage changes to diet and health in very practical terms: "Smoking has been shown to increase the chance that HPV might move from mild to more severe, and increase the chance of damaged cells being more susceptible to HPV even after treatment" (Down There Health Collective 2007, 29). Although the research and preparation for the production of this zine was intensive and wide-ranging, the authors conform to an approachable style of writing that calls the reader toward action through direct information, while simultaneously acknowledging the limits and the confusion that swirls around the virus.

When considering the effects of a vaccine marketed to protect girls from cervical cancer, one cannot take lightly or ignore the resulting changes to sexual health behavior based on a falsely advertised promise. As several stories submitted to the Tune in HPV site demonstrate, there is a popular misunderstanding that the choice to accept a vaccine negates or reduces a need for safe-sex practice. Several stories reveal confusion about the HPV virus on the part of those infected. It is not the responsibility of a pharmaceutical ad campaign to clarify misconceptions and to simplify the complexity of sexually transmitted infections. But the effects of this advertising campaign on sexual health behavior deserve attention. Perhaps in recognition of these effects, Merck launched another site in the summer of 2009: www.hpv.com. Coupled with a similarly designed print campaign, this basic black-text-on-yellow-background interface calls attention to some facts about HPV. Minuscule font indicates that "educational content is brought to you by Merck & Co. Inc." At the bottom of the numbered facts list, in large red text, the site encourages the visitor, "Learn how to protect yourself." But in clicking on these words, a visitor is redirected to the Gardasil site. While alternative spaces like the Down There *HPV Zine* and Tune in HPV build from facts, they use alternative means of presenting information, so as not to mimic the Centers for Disease Control and Prevention or other groups that provide resources. In the latest installment of the Gardasil campaign, www.hpv.com leaves little room between the pharmaceutical brand and the HPV virus. With this in mind, counternarratives, mobilization of publics, and artistic reinterpretations of the sexual-health narratives popularized by pharmaceuticals are imperatives. The stories and videos of Tune in HPV produce one countersite in a nascent, experimental form. More efforts must disperse and add complexity to this cultural moment in which a virus continues to be reimagined by a pharmaceutical branding strategy.

NOTES

1. This is the not the first time that feminist mantras of choice and empowerment have been employed in the marketing of pharmaceuticals intended for female consumers. In the promotion of Seasonale, a birth-control method promoted as a menstrual suppressant, the president and Chief Operating Officer of Barr Labs spoke of her latest product with this rhetoric: "With today's approval of Seasonale, women have a new *choice* when deciding on oral contraception" (George 2005; our emphasis). Depo-Provera, a hormonal birth-control shot produced by Pfizer that also eliminates the monthly period is advertised through the slogan "Freedom from the everyday."

2. The three videos in the *HPV Boredom* series were posted on YouTube in October 2006 but had to be reposted in May 2007 (their current date of publication online). These performances began shortly after Merck's "Tell Someone" campaign for Gardasil was launched. Although the videos were made in an effort to perform illness in the space of transmission of the virus (a bed) they also served to "tell someone" about the experience of recovering from an HPV-related surgery.

3. *R.H.K.*'s writing implicitly references Gloria Steinem's Oct. 1978 *Ms.* magazine piece "If Men Could Menstruate."

4. For more examples of alternative media responses, see Kessler, Bree, and Wood 2008.

REFERENCES

Brown, Larry K., Barbara Silver, William E. Schlenger, Ralph DiClemente, Richard Crosby, M. Isabel Fernandez, David Pugatch, Sylvia Cohn, Celia Lescano, Scott Royal, and Jacqueline R. Murphy. 2008. "Condom Use among High-Risk Adolescents: Anticipation of Partner Disapproval and Less Pleasure Associated with Not Using Condoms." *Public Health Reports*, Sept.–Oct., 601–607.

Clark, Jennifer, and Patricia Aufderheide. 2009. "Public Media 2.0: Dynamic, Engaged Publics," Center for Social Media, American University, Feb.

Down There Health Collective. 2007. *HPV Zine*. www.myspace.com/downtherehealth.

Foucault, Michel. 1998 [1967], "Different spaces", in Aesthetics, Method, and Epistemology: Essential Works of Foucault 1954–1984 Paul Rabinow, Series Ed., Volume 2, Ed. James D. Faubion. New York: The New Press.

Foucault, Michel. 1986. "Of Other Spaces." *Diacritics* 16, no. 1: 22–27.

George, Lianne. 2005. "The End of the Period." Macleans.ca, www.macleans.ca/article .jsp?content=20051213_117621_117621, Dec. 12.

Kessler, Bree, and Summer Wood. 2008. "At Your Cervix: Talking Back to Merck." *Bitch Magazine* 41 (Fall): 12.

McFarlane, M, M. W. Ross, and J. Elford. 2004. "The Internet and HIV/STD Prevention." *AIDS Care* 15, no. 8 (Nov.): 929–930.

Pew Internet and American Life Project. 2006. "Online Health Search 2006." www. pewinternet.org.

———. 2008. "Home Broadband 2008." www.pewinternet.org.

Rietmeijer, Cornelis A., and Mary McFarlane. 2008. "STI Prevention Services Online: Moving beyond the Proof of Concept." *Sexually Transmitted Diseases* 34, no. 8 (Aug.): 770–771.

Serlin, David. 2010. "Toward a Visual Culture of Public Health: From Broadside to YouTube." In *Imagining Illness: Histories of Public Health and Visual Culture*, ed. David Serlin. Minneapolis: University of Minnesota Press.

Part III / Focus on the Family

Parents Assessing Morality, Risk, and Opting Out

CHAPTER NINE

Parenting and Prevention

Views of HPV Vaccines among Parents Challenging
Childhood Immunizations

Jennifer A. Reich

Parents employ various strategies to make decisions they believe are in their children's best interests. Their decisions are informed by and reflect a complex web of meaning made up of interpretations of culture, experience, tradition, media, peers, expert advice, and their own sense of morality (Bobel 2001; Hulbert 2003; Lareau 2003). The interplay between parents' interpretations of cultural meaning and parenting strategies can be seen most clearly in parents' decisions whether to consent to childhood immunizations. Although all states require certain vaccinations to attend schools, the safety and necessity of vaccines remain controversial (Colgrove 2006). The availability of a vaccine against human papillomavirus raises additional issues as well (Casper and Carpenter 2008). As an elective vaccine not marketed for the collective good (an ethos that underscores compulsory childhood vaccine laws) but for individual benefit against a virus that is transmitted through sexual contact, it prompts new questions. Using multiple sources of qualitative data from interviews with parents and from Web sites aiming to support parents who choose to refuse vaccinations, I explicate this complex interplay in the lives of families as they contemplate vaccine choices more generally and the HPV vaccine specifically. In this chapter, I show that parents—even those generally supportive of vaccine policy—are resistant to the HPV vaccine for reasons not often discussed.

Perceiving Risk, Seeking Prevention

We know that before individuals become motivated to seek out health interventions, they must perceive that they are susceptible to a particular health problem, that the health problem is a serious one, that utilization of a treatment or medication will reduce risk of that condition, and that there are no serious barriers preventing access to the treatment (Rosenstock 1966). Consumers (or patients) must also find ways that the prescribed treatment fits their lives and matches their own goals and lifestyle (Conrad 1985).

Risk of disease—for example, cervical cancer—is often framed in terms of what Deborah Lupton calls "lifestyle risk discourse." As she explains, "lifestyle risk discourse is the responsibility of the individual to avoid health risks for the sake of his or her own health as well as the greater good of society" (2009:463). Girls and their parents are told abut the risks of HPV with data showing the high rates of HPV infection in the population, suggesting a high likelihood of infection upon initiation of sexual activity. They are also given information on how cervical cancer is caused by the virus (even though the limited number of strains that are carcinogenic is not always clearly communicated). To want the vaccine, parents need to be convinced that their daughters are at risk of cervical cancer from a nearly ubiquitous virus that is sexually transmitted, that the vaccine—given in early adolescence and before sexual activity begins—is the best prevention for that disease, and that despite the significant costs for the three-part inoculation, the vaccine mitigates risk. As Lupton notes, the discourse of risk is weighted toward the high probability of disaster, in this case, the development of a preventable illness. Yet to persuade parents to seek out the vaccine, promoters must also convince parents that risk and protection lie outside of the parent-child relationship and are best met by pharmaceutical companies.

Methods

This chapter uses two sources of qualitative data, collected as part of a larger ongoing study of parents' health care and vaccine decisions. First, data were collected during in-depth interviews with twenty-one parents who are modifying existing compulsory vaccination schedules for their children, by opting out completely, by providing consent to select vaccines on a schedule of their own choosing, or by negotiating alternative schedules in some way with their health providers. These parents do not represent the views of all parents and are not representative of the larger world of parental decision making. Yet these parents are already engaged in complex efforts to challenge physician knowledge and public health prescription and are strategizing these health care decisions based on their own understandings of health, illness, risk, corporate power, and trust. As such, they are an important group ready to discuss their concerns about new HPV vaccines.

These parents (twenty mothers and one father) all reside in Colorado. Colorado had until recently the lowest rates of vaccination in the country and remains one of the states with the highest numbers of parents exercising personal-belief exemptions from mandatory vaccines. Although all but two states allow parents to waive out of compulsory vaccines if they have a religious objection, only eighteen states, of which Colorado is one, allow parents to claim an exemption from state-required vaccines based solely on personal belief (NVIC 2004). No vaccine against HPV is currently required by law, but the intense marketing, the high cost, the newness of the vaccine, and the promise to protect girls from potential illness brought on by sexual contact have left many parents questioning its importance, safety, and necessity.

All parents in this study are white. Seven parents have bachelor-level degrees, seven have graduate degrees, and the remaining seven have either not attended college or have attended but without earning bachelor's degrees. The parents are between 30 and 55 years old. Eight parents have one child, seven have two children, four have three children, one has four children, and one has eight children. All but four parents (all women) in this study have at least one daughter. Many of the mothers have strong opinions about the HPV vaccine because they themselves have experienced cervical cancer screenings and issues of women's health. They are also engaged in the lives of other children, including nieces, friends' daughters, and girls they know socially or professionally. Although this sample is not representative of all parents who alter vaccine schedules or opt out completely, this sample is similar to samples described in other research on families who opt out (Smith, Chu, and Barker 2004). In religion, these twenty-one parents range from conservative Christians to a self-identified pagan.

Although all of them have made nontraditional choices about childhood vaccinations, their choices about childhood immunizations and the reasons for their choices differ from one another. Five parents have fully immunized their children but have done so on a schedule of their own choosing; one of the five expedited infant vaccines against hepatitis B because her husband was infected. Of those five, three voiced their discomfort with the HPV vaccine and could not imagine giving it to their children. One parent had refused all vaccines until she began planning a trip to India, at which point her elementary-school-aged daughters received all of the missing mandatory inoculations. One mother objects to vaccines and had refused all, but her child was fully vaccinated at her husband's insistence. Fourteen parents have given either no vaccines or only select vaccines to their children. It is worth noting that some parents may have consented to vaccinate their first child but decided not to vaccinate later children or might have conscientiously chosen vaccines against illnesses they thought represented risk to their children (like tetanus) while rejecting others. In all cases, parents were interpreting medical information and using their assessment of risk and benefit in making decisions about their children's health.

All parents were asked about the HPV vaccine, even though it was not equally relevant to all parents. Some had children past the ages for which the vaccine is recommended, and several parents' children were still several years younger than the recommended ages. Nonetheless, all had considered the vaccine and had formed opinions about it. Many then also anticipated how they would manage an HPV vaccine decision for their own children in the future.

Parents were recruited through convenience sampling. Some participants were referred by others familiar with the research study, and others were contacted via requests circulated on listserves, including those involving homeschooling families or families whose children attended private preschools and elementary schools that touted an alternative philosophy of learning. Many invitations were sent by e-mail through social networks. Interviews lasted between one and three hours. All interviews were tape-recorded and transcribed verbatim for analysis. Transcripts have been coded and analyzed thematically (Braun and Clarke 2006).

In addition, I collected and analyzed materials, e-mails, and Web sites produced by opponents of mandatory vaccination laws and those aiming to support parents in nonnormative child-rearing. Among these is correspondence sent by the National Vaccine Information Center (NVIC), a leading national advocacy organization committed to challenging state efforts to require vaccines. Between January 2005 and September 2008, NVIC has sent out 531 action alerts, news briefs, and essays by e-mail, 69 of which specifically address the HPV vaccine. The periodical *Mothering* and the related Web site, mothering.com, a virtual community committed to "natural family living," provide information critical of vaccines. Focus on the Family, a conservative faith-based organization headquartered in Colorado, has also published informational brochures and a policy statement about HPV; the group supports making HPV vaccines available but opposes efforts to require them. This correspondence and these resources offer insight into ways that parental distrust is shaped by forms of information that are outside of mainstream medicine or governmental agencies.

Parents' Concerns and Strategies

According to most news accounts, critics of the HPV vaccine believe that vaccinating girls will encourage them to become sexually active (Gyapong 2007; Stein 2005). In fact, no parents in this study have expressed this concern. These parents believe they know their children and trust their parenting and their ability to communicate the family's morality on this issue. Instead, parents' concerns overwhelmingly focus on three issues. First, the parents question the necessity of the vaccine, particularly since their children are not sexually active or because routine screening can detect pre-cervical-cancer cells, making the vaccine unnecessary. Second, the parents view the HPV vaccine development, review, and market process as corrupted by politics and through lobbying efforts funded by pharmaceutical companies. So they distrust the science that is used to tout the vaccine's safety and efficacy. Third, the parents consider the vaccine to interfere with parental autonomy; they envision strategies they will employ to address their children's future sexual interest or activity that will allow them to promote HPV prevention without vaccination, keeping risk reduction within the family. The following sections explore these objections.

Questions of Necessity

First, parents frequently questioned the necessity of the vaccine, for varying reasons. Some questioned its efficacy or necessity, particularly since routine screening can detect and treat pre-cervical-cancer cells. Tammy, a mother of one 2-year-old girl, explains her concerns with both the limited number of strains the vaccine covers and the availability of other methods of prevention.

[This is] for a vaccine that covers four papilloma viruses. Yeah, that can cause upwards of 70 percent, but the reality is, is that if you get routine Pap smears . . . that it is extraordinarily rare for HPV to turn into cervical cancer if you have Pap smears to detect it and then treat it. Because they can treat it. They can burn the cells, they can do a lot of, you know. And it's—with preventative treatments, you know cervical cancer is completely preventable. I'd rather—I'd rather teach her safe sex practices. Though granted these days if you just cough on each other you can get—it seems you can get things like HPV, but yeah, I'm just not comfortable with the vaccine.

Faith in prevention through Pap smears and exams was a strong theme. Heather, a mother of two boys, expresses her frustration with how the vaccine might undermine other methods of prevention or detection: "I just feel like it's marketed as if it's a completely safe, completely—you'd-be-silly-if-you-didn't-get-it kind of thing, when it really is, I think, a serious thing to consider—if it's the right thing for your child or not. Um, or that it would prevent girls from getting the right kind of healthcare, because they think, you know, they're vaccinated so they don't need to go in for um, you know, Pap smears or whatever."

Kristin, a mother of two girls, contemplates how prevention might be enough without the vaccine: "I think also, you know, cervical cancer isn't [passed] hereditarily. [So we can prevent it] from, you know, giving them annual exams or you know—I don't know about that one." Although she acknowledges her own uncertainty about the etiology of HPV-related disease, she is still unsure of her willingness to vaccinate against it. Kristin's discomfort, even though she lacks personal knowledge, mirrors that of other parents facing the new HPV vaccines (Noakes, Yarwood, and Salisbury 2006).

Some parents recognized the general relevance of the vaccine but viewed it as unnecessary since their daughters were not yet interested in sex. Jane, a mother of one daughter, imagined seeking out the vaccine for her 11-year-old daughter eventually but felt that it was currently unnecessary. "She's a very immature eleven; she has no crushes on boys, you know, and so I think we'll do—I understand it needs to be given before sexual activity begins and I understand that mothers are prone to always underestimate, you know, you know, how advanced their children might be, but I don't think I'm gonna do it this year. I think I'm gonna wait I think til she's thirteen."

As Jane considers the issue, she recalls the interaction she and her daughter had with her daughter's pediatrician when the vaccine was offered. "I was okay with the idea that she just turned eleven [and] didn't want a shot. It seemed to me that for her—I can see that—she knows some girls for whom it should, you know? Because you're not gonna be able to predict when they become sexually active and it could happen sooner, but not for her."

As demonstrated by Jane's comments, unless parents believe that their children face a reasonable risk of HPV infection in the future, they will not seek the vaccine in their children's youth. This parental perspective points to the way the HPV vaccine is different from others. Many mandatory vaccines provide benefit to others not receiving the vaccine, including pregnant women, immune-compromised individuals, or infants, and thus, the rationale to mandate those vaccines is underscored by a broader concern for public health (Salmon and Omer 2006). The HPV vaccine is still largely a vaccine for individual benefit, much as inoculations against tetanus are, since tetanus is not contagious but is ubiquitous in the environment. Recognizing this issue, parents' willingness to vaccinate was based on whether their child needed the protection. For example, Tammy explains her continued opposition to HPV: "It's for her in the end. Yeah, kinda selfish, but yeah it's for her in the end. . . . I mean, you do hear the—that you hear the one in a million stories and they're heart-wrenching, but you can hear the one in a million stories about, you know the kid who got stung by the ten bees and died. Are you gonna not let them play outside?"

Billy has a 16-year-old son who, likely as a result of the since-modified diphtheria-tetanus-pertussis vaccine he was given, developed a seizure disorder as an infant that led to profound mental retardation. Billy has chosen to refuse all vaccines for his daughter and is greatly concerned with vaccine safety: "My question is again how do you know that your daughter's not gonna react to that, to the shot? So you know that's the biggest thing. How do you know she's not gonna react? How do you know there's not something in that shot that's gonna cause her to either be brain damaged or she's gonna die from one little shot? So what are the odds of that and the odds of her getting something?"

Parents collect information to gauge the risk of the disease against the risk of the vaccine. They reflect on the meaning of a vaccine that covers only four strains of the virus, the meanings of prevention and screening, and the risk of cervical disease. They also weigh the timing of the vaccine against their estimations of when their daughters might be exposed to the virus through sexual contact (always assumed to be consensual). Given the relative rarity of cervical cancer in the United States for middle-class white women—approximately 10,000 cases per year, resulting in 3,700 deaths (Sonfield 2008)—these questions are reasonable. Taken together, parents grapple with their perceptions of risk and their evaluation of the necessity of the vaccine.

HPV Vaccine Approval and the Corruption of Process

A significant proportion of parents objected to the HPV vaccine because they believed its approval and marketing resulted from political corruption, powerful lobbying efforts, and profit motives of large pharmaceutical companies. Many parents who object to most or all vaccines, beyond HPV, believe these issues proliferate; they cite the very existence of the federal Vaccine Injury Compensation Program, "a no-fault alternative to the traditional tort system for resolving vaccine injury claims that provides compensation to people found to be injured by certain vaccines" (HRSA 2009), as evidence. In the HPV case, some parents consider the rapidity with which Texas mandated the vaccine to be evidence of this issue (Blumenthal 2007). NVIC has the same suspicions: "If anyone doubts why drug companies making and selling vaccines are lobbying so hard to convince California legislators to pass a precedent-setting law that would automatically mandate every new vaccine the CDC recommends for 'universal use' by all children, the recent market analysis by Global Industry Analysts gives the answer: forcing vaccine use is a multi-billion dollar business" (NVIC, e-mail correspondence, July 19, 2007).

As many NVIC alerts point out, the role of politics in pharmaceutical regulation affects both daily issues such as drug pricing and more intricate ones such as requirements for expensive new vaccines. Illustrating this framing of pharmaceutical companies, a May 8, 2008, update sent by e-mail explains:

> It has been nearly two years since the FDA fast-tracked the licensure of Merck's GARDASIL vaccine and the CDC's Advisory Committee on Immunization Practices (ACIP) recommended it for universal use by all 11 year old girls. Calls for state mandates for sixth grade entry soon followed with Merck leading an aggressive lobbying campaign with pro-forced vaccination proponents talking about 12 year old girls having sex. The National Vaccine Information Center countered by publicly framing the debate about GARDASIL mandates as a product safety and informed consent issue: the vaccine had only been tested in fewer than 1200 girls under age 16 followed up for less than two years.

Beyond issues of corruption, most parents voiced their distrust of the testing of pharmaceutical products; they expressed doubt that testing in clinical trials is adequate to ensure a product's safety. Tammy clarifies why she would not seek out the HPV vaccine for her daughter, even as she remains supportive of many other vaccines: "It's too new. I don't care if it's FDA approved. I don't necessarily trust the FDA. You know, EPA, FDA, they all have their interests that they protect."

Heather echoes this sentiment in explaining her disapproval of the vaccine, and she repeats a rumor suggesting that the vaccine's safety was uncertain: "I don't know if this is true, but I heard that the guy who invented the vaccine, or came up with the vaccine, has not vaccinated his own child for that. That could tell you something if that's true." Although Heather admits that "that's kind of secondhand hearsay," she reiterates her broader discomfort with the vaccine as originating in the inadequacy of the testing process. She confesses, "I guess I just feel like things are rushed through so fast to get on the market without really due diligence done to safe—safety testing. And I feel it's a money thing. You know, they want to get out—go out there and make money. It's a business. And they'll find out later. Wait and see what happens and they've tested us as we go, and I don't feel like that's the right approach."

Marlene, a mother of three with two adult sons and one preadolescent daughter, also describes her discomfort with the vaccine's safety. Explaining how her rejection of all required vaccines led her to reject HPV, she notes, "[It's] fairly untested on a big range of population and they want to give it to so many millions of young women." As she thinks through its likely efficacy and safety, she continues, "Well, it only covers for four out of the hundred diseases and it's so untested and some of the side effects that have more anecdotes because they haven't done the long term testing sounds like they're going to cause more problems than they'll solve."

Katie, a mother of two, points to the limitations of safety studies: "My thinking is I would never want to be the early adopter of any vaccine because you never—there have been another they've pulled off the market, like the rotovirus one and stuff because they ended up not being safe." Here she points to the ways vaccines approved for market use can often have deleterious results not discovered until they are widely used, as was the case with the rotavirus vaccine. Absorbing this possible risk does not make sense for her, given her overarching view that the vaccine's safety is unknown. This concern is magnified by the reality that her 2-year-old daughter will not be sexually active for a long time.

Many parents were familiar with the media coverage of the HPV vaccine and were following the discussions about whether it should be required, and if so, for whom. This popular discourse shaped their views of vaccines as well. Patricia, a 42-year-old mother of eight, objects to efforts to broaden the distribution of HPV vaccine and cites her existing concerns about the flu shot as evidence:

> It's crazy. I just read today they want to give it to boys now. So um I just—and the flu shot, too—my mother-in-law, she has dementia . . . I said, "you know we don't do flu shots. There's mercury still in the flu shots from what I understand, and that can be a contributor to dementia." Before she moved out here, she was getting the flu shot every year. That may have, you know, that may have been a big contributor to her problem. Maybe not, but also there's—I wish I had more time to research the whole science of, um, vaccination to begin with, because from what little I've read on it, it's kind of disturbing. So I'm not even convinced that if it worked, it was all a good thing, if it's even, you know a good thing to do.

Like Patricia, Katie has concerns about the safety of vaccines and is cognizant of how vaccines are marketed and promoted in ways that ignore how they might be problematic. She explains, "I know I've seen something in some parenting magazine and they're like, 'isn't that great?' This new vaccine that's coming out where it's like six—like all—you can get all your vaccines in one day, in one shot. And I'm like, what? But it's like, 'isn't this great?' you know. And I thought—wait a minute, is anybody—where is the consideration of, is it a good idea?"

Although motivated by different specific concerns about the HPV vaccine, these parents articulate their distrust of the review system, their belief that regulatory agencies charged with evaluating vaccine safety are not objective, and their suspicions that lobbying efforts of pharmaceutical companies can alter state policy. Many also point to the significant profits Merck stands to make on this vaccine. In fact, sales of Merck vaccines, most specifically Gardasil and RotaTeq (a new rotavirus vaccine for children), were credited with helping raise stock prices at Merck, particularly after the expensive settlements from their troubled anti-inflammatory drug Vioxx (Pettypiece 2007). In 2007 Merck saw a sales jump for Gardasil alone to $418 million (Kennedy 2007). Parents are not misguided in recognizing the profit motives of pharmaceutical companies.

It is easy to dismiss such parental concerns as paranoid or conspiratorial, but these parents are finding increasing examples to support their suspicions. For example, the U.S. Food and Drug Administration (FDA) rejected an application to make the Plan B emergency contraceptive available over the counter despite strong support for doing so from its own advisory panel (Davidoff 2006), drugs like Vioxx have been removed from the market for unknown dangers, and ingredients for Heparin have been counterfeited during their manufacture in unregulated China (Wood 2008); so the parents' concerns seem well supported. Furthermore, the Vaccine Injury Compensation Program and the U.S. Supreme Court protect pharmaceutical companies and medical device manufacturers from liability (Schweitzer 2008). Thus, parents' concerns are informed by real conditions, whether or not they necessarily apply to the review of HPV vaccines specifically.

Parent Autonomy and Children's Health

The parents are also significantly concerned about how efforts to mandate the HPV vaccine represent state effort to ignore parental autonomy and influence. These parents understand that the vaccine's purpose is to protect their children from future risk presented by sexual activity, but they perceive this issue as one among many on a landscape full of challenges that pertain to parenting adolescents. Parents who reject (or plan to reject) the HPV vaccine for their children see themselves as needing to address broader issues of youth choices and behavior. These parents envision working with their children within the family on issues of sexuality as an alternative to an inoculation against a sexually transmitted infection.

As Katie contemplates the HPV vaccine for her now 2-year old daughter, she explains,

> I don't know. I haven't had to make the decision on that per se, because I don't have a twelve year old. But, I just think, I think with that I have more of an issue with how it's being marketed where it's like—I don't know, it just seems like such a huge marketing push. . . . I think it's a decision that you have to make between yourself and your child and your doctor and you know it needs to be part of the bigger discussion, I guess, with your child. . . . I mean this is not totally like this, but to me in a way, it seems, just, you know, like putting like birth control pills in your child's orange juice, you know, just in case.

For some parents, the largest issue is indeed the assumption that their children will become sexually active before adulthood. Kristin explains how the question of the HPV vaccine gets wrapped around these other issues: "Well, I was thinking, since I have two girls, that I hope that they don't get involved in, you know, being interested in boys, um, as much as—or as—I guess as much as you know the average girl would. And I'm just going to try to do anything and everything I can to avoid that and once I see that maybe they have a crush or something like that, then maybe I'll start thinking about it." Kristin is aware that her oldest daughter, now 8 years old, will not eligible for the vaccine for four more years. She also considers at what point her daughters will become interested in sexual behavior. "Yeah I have—have some time, so but I'll keep track of it and see what—my plan is trying, you know, to try to keep them interested in other things."

While Kristin believes she could simply redirect her daughters from any future emerging interest in sexual activity, some parents reiterate the importance of firmly encouraging abstinence. Billy, a father of one disabled teenage boy and a healthy 7-year-old girl, is torn about the HPV vaccine. Having dealt with his wife's health problems caused by cervical dysplasia, Billy recognizes the significant impact cervical disease can have. Nonetheless, he is not convinced he would consent to the HPV vaccine for his daughter, although he has considered it. "I went through that part, but I mean abstinence is the best key. But you know, [my wife] and I both—we were virgins when we married, you know. We both were together and we've only been together, you know, just us so are you really gonna have your 14 year old out there having sex?" Billy does not explain how his wife's illness occurred even in the face of their own reported abstinence and monogamy. But he anticipates protecting his child and reiterates that a core parental responsibility is to know where your child is and in what kind of activities she is involved.

> And if you're the parent that you're supposed to be to begin with, you're gonna know what your child's doing from then on. Or at least I do know about mine. So you know, I know where she's at every minute of the day, except when she's at school and she's at school, so if she's having sex there, then we've got some problems . . . and it's talking to our kids, it's what we need to be doing as parents. We need to be talking to our kids. Sure, I know you're gonna have those feelings, I know you're going have—be curious as to what's going on—[she] is 7 years old and she knows where children come from, where babies come from. You know

she knows all of that and she also knows to, um, you know, if she wants to save that for her marriage and be the most special moment of her life, then she's gonna do that. But there's no way for us to stop her if she does, unless we're with her, you know, all the time. So, I mean you got 14 year-old girls getting out of the windows and running off with their boyfriend down the street or what not. Where's the parents? I mean they should know if they're getting in and out of their house or not. So that's the way I look at that. It's better to do that than it is to give them the shot wondering what's gonna happen 'cuz they really don't know.

Billy believes the responsibilities of parenting include protecting his children. He envisions doing so by presenting options of behavior and structuring choices, while also maintaining awareness of his children's location and behavior. He summarizes his position, saying, "I thoroughly believe that it all starts here. . . . If we try to raise her up as best we can and keep her out of—you know, bad things—if you're gonna be a good parent, then you're gonna know. You'll know if your daughter or son is getting into things they shouldn't be into and then it's time to jump in and say, well, which two things will it be?"

For other parents, giving the HPV vaccine to children before they are sexually active removes a potential tool for teaching responsibility. Astrid, a mother of one 6-year-old daughter, explains her choice to reject both the hepatitis B vaccine, given to infants to protect them against future exposure to a disease transmitted in blood and through sexual activity, and the HPV vaccine, which would be offered to her daughter as a preteen. Unlike parents who believe they can teach abstinence or redirect their children's sexual interest, Astrid sees value in allowing her daughter to make choices about prevention as she is making choices about her own behavior. She notes, "She needs to be a voice for herself. How is she going to learn that process?" As Astrid imagines it, her daughter will have an opportunity to claim responsibility for her own behaviors and make appropriate choices about prevention. Astrid hopes to communicate the importance of taking responsibility for one's own decisions and seeking out prevention for one's own risk behaviors.

To be clear, although many of the parents are cognizant that their children will likely develop interest in sexual activity, none believe that the vaccine will encourage sexual initiation. In fact, some parents resent the presumption, encoded in vaccine marketing, that sexual activity is inevitable for teens. Nonetheless, many have confidence in their abilities to work with their children within their families on these issues. It is worth noting that most states require minors to receive parental consent for health care. Should a teen want to obtain the HPV vaccine (or even the hepatitis B vaccine), she or he would be unlikely to succeed in light of these rules (Gordon et al. 1997), which match other laws that require parental consent or notification for reproductive health measures (Ehrlich 2003; Henshaw and Kost 1992). In this way, parents appropriately expect to remain central in their children's access to health care, even as they recognize that their children may make future decisions without them on issues of sexuality.

Implications

Parents aim to make decisions in their children's best interests. Health care decisions and strategies clearly cause them to grapple with how to do so most effectively. They must decide how they can best protect their children from future harm while remaining true to their (the parents') values, beliefs, and understandings of risk. In facing public policy that touts new medicines, technologies, and vaccines as safe and sometimes even legally mandates their consumption, parents also try to find ways to maintain the centrality of their parenting decisions in the context of their private families, away from the gaze of the state. As they see it, they know their own children best, are best qualified and most motivated to protect their children, and find that political corruption or negligence can inadvertently cause harm or undermine their goals.

Newly released vaccines against HPV provide an exciting opportunity to explore how daily parental decisions are framed by law and culture and informed by parents' own experiences. HPV vaccines differ from other vaccination decisions because, unlike vaccines that protect others (including infants, pregnant women, the immune-compromised, or the disabled), the HPV vaccines primarily protect the individual who receives them. Although the vaccines may eventually provide some herd immunity, such a benefit is not likely to occur for many decades. And since HPV is not transmitted through casual contact, the importance of herd immunity in this case is different than for other diseases. For many parents, this understanding changes their risk-benefit calculation. For example, many of the parents opposed to compulsory childhood vaccines against infectious diseases have sought out tetanus vaccines for their children. This is remarkable because tetanus is not contagious but is ubiquitous in the environment. Parents recognize that children can be infected by playing barefoot in the yard or in other invisible and innocent ways and can suffer tremendously from infection. Tetanus is also a vaccine that promises individual benefit, rather than collective benefit. What health providers and public health policymakers refer to as the "free-rider" problem—in which unvaccinated children remain safe because of the herd immunity provided by the immunized (May and Silverman 2005)—will not likely be the issue with refusal to obtain a vaccine against HPV in their children's lifetime.

Some parental concern might be alleviated if a more transparent accounting of the FDA's review process were made available and if parents had a better understanding of the science behind vaccines. Additional information about the lack of profit in manufacturing older vaccines for which patents no longer exist might also help diminish suspicion of corporate profit motives, even as companies do profit from new vaccines and medicines. In these ways, health providers and public health advocates could address the actual issues parents consider when making vaccine decisions.

By showing how parents' decisions regarding the new HPV vaccines exist within a web of complex parenting decisions, this chapter offers a new way of thinking about the public health implications of new technologies, particularly those that target young people. In this light, we can begin to consider more closely the ways that parental decisions are situated in and bound by social forces. Indeed, parents make decisions they see as in their children's best interests, in dialogue with law, policy, providers, and cultural understandings of risk and trust.

REFERENCES

Blumenthal, Ralph. 2007. "Texas Is First to Require Cancer Shots for Schoolgirls." Online in *New York Times*, Feb. 3.
Bobel, Chris. 2001. *The Paradox of Natural Mothering*. Philadelphia: Temple University Press.
Braun, Virginia, and Victoria Clarke. 2006. "Using Thematic Analysis in Psychology." *Qualitative Research in Psychology* 3:77–101.
Casper, Monica J., and Laura M. Carpenter. 2008. "Sex, Drugs, and Politics: The HPV Vaccine for Cervical Cancer." *Sociology of Health and Illness* 30:886–899.
Colgrove, James. 2006. *State of Immunity: The Politics of Vaccination in Twentieth-Century America*. Berkeley: University of California Press.
Conrad, Peter. 1985. "The Meaning of Medications: Another Look at Compliance." *Social Science in Medicine* 20:29–37.

Davidoff, Frank. 2006. "Sex, Politics, Morality at the FDA." *Hastings Center Report* 36:20–25.

Ehrlich, J. Shoshanna. 2003. "Choosing Abortion: Teens Who Make the Decision without Parental Involvement." *Gender Issues* 21:3–39.

Gordon, T. E., E. G. Zook, F. M. Averhoff, and W. W. Williams. 1997. "Consent for Adolescent Vaccination: Issues and Current Practices." *Journal of School Health* 67:259–264.

Gyapong, Deborah Waters. 2007. "HPV Vaccine Could Encourage Sexual Activity." *Catholic Register*, www.catholicregister.org/content/view/991/856/, Aug. 27.

Health Resources and Services Administration (HRSA). 2009. "National Vaccine Injury Compensation Program." U.S. Department of Health and Human Services, Washington, DC., www.hrsa.gov/vaccinecompensation/.

Henshaw, Stanley K., and Kathryn Kost. 1992. "Parental Involvement in Minors' Abortion Decisions." *Family Planning Perspectives* 24:196.

Hulbert, Ann. 2003. *Raising America: Experts, Parents, and a Century of Advice about Children.* New York: Knopf.

Kennedy, Val Brickates. 2007. "Merck's Profit Rises on 12% Jump in Sales." Online in *MarketWatch*, www.marketwatch.com/story/mercks-profit-rises-on-12-jump-in-sales?print=true&dist=printMidSection, Oct. 22.

Lareau, Annette. 2003. *Unequal Childhoods: Class, Race, and Family Life.* Berkeley: University of California Press.

Lupton, Deborah. 2009. "Risk as Moral Danger: The Social and Poltical Function of Risk Discourse in Public Health." In *Sociology of Health and Illness*, ed. P. Conrad, 460–467. New York: Worth.

May, Thomas, and Ross D. Silverman. 2005. "Free-Riding, Fairness, and the Rights of Minority Groups in Exemption from Mandatory Childhood Vaccination." *Human Vaccines* 1:12–15.

National Vaccine Information Center (NVIC). 2004. "State Information." Vol. 2004. NVIC, Vienna, VA.

Noakes, Karen, Joanne Yarwood, and David Salisbury. 2006. "Parental Response to the Introduction of a Vaccine against Human Papilloma Virus." *Human Vaccines* 2:243–248.

Pettypiece, Shannon. 2007. "Merck Earnings Rise on Vaccines, Cholesterol Drugs (Update5)." Bloomberg.com, www.bloomberg.com/apps/news?pid=20601087&sid=aG9S9ghJY4pk&refer=home, April 19.

Rosenstock, Irwin M. 1966. "Why People Use Health Services." *Milbank Memorial Fund Quarterly* 44:94–127.

Salmon, Daniel A., and Saad B. Omer. 2006. "Individual Freedoms versus Collective Responsibility: Immunization Decision-Making in the Face of Occasionally Competing Values." *Emergent Themes in Epidemiology* 3:13.

Schweitzer, Stuart. 2008. "Trying Times at the FDA—The Challenge of Ensuring the Safety of Imported Pharmaceuticals." *New England Journal of Medicine* 358:1773–1777.

Smith, P. J., S. Y. Chu, and L. E. Barker. 2004. "Children Who Have Received No Vaccines: Who Are They and Where Do They Live?" *Pediatrics* 114:187–195.

Sonfield, Adam. 2008. "New Study Points to High STI Rates among Teens, Major Disparities among Population Groups." *Guttmacher Policy Review* 11, no. 2: 20.

Stein, Rob. 2005. "Cervical Cancer Vaccine Gets Injected with a Social Issue; Some Fear a Shot for Teens Could Encourage Sex." *Washington Post*, Oct. 31, A03.

Wood, Alastair J. J. 2008. "Playing "Kick the FDA"—Risk-Free to Players but Hazardous to Public Health." *New England Journal of Medicine* 358:1774–1775.

CHAPTER TEN

Decision Psychology and the HPV Vaccine

Gretchen Chapman

New medical advances pose new decision-making challenges for patients and their families. When a medical intervention is made available, patients must decide what information to gather about this new option and under what circumstances they would want to receive it. The introduction of a vaccine for human papillomavirus gives adolescents and their parents the responsibility of making decisions that have implications for the health of the entire population. This chapter views the HPV vaccine through the lens of theory and research on the psychology of decision making. I examine the vaccine as a case study in making decisions on behalf of others, where concern for others must include a consideration of the delayed consequences and risky outcomes of the vaccine. These issues, I argue, illuminate the complexity of the HPV debate at the level of families and society. However, my primary focus is not to analyze the HPV vaccine debate per se; rather, I use this vaccine as an important opportunity to think more broadly about vaccination behaviors, drawing on the theoretical insights of decision psychology.

Chapter 9, by Jennifer Reich, examines in depth the decision processes of the primary decision makers in the cancer vaccine controversies—the parents of adolescent girls. The parents are the people who say the ultimate yea or nay regarding whether a particular girl will be vaccinated. Here, I continue Reich's exploration of decision making but in a different framework. Rather than employing qualitative methods to answer sociological questions, I review several quantitative studies that address questions emerging out of the framework of decision theory.

The Decision Theory Framework

Decision theory views individual decision makers as comparing choice options based on the expected costs and benefits, weighted by the likelihood that those outcomes will occur. Four key issues about vaccination decisions generally, and about the HPV vaccine in particular, are of interest to researchers focusing on the psychology of decision making. First, the benefits of vaccination are delayed: one vaccinates now to prevent the diseases of tomorrow. In this respect, the HPV vaccine is of particular interest for decision psychology because, unlike the flu shot, for example, which accrues benefit several months after vaccination, the HPV vaccine accrues benefit several *decades* after vaccination. Thus, it provides a rare window onto decision making about very long-term consequences. Second, vaccination entails uncertain outcomes: one may escape the disease even if unvaccinated, and vaccination only partially reduces risk. The HPV vaccine also has uncertain consequences, but the uncertainties of effectiveness are complicated because the vaccine acts on a virus rather than directly on the cancers to which it has been linked, and, moreover, most women could avoid cervical cancer even if unvaccinated. Third, vaccination entails social dynamics: because of herd immunity, your vaccination decision benefits not only you but also me, and consequently, I have the opportunity to benefit from others' vaccination without vaccinating myself. The HPV vaccine provides a unique opportunity to examine the role that others' outcomes play in personal vaccination decisions. The vaccine has only recently been approved for boys and men, giving them the opportunity to protect future female partners from cancer by vaccinating themselves. Fourth, vaccination frequently entails surrogate decision making, as when a parent makes vaccination decisions on behalf of a child. Like other pediatric vaccinations, the HPV vaccine requires surrogate decision making. However, unlike other pediatric vaccinations, the HPV vaccine is delivered to adolescents, who are old enough to have views and opinions of their own. This vaccination decision thus affords a special opportunity to examine the extent of agreement between the surrogate decision maker and the beneficiary.

Decisions with Delayed Consequences

The concept of *intertemporal choice*, or decision making among options with consequences that occur at different times, is a useful one in exploring decision making about delayed consequences. A standard finding among researchers is that decision makers *discount* future outcomes; that is, they value delayed outcomes less than immediate outcomes with the same nominal value. Furthermore, the rate at which they discount future outcomes varies with factors that normatively should be irrelevant. This behavior demonstrates that a decision requiring the calculation of a delayed outcome is a complex process resulting in choices that deviate systematically from those described by normative decision theory. For example, descriptively, positive outcomes (or gains) are discounted at a much steeper rate than negative outcomes (or losses), even when the gains and losses are two equivalent descriptions of the same outcomes. Thus, someone might view a gain of $500 in ten years to be just as attractive as, say, $200 immediately but also view paying a $500 fine ten years from now to be just as unattractive as paying a $500 fine today. This pattern would indicate steep discounting of the monetary gain but no discounting at all of the monetary fine.

Most vaccination decisions entail a choice between a small cost now (the pain and inconvenience of the shot, the time spent at the doctor's office, the small risk of immediate side effects) and the risk of a large cost much later (the disease). If decision makers discount the delayed outcome sufficiently, they might view the future cost as less extreme than the immediate cost and choose not to vaccinate. Thus, discounting of delayed outcomes could help to explain why many patients fail to get vaccinated or to take other preventive health measures. The fact that vaccination consequences are often framed as negative outcomes (cost and inconvenience now versus disease later), however, may serve to mitigate this discounting of the future.

The HPV vaccine is of particular interest for the study of intertemporal choice because the disease it is designed to prevent occurs many decades after vaccination—a much longer delay than that entailed by other vaccines. For example, the pneumococcal vaccine and many pediatric vaccinations are designed to prevent diseases that could occur anytime over the following years and decades. In contrast, a 12-year-old girl who is vaccinated against HPV is reducing her risk of cervical cancer, a disease that she almost assuredly would not get for at least twenty years and probably would not get for several decades beyond that. Given that the HPV vaccine represents a case of a decision about an extremely delayed outcome, one might expect temporal discounting to have an even larger effect on this decision than it does on other health decisions.

A curious feature of the HPV vaccine is that it is cancer (and also genital warts) that is the outcome of interest; in other words, HPV infection in and of itself is of little concern apart from the disease states that result from it. Yet, the fact that the vaccine prevents an event (a virus-related infection) that could occur in the near future could potentially make the vaccine seem more similar to traditional vaccines, which prevent near-term infections. That is, the HPV vaccine may be viewed more favorably when it is seen as preventing a near-term event (virus infection), even though that event is significant only because of the longer-term diseases to which it leads. In addition, the genital warts prevented by Gardasil have a much shorter time horizon than cervical cancer. Gardasil has the obvious advantage over Cervarix that it prevents not only cancer but also genital warts, and a key aspect of this advantage may be that Gardasil prevents something that might occur only a few years in the future, whereas Cervarix prevents only outcomes that are decades away.

One might expect that Merck would seek to capitalize in its advertising on the shorter time frame of the "other HPV diseases" that it prevents as a way of emphasizing that the vaccine has both short-term and long-term benefits. The fact that it has not already done this may have less to do with how consumers think about delayed benefits and more to do with the emotional impact of the fear of cancer and with the desexualizing of Gardasil in its presentation to consumers (see chapter 7). Indeed, as Robert Aronowitz (chapter 2) argues, the HPV vaccine has been constructed and marketed as a device to reduce individual risk (including the emotional experiential aspect of risk) rather than to prevent disease. Presenting Gardasil as a source of "peace of mind" is likely meant to convey that the vaccine does not simply reduce the likelihood of a severe disease decades in the future, but also prevents a near-term viral infection that, once contracted, could lie in wait, undetected, for decades in one's body.

A final note about the benefits and limits of the concept of temporal discounting: although the claim that decision makers discount future outcomes is well supported, there is far less empirical support for the idea that individual variation in the rate of discounting future outcomes is associated with individual variation in health behaviors more generally. Chapman (2005) conducted a meta-analysis of studies that examined the association between subjective discount rate and health behaviors and found no reliable association between the two except when the health behaviors concerned addiction (drug or alcohol use, gambling, etc.). Subjective discount rate predicts addiction status, but it does not predict health behaviors such as vaccination for influenza or other diseases. What this research suggests is that although we know that people generally tend to discount future outcomes, it is not entirely clear why some choose to vaccinate and others do not. In other words, temporal discounting may be more useful as a way to understand why vaccines are *generally* underutilized than as a way to understand individual variation—why some people get vaccinated and others do not.

Decisions with Risky Outcomes

Almost all difficult decisions have to do with risky or uncertain outcomes. Perceived risk is a robust predictor of vaccination behavior (e.g., see chapter 9). Brewer and co-workers (2007) conducted a meta-analysis and found that perceived risk likelihood and perceived risk severity (e.g., the severity of disease) are consistently associated with vaccination behavior. This relationship is also present for the HPV vaccine specifically (Brewer and Fazekas, 2007; Fazekas, Brewer, and Smith 2008) and is in accord with *Expected Utility Theory*, a normative theory of decision making that prescribes that decisions should be driven by the value or utility of potential outcomes weighted by the probability that they will occur.

In many decisions the risk likelihood can be quantified as a probability. For example, an unvaccinated woman in the United States has about a 1 percent lifetime risk of developing cervical cancer, given current screening prevalence (Saraiya et al. 2007). The Gardasil vaccine is purportedly fully effective against the two strains of HPV that account for 70 percent of cervical cancers. Thus, given certain assumptions, vaccination should reduce the risk of cervical cancer by 70 percent. Importantly, however, the large majority of women will escape cervical cancer even if unvaccinated, and there is a chance of getting cervical cancer even if vaccinated.

Although perceived risk is a reliable predictor of vaccination decisions, perceived risk can differ drastically from actual risk. For example, a recent questionnaire study found that both adolescent children and their parents were poorly informed about the objective risks of HPV. Parents of girls estimated, on average, that the lifetime risk that their daughter would develop cervical cancer if not vaccinated was 30 percent (Basu, Chapman, and Galvani 2008). Girls estimated this risk at 36 percent on average (Chapman et al. 2008). Both mean estimates are far above the actual statistical risk of 1 percent (Saraiya et al. 2007). Children estimated the lifetime risk of contracting HPV if not vaccinated at 34 percent on average (Chapman et al. 2008), whereas the actual risk is 50 to 80 percent.

Risk-estimation errors, even errors of that magnitude, might be attributed to lack of information about the disease and lack of familiarity with the probability scale, such that low risks are overestimated and high risks are underestimated. Risk perceptions are not inaccurate just because of lack of knowledge, however; they also display systematic errors. One such error is the *optimistic bias*, which is the tendency for people to think their risk of adverse events is below average (Weinstein 1980). Although certain individuals legitimately do have below-average risk, not everyone can have below-average risk. The optimistic bias is demonstrated when the average of individual risk estimates from a group of participants is significantly below the perceived average risk for that group. For example, Li and colleagues (unpublished) asked a group of college women to rate their lifetime risk of cervical cancer on a five-point scale from much below average to much above average, where "average" refers to the risk of the average college woman. The mean rating was below average. When asked to estimate their risk on a percentage scale, the women rated their risk at about 25 percent, on average, well above the national average for cervical cancer incidence. The women were then told that the average American woman has a 4 percent lifetime risk,[1] and they were asked to make a revised estimate of their own lifetime risk. The mean revised perceived risk was just under 4 percent. This result suggests that lay women do not know what the average prevalence of cervical cancer is, but they think their own risk is less than whatever that average is. Not knowing the prevalence is an understandable lack of knowledge about an unfamiliar topic. But the tendency to conclude that one's own risk is less than whatever the average is represents a systematic bias in risk perception.

Not only are vaccination decisions based on biased and inaccurate risk perceptions, but the way in which risks are presented can affect the choice made, including vaccination decisions. Li and Chapman (2009) presented college students with one of two descriptions of a hypothetical vaccine intended to mimic the HPV vaccine. One group was told that the vaccine was 100 percent effective against the viruses responsible for 70 percent of cases of a particular type of cancer. The other group was told that the vaccine was 70 percent effective against the viruses responsible for all cases of a particular type of cancer. The two descriptions, of course, could be applied to the HPV vaccine and are equivalent in the sense that both describe a vaccine that prevents 70 percent of the cancer cases. Participants gave higher ratings of willingness to vaccinate in the first condition than in the second, however. This effect illustrates a *"certainty effect,"* where certainty is overweighted relative to other probabilities. Merck describes Gardasil as a "vaccine that helps protect against . . . 2 types that cause 70% of cervical cancer cases."[2] Although the firm does not describe the vaccine as 100 percent effective against a subset of viruses, it is notable that it also does not describe the vaccine as 70 percent effective against cervical cancer, which might be viewed as a low efficacy.

Recent research on risk perception (e.g., Finucane et al. 2000; Mellers, Schwartz, and Ritov 1999; Slovic 1987) has emphasized not only the cognitive aspects of risk perception (e.g., estimates of risk likelihood and risk severity) but also the emotional aspects of risk perception or "risk as feelings" (Loewenstein et al. 2001). Perceiving risk entails feeling worry and dread or anticipating later regret, as well as estimating a probability. Research has demonstrated that worry is correlated with risk estimates (Cameron and Diefenbach 2001; Chapman and Coups 2006; Constans 2001; Easterling and Leventhal 1989; McCaul et al. 2003; Sjöberg 1998) and that worry is related to preventive health behaviors such as cancer screening (Diefenbach, Miller, and Daly 1999; McCaul et al. 1998; McCaul, Schroeder, and Reid 1996), genetic testing (Cameron and Diefenbach 2001), and vaccination (Chapman and Coups 2006). Anticipated regret also drives various preventive behaviors such as condom use (Richard, de Vries, and van der Pligt 1998; van der Pligt and Richard 1994), exercise (Abraham and Sheeran 2003), and immunization (Chapman and Coups 2006; Connolly and Reb 2003; Wroe, Turner, and Salkovskis 2004).

Thus, it appears that vaccination decisions and other choices about preventive health are driven more by emotions surrounding perceived risk than by the perception of risk likelihood itself. Indeed, as both Robert Aronowitz (chapter 2) and Laura Mamo and coauthors (chapter 7) have argued, the marketing of Gardasil has certainly not relied on the explicit quantification of risk likelihood but rather on appealing to the emotionally rich themes of girlhood, innocence, agency, and health. Merck describes Gardasil as "the only cervical cancer vaccine that helps protect against 4 types of HPV: two types that cause 70% of cervical cancer cases, and two more types that cause 90% of genital warts cases."[3]

Although this description does reference the efficacy percentages explicitly, it is qualified by the "helps protect" phrase, which undermines the implied efficacy. Rather than providing precisely quantified risks, then, advertisements for Gardasil promote it as "the only" vaccine offering the possibility to be "One Less" and to "choose" to protect one's health—all appeals to the feeling of being at risk rather than reflecting the actual likelihood of disease states.

Concern for Others

Vaccination has implications not only for the person getting vaccinated but also for others in the population. Vaccination decreases the likelihood that the individual will become infected and therefore decreases the likelihood that the individual will spread the virus to others in the population. If a large enough fraction of the population gets vaccinated (and reaches what is known as *"the herd immunity threshold"*), an epidemic cannot persist, even if the remainder of the population is not vaccinated. Consequently, individuals can benefit not only themselves but others by being vaccinated. Likewise, individuals have the opportunity to free-ride on the vaccination of others and reap the benefits of herd immunity even though they themselves were not vaccinated.

Vaccination decisions are therefore somewhat parallel to what decision psychology calls a *"social dilemma."* In a social dilemma, each of several players is faced with the same choice options: to "cooperate" or to "defect." Each individual is better off defecting, regardless of what the others do. But if everyone defects, everyone will be worse off than if everyone cooperates. Everyday life presents numerous examples of social dilemmas, such as recycling, donating blood, voting, and avoiding overfishing. Hundreds of psychology experiments have presented laboratory participants with social dilemmas. A striking finding is that a sizable proportion of participants cooperate, even if they are playing only one round of the game against players whom they cannot identify (Dawes and Messick 2000; Fehr and Gächter 2002). This result strongly suggests that decision makers take other people's outcomes into account.

Vaccination decisions deviate somewhat from a classic social dilemma in that vaccination behavior very often is in the individual's self-interest. That is, it is not the case that one option benefits the individual and the alternative benefits the group. Instead, vaccination very often benefits both the individual and the group. There is a discrepancy between group and individual welfare only when a sufficiently high proportion of the population has vaccinated such than an individual's self-interest would be maximized by not being vaccinated and "free-riding" on others' vaccination, whereas the population outcome would be improved if the individual was vaccinated. Galvani, Reluga, and Chapman (2007) present modeling results indicating that the level of influenza vaccination that would occur if individuals were motivated only by self-interest is lower than that needed to achieve the optimal population outcome. Thus, the model predicts a substantial amount of free-riding. This result is analogous to the less-than-100-percent cooperation found in numerous psychology experiments on social dilemmas.

Yet another way that vaccination decisions differ from a classic social dilemma is that not all individuals face the same consequences from vaccination. For example, in the case of influenza vaccination, elderly individuals face the highest risk of dying from the virus, but young individuals are responsible for most of the transmission. Indeed, the population outcome would be maximized by vaccinating many of the young and none of the elderly, thereby decreasing transmission so much that the elderly are protected. If individuals act purely out of self-interest, however, just the opposite will occur: all of the elderly and few of the young will get vaccinated (Galvani, Reluga, and Chapman 2007).

Aronowitz and Reich (chapters 2 and 9) make the case that Gardasil has been presented, and potential consumers have perceived it, as an individual-consumer commodity rather than as a public good. The population-level herd-immunity benefits of Gardasil vaccination have not been emphasized in public discourse, in large part because the vaccine was, until very recently, targeted only to girls and women in a narrow age range (not to men) and because it is priced and marketed to appeal to families of higher income (not the broader population). In addition, because HPV is spread through intimate contact, not casual contact, an appeal to vaccinate to protect the population seems less relevant than an appeal to vaccinate to protect one's future sexual partners. Indeed, the Gardasil ad that says, "She won't have to tell him she has HPV . . . because she doesn't" (see chapter 2) touches on this motivation, although the ad seems to be more about preventing shame or embarrassment than about protecting a future partner from disease. Thus, for various reasons, herd immunity is not part of the current dialogue about HPV, but it is possible that this may change in the future.

A parallel issue is that the HPV vaccine presents an interesting trade-off between men and women. In 2006 Gardasil was approved only for girls and women, but in 2009 it was approved for boys and men as well. Vaccinating boys would decrease the likelihood that they would spread the virus to later female sexual partners. Thus, the population outcome could be enhanced by vaccinating boys, especially when vaccination levels among girls are modest (Basu, Chapman, and Galvani 2008). Boys, however, have little to gain from the vaccine in terms of self-interest. Boys could benefit directly from the Gardasil vaccine because it protects against HPV strains that cause genital warts in both men and women. But boys would receive no direct benefit from Cervarix, which protects only against cervical cancer. (Boys would also benefit directly through the prevention of anal cancer, a little-discussed benefit of the vaccine [see chapter 4].) Both Cervarix and Gardasil can prevent boys from transmitting HPV to future female partners. The uptake of either vaccine, but especially Cervarix, among boys would provide compelling evidence of vaccination behavior motivated by the benefits to others. A survey study by Vietri and colleagues (under review) found little evidence of such motivation. Boys gave low preference ratings for Cervarix, lower than they gave for Gardasil, and lower than girls gave for Cervarix. Interesting research questions raised by the HPV vaccine include, When is vaccination behavior driven by a concern for benefits to others, and How can such motivation be augmented among decision makers.

Surrogate Decision Making

Many medical decisions are made on behalf of someone else. For example, doctors decide on behalf of patients, and family members make choices on behalf of incapacitated loved ones. Decisions about pediatric vaccinations are made by parents on behalf of their children. Because the HPV vaccine is primarily targeted to minors, it is the parent who makes the decision about vaccination. The adolescent girls who are the beneficiaries of those decisions are, however, old enough to have preferences of their own about vaccination. Thus, HPV vaccination decisions provide an opportunity to examine the accuracy of surrogate decisions. That is, can parents accurately predict the vaccination preferences of their adolescent daughters, and do they make decisions that are in line with their daughters' preferences?

Several studies have examined the accuracy of surrogate medical judgments. In the Advance Directives Values Assessment and Communication Enhancement study by Pete Ditto and colleagues (Coppola, Ditto, Danks, and Smucker 2001; Ditto et al. 2001), the preferences of patients for various medical interventions were compared with family members' predictions of the patient's preferences. These surrogate judgments were modestly accurate, but providing an advance directive or a living will did not improve accuracy. When a family member makes end-of-life medical treatment decisions on behalf of a loved one, the goal is arguably to predict what the loved one would have chosen if the loved one were able to make his or her own choices. In contrast, when parents make medical decisions on behalf of a child, the goal is not simply to predict what the child wants or what the child's future preferences will be. Instead, the goal is to act in the best interest of the child, without regard for the child's actual preference (hence the high frequency with which parents cajole children to eat more vegetables and watch less TV). Thus, it is reasonable to predict that a parent's decision on behalf of the child might differ from the parent's prediction about the child's preference.

Indeed, Vietri and co-workers (under review) found just this. Parents of adolescent children had a stronger preference for giving their children the HPV vaccine than the children had for getting the vaccine. But parents knew about this disparity in preferences: when predicting the child's preference, parents were very accurate. Likewise, children were accurate in predicting their parents' preferences. Thus, parents and children had different preferences, but each party knew the other's preference, demonstrating high surrogate judgment accuracy.

In this study, the parents' preferences for the vaccine were associated with their perceptions of the risks and benefits of the vaccine (Chapman et al. 2008). These findings are in line with those of Reich (chapter 9), even though her smaller qualitative study included only parents opposed to vaccines, whereas our larger questionnaire study included a broader sample of parents of adolescents. Our study did not examine, as Reich's did, parental views about parental autonomy or the role of corrupt politics in the promotion of Gardasil. The Reich study, in turn, examined only parental views and not those of the children. Her results therefore may not reflect surrogate decision making specifically, but rather medical decision making more generally (i.e., the parents may have had similar views about vaccinating themselves). An interesting topic for future research is whether being in the surrogate or the parental role changes the decision process, for example, by causing the decision maker to be more cautious, scrutinize evidence more skeptically, or consider long-term effects more thoroughly.

Conclusion

Decision theory views decision making as a process of comparing options based on the outcomes that might result from each option, the value of each outcome, and the likelihood and timing of each outcome. The HPV vaccine represents a real-life decision with high-stakes outcomes and extremes in the likelihood and timing of the outcomes. Unlike medical decisions that are essentially individual choices, decisions about the HPV vaccine are made within a family context and have outcomes that impact the entire population. Thus, decisions about the HPV vaccine represent a nexus of some of the most interesting issues being explored by decision psychology.

NOTES

1. Four percent is the estimated lifetime incidence conditional on no Pap smear screening.
2. Gardasil, www.gardasil.com, accessed November 26, 2008.
3. Ibid.

REFERENCES

Abraham, C., and P. Sheeran. (2003). Acting on intentions: The role of anticipated regret. *British Journal of Social Psychology* 42:495–511.
Basu, S., G. B. Chapman, and A. P. Galvani. (2008). Integrating epidemiology, psychology, and economics to achieve HPV vaccination targets. *Proceedings of the National Academy of Sciences* 105, no. 48: 19018–19023.
Brewer, N. T., G. B. Chapman, F. X. Gibbons, M. Gerard, K. D. McCaul, and N. D. Weinstein. (2007). A meta-analysis of the relationship between risk perception and health behavior: The example of vaccination. *Health Psychology* 26:136–145.
Brewer, N. T., and K. I. Fazekas. (2007). Predictors of HPV vaccine acceptability: A theory-informed systematic review. *Preventive Medicine* 45:107–114.
Cameron, L. D., and M. A. Diefenbach. (2001). Responses to information about psychosocial consequences of genetic testing for breast cancer susceptibility: Influences of cancer worry and risk perceptions. *Journal of Health Psychology* 6:47–59.
Chapman, G. B. (2005). Short-term cost for long-term benefit: Time preference and cancer control. *Health Psychology* 24:S41-S48.
Chapman, G. B., and E. J. Coups. (2006). Worry, regret, and preventive health behavior. *Health Psychology* 25:82–90.
Chapman, G. B., J. Vietri, M. Li, S. Basu, and A. Galvani. (2008). Surrogate decision making and the human papillomavirus vaccine. Paper presented at the annual meeting of the Society for Medical Decision Making, Oct. 2008, Philadelphia.
Connolly, T., and J. Reb. (2003). Omission bias in vaccination decisions: Where's the "omission"? Where's the "bias"? *Organizational Behavior and Human Decision Processes* 91:186–202.
Constans, J. I. (2001). Worry propensity and the perception of risk. *Behaviour Research and Therapy* 39:721–729.

Coppola, K. M., P. H. Ditto, J. H. Danks, and W. D. Smucker. (2001). Accuracy of primary care and hosptial-based physicians' predictions of elderly outpatients' treatment preferences with and without advance directives. *Archives of Internal Medicine* 161:431–440.

Dawes, R. M., and D. M. Messick. (2000). Social dilemmas. *International Journal of Psychology* 35 (2): 111–116.

Diefenbach, M. A., S. M. Miller, and M. B. Daly. (1999). Specific worry about breast cancer predicts mammography use in women at risk for breast and ovarian cancer. *Health Psychology* 18:532–536.

Ditto, P. H., J. H. Danks, W. D. Smucker, J. Bookwala, K. M. Coppola, R. Dresser, A. Fagerlin, R. M. Gready, R. M. Houts, L. K. Lockhard, and S. Zyganski. (2001). Advance directives as acts of communication: A randomized controlled trial. *Archives of Internal Medicine* 161:421–430.

Easterling, D. V., and H. Leventhal. (1989). Contribution of concrete cognition to emotion: Neutral symptoms as elicitors of worry about cancer. *Journal of Applied Psychology* 74:787–796.

Fazekas, K. I., N. T. Brewer, and J. S. Smith. (2008). HPV vaccine acceptability in a rural, southern area. *Journal of Women's Health* 17:1–10.

Fehr, E., and S. Gächter. (2002). Altruistic punishment in humans. *Nature* 415:137–140.

Finucane, M. L., A. Alhakami, P. Slovic, and S. M. Johnson. (2000). The affect heuristic in judgments of risks and benefits. *Journal of Behavioral Decision Making* 13:1–17.

Galvani, A., T. Reluga, and G. B. Chapman. (2007). Long-standing influenza vaccination policy is in accord with individual self-interest but not with the utilitarian optimum. *Proceedings of the National Academy of Sciences* 104:5692–5697.

Li, M., and G. B. Chapman. (2009). 100% of anything looks good: The appeal of one hundred percent. *Psychonomic Bulletin and Review* 16:156–162.

Li, M., M. D. DiBonaventura, R. L. Stuart, and G. B. Chapman. (unpublished). Risk perception biases at the group and individual level. Available from Chapman.

Loewenstein, G. F, E. U. Weber, C. K. Hsee, and N. Welch. (2001). Risk as feelings. *Psychological Bulletin* 127:267–286.

McCaul, K. D., A. D. Branstetter, S. M. O'Donnell, K. Jacobson, and K. B. Quinlan. (1998). A descriptive study of breast cancer worry. *Journal of Behavioral Medicine* 21:565–579.

McCaul, K. D., A. B. Canevello, J. L. Mathwig, and W. M. P. Klein. (2003). Risk communication and worry about breast cancer. *Psychology, Health, and Medicine* 8:379–389.

McCaul, K. D., D. M. Schroeder, and P. A. Reid. (1996). Breast cancer worry and screening: Some prospective data. *Health Psychology* 15:430–433.

Mellers, B., A. Schwartz, and I. Ritov. (1999). Emotion-based choice. *Journal of Experimental Psychology: General* 128:332–345.

Richard, R., N. K. de Vries, and J. van der Pligt. (1998). Anticipated regret and precautionary sexual behavior. *Journal of Applied Social Psychology* 28:1411–1428.

Saraiya, M., F. Ahmed, S. Krishan, T. B. Richards, E. R. Unger, and H. W. Lawson. (2007). Cervical cancer incidence in a prevaccine era in the United States, 1998–2002. *Obstetrics and Gynecology* 109:360.

Sjöberg, L. (1998). Worry and risk perception. *Risk Analysis* 18:85–93.

Slovic, P. (1987). Perception of risk. *Science* 236:280–285.

van der Pligt, J., and R. Richard. (1994). Changing adolescents' sexual behaviour: Perceived risk, self-efficacy and anticipated regret. *Patient Education & Counseling* 23:187–196.

Vietri, J., M. Li, G. B. Chapman, and A. Galvani. (under review). Attitudes towards HPV vaccination in parent-child dyads: Similarities and acknowledged differences. Available from Chapman.

Weinstein, N. D. (1980). Unrealistic optimism about future life events. *Journal of Personality and Social Psychology* 39 (5): 806–820.

Wroe, A. L., N. Turner, and P. M. Salkovskis. (2004). Understanding and predicting parental decisions about early childhood immunizations. *Health Psychology* 23:33–41.

CHAPTER ELEVEN

Nonmedical Exemptions to Mandatory Vaccination

Personal Belief, Public Policy, and the Ethics of Refusal

Nancy Berlinger and Alison Jost

Do "good" parents refuse to vaccinate their children? Should policymakers allow parents to opt out of having their children vaccinated based on the parents' personal beliefs about the safety of vaccination, even if there is no scientific evidence to support these beliefs?

Continuing battles between public health experts and parents who question the safety of childhood vaccinations have raised concern over the growing practice of invoking nonmedical exemptions to mandatory vaccinations. These opt-out provisions, like "conscience clauses" for health care providers, are written into the laws of most states. Several state legislatures are now considering separate nonmedical exemption policies for the human papillomavirus vaccine. The introduction of the HPV vaccine offers an opportunity to examine the tension between the public health goals of mandatory childhood vaccinations and the argument that individuals' beliefs about science are valid grounds for exemption from public health obligations. Because the HPV vaccine serves different goals than other mandatory vaccinations—it aims to protect the individual who receives it, rather than the communal "herd," and one of its benefits is cancer prevention—the story of its introduction and uptake also offers an opportunity to examine the ethics of responsibility from a different perspective. When "strengthening the herd" is not the primary aim of the social contract associated with participation in vaccination, do parents still have an obligation to protect their adolescent daughters from a sexually transmitted disease they may someday be exposed to? (And if so, what about protecting their sons from HPV-related cancers and other diseases affecting men?) Has the HPV-vaccination story, as it has played out in the American media over the past few years, had a different ending than we expected, based on what parents decided to believe, and to do, once this vaccine was introduced? Did "cancer" trump "conscience"?

Medical and Nonmedical Exemptions to Routine Childhood Vaccinations

School immunization laws in every state permit medical exemptions for children whose underlying health conditions, such as HIV infection, cancer, or immunosuppressive therapies, place them at undue risk from one or more routine immunizations.[1] Forty-eight states also permit "nonmedical" exemptions based on religious convictions, although most religions do not take a position on vaccination: Christian Science and other faith-healing traditions are among the few recognized religions in which immunization refusal might be construed as the expression of a religious belief. Twenty of these forty-eight states also permit nonmedical exemptions based on nonreligious personal beliefs.[2] States vary in the language used to describe the basis for a nonmedical exemption and in the standards of proof required of parents. (In the interest of clarity and consistency, this essay uses the phrase *religious conviction* when referring to laws, policies, and practices that explicitly concern adherence to religious doctrine, and *personal belief* when referring to laws, policies, and practices that concern beliefs about science that are not necessarily religious in nature.)

In some states, parents have been required to defend their requests at a "religious sincerity" hearing or to provide written documentation that their religious convictions or personal beliefs are incompatible with state law. In other states, they simply sign a form or check off a box to claim a nonmedical exemption. This is an active area of legislation, with recent legislative trends moving toward adding nonmedical exemption categories or making it easier for parents to obtain such exemptions.[3] States where it is easy to get a nonmedical exemption tend to have the largest numbers of such exemptions.[4]

The HPV Vaccine in the Context
of Nonmedical Exemptions

The introduction of the HPV vaccine Gardasil, which reduces the risk of cervical cancer by blocking cancer-causing viral strains, has added a new dimension to debates about nonmedical exemptions.[5] Since 2006, when the national Advisory Committee on Immunization Practices recommended that the HPV vaccine should be routine for girls by age 12, virtually every state legislature has proposed bills addressing whether Gardasil should become a mandatory vaccine, who should pay for it, how parents should be educated about the risks and benefits of Gardasil, and how exemptions should be handled. Unlike other routine childhood immunizations, which protect against diseases that are communicable through everyday contact, Gardasil provides protection against a common sexually transmitted disease. Even though all but two states already allow some form of nonmedical exemption, some state legislatures are drafting HPV-specific opt-out policies, perhaps owing to the additional issue of sexual morality or scrutiny from socially conservative groups.[6] The HPV-vaccine-specific policies may also reflect a somewhat different handling of the nonmedical exemption issue: this vaccine is aimed at adolescents, who may be more aware of health care decisions involving themselves than younger children are and may participate in these decisions.

Why Do Some Parents Opt Out?
Religious Convictions

Some parents may seek or consider nonmedical exemptions because of strongly held religious convictions about faith healing.[7] The Supreme Court's landmark 1944 decision in *Prince v. Commonwealth of Massachusetts*, on the limits of parents' religious freedom, states that the "right to practice religion freely does not include the liberty to expose the community or the child to communicable disease or the latter to ill-health or death."[8] While *Prince v. Massachusetts* upheld the state's authority to protect children under child labor laws when parents or guardians claimed that their own religious rights extended to the activities of the children in their care, this ruling is frequently cited in cases where parents claim a religious right to forgo medical treatment for a child. When such cases arise, courts may intervene to protect the health and welfare of the affected children and, if there is a risk of disease outbreak, to protect the health of the community in general. In 1991, a measles epidemic in Philadelphia that resulted in more than 500 cases and 7 fatalities was traced to unvaccinated children whose families were members of two faith-healing churches.[9] In an effort to halt the spread of the disease, public health officials were granted a court order to vaccinate six children of church members.

Beliefs about the Dangers of Vaccination

Parental resistance based on personal beliefs about immunization present a different clinical, ethical, and policy challenge. While some of these parents join mail-order or Internet "churches" to bolster their case for a religious exemption—available in nearly every state—they may have little common ground with those whose resistance to vaccination is the product of their religious convictions concerning disease and medicine.[10]

Many of these parents have strong personal beliefs about the dangers of vaccines, in particular, the belief that certain childhood vaccines are linked to rising rates of autism or other disorders. The 2008 case of Hannah Poling, whose family was deemed eligible for compensation from a federal vaccine-injury fund, has reinforced these families' fears. The Polings contended that the onset of their daughter's autism was triggered by a series of vaccines she received at the age of 18 months. The National Vaccine Injury Compensation Program agreed but noted that Hannah Poling's case was both unusual and medically complex: at the time she received the vaccines, she was likely suffering from an undetected and rare mitochondrial disorder. The court concluded that the vaccines exacerbated this disorder, resulting in symptoms consistent with autism. Although the Polings are not personally opposed to childhood vaccinations and have stated that they "support a safe vaccination program against critical infectious diseases," they remain unconvinced that Hannah's mitochondrial disorder preceded the onset of her autism.[11]

Vaccine skeptics heralded the outcome of the Poling case as an admission by the federal government that vaccines cause autism.[12] But physicians, scientists, and officials at the National Vaccine Injury Compensation Program itself maintained that the Poling case was an outlier and that there is no proven link between vaccines and autism: Julie Gerberding, then director of the Centers for Disease Control and Prevention (CDC), stated with reference to this case that "the government has made absolutely no statement indicating that vaccines are a cause of autism."[13] Several recent Institute of Medicine reports have also found no credible scientific evidence of a causal link between vaccines—in particular, the preservative thimerosal—and neurological damage.[14] Beginning in 1999, thimerosal was removed from most childhood vaccines used in the United States, and diagnoses of autism have continued to rise among the generation of children who were never exposed to thimerosal through vaccination. A case-control study released in 2008 by Columbia University's Mailman School of Public Health concluded that there was no evidence of a link between autism and the measles-mumps-rubella vaccine, another focus of speculation and anxiety among vaccine-skeptic parents.[15] However, the decision in the Poling case—and the tenacious nature of beliefs about "science," which may include the rejection of any credible scientific evidence that challenges these beliefs—has ensured that some parents will continue to seek nonmedical exemptions from childhood vaccinations based on their belief that vaccines are unsafe, rather than from any core conviction that vaccines are morally wrong.

Beliefs about "Natural Parenting"

Other parents who seek personal-belief exemptions from mandatory childhood vaccinations do not believe in the need for vaccines, or for certain vaccines. This belief may coexist with strong beliefs in alternative medicine and "natural-health oriented" parenting. In early 2008, the *New York Times* reported a natural-parenting trend: hosting "measles parties" to expose unvaccinated children to children in the contagious stage of that disease, with the goal of avoiding vaccination through deliberate infection.[16] While measles can lead to ear infections, pneumonia, encephalitis, or death, Americans of childbearing age have little firsthand knowledge of this or other serious communicable childhood diseases, because of their own childhood vaccinations.[17] The introduction of the measles vaccine in the early 1960s sharply decreased not only the incidence of this disease in the United States, but also common knowledge of its perils.

Libertarian Values

Fear of government intrusion into the lives of families and the rights of parents may also fuel skepticism concerning and resistance to mandatory childhood vaccinations. Communities defined by belief systems that value separation from mainstream society or alternative, "natural" approaches to parenting may exhibit these characteristics, as may families and communities whose values are libertarian. In January 2009, the Department of Health and Human Services National Vaccine Program, which coordinates the vaccination-related activities of various federal agencies, organized a community meeting in Ashland, Oregon, to find out why so many local parents were opting out of vaccinating their children. In this town, 28 percent of parents get exemptions, far above Oregon's statewide average of 4 percent. Among students at a local "alternative" school, the exemption rate was 67 percent. One Ashland parent told researchers: "One of the basic tenets of my decision-making is mistrust of the government, a mistrust of the pharmaceutical companies, and mistrust of the big blanket thing that says this is what everybody has to do. . . . I get the public health standpoint. . . . I am still questioning (vaccines') safety."[18] The Internet is integral to building and sustaining virtual communities of vaccination skeptics and resisters—and to showing parents exactly how to claim a nonmedical exemption in their state.[19] Googling a phrase such as "vaccine information" will bring up authoritative-looking sites maintained by vaccine skeptics as well as Web sites maintained by the CDC and by public health programs.

Public Health Consequences of Nonmedical Exemptions

Nationwide, only a tiny percentage of parents—2.54 percent—will ever invoke any nonmedical exemption.[20] However, until recently, the percentage was just 1 percent, according to the pediatricians' rule of thumb. And, as the example of Ashland, Oregon, demonstrates, because families with similar beliefs and values may choose to live together, worship together, send their children to the same schools, or take part in the same homeschooling networks, local rates of immunization refusal, when coupled with lenient exemption policies, may be much higher than national averages.[21] These high local rates of refusal have public health consequences, as they compromise the goal of mandatory vaccination: herd immunity.

There is no community—apart from a "closed" community such as the active-duty military—in which every single member can be protected from disease through mandatory vaccination. For infants and persons with medical contraindications, the risks of vaccination outweigh the benefits. An American community may include travelers and new immigrants who were not vaccinated in their home countries. Undocumented residents may remain unvaccinated because they lack access to health care or because they desire to avoid authorities. A community also includes children who have not yet completed their full vaccinations and so remain vulnerable to infection. Others in the same community may have been vaccinated or rendered immune to a particular disease through past infection but later become vulnerable to disease during autoimmune dysfunction: adults with suppressed immune systems can get chicken pox twice. All of these community members are protected by herd immunity, the protective ring created by the majority of community members who are immune from disease. Among public health experts, the standard for maintaining herd immunity is to have 83–94 percent of total community members immune through vaccination or previous infection.[22] As the cohort of adults who acquired immunity through infection rather than vaccination ages, maintaining herd immunity will depend more and more on vaccination.

The CDC has determined that nonmedical exemptions to vaccinations are a factor in the development of "hot spots": locations where the herd immunity provided by compulsory vaccination has been weakened sufficiently for disease outbreaks to occur. In the spring of 2008, the CDC was tracking measles outbreaks in New York City, San Diego, and communities in other states. These outbreaks involved very young children, older children whose parents had refused vaccination, and travelers from countries where vaccination refusals are also contributing to outbreaks.[23] By August 2008, the CDC reported 131 diagnosed cases of measles for the year to date, compared to 42 cases for all of 2007. Of these 131 individuals, most (112) were unvaccinated or had an unknown vaccination status; 63 of the 112 had never been vaccinated because of parental refusal on grounds of religious conviction or personal belief. One outbreak, in Washington state, involved 16 unvaccinated children attending a church conference.[24]

Herd immunity can also be weakened in communities where large numbers of children are "undervaccinated": these children have gaps in their vaccination coverage, or their program of vaccinations was delayed at the beginning owing to factors like lack of access to health care, incomplete records that resulted in dosage errors, or frequent family relocations that disrupted health care relationships.[25] A CDC report released in April 2008 found that rates of undervaccination are even greater than previous studies showed and have a measurable impact on herd immunity.[26] While undervaccination is a different public health problem than vaccination refusal, laws that ease nonmedical exemptions are of special concern to physicians and public health officials in communities where undervaccination is already a persistent problem.[27]

Nonmedical Exemptions as "Conscience Clauses"

A little-noticed fact about nonmedical exemptions to public health mandates is that they look a lot like "conscience clauses": the statutes that, in forty-seven states, allow physicians, and sometimes other health care professionals or institutions, to opt out of providing or participating in health care procedures on the grounds of religious or moral conviction. Most of these state laws, as well as similar federal laws and provisions in professional and institutional codes of ethics, were enacted after the passage of *Roe v. Wade* in 1973 to permit physicians to opt out of performing or participating in legalized abortions. Some conscience clauses cover certain procedures, while others acknowledge a general right of moral refusal. A recent spike in conscientious-objection cases and related policymaking, centering on access to emergency contraception, has been fueled by the recent political climate in the United States, where value conflicts over the validity or interpretation of scientific data are often characterized by policymakers, activists, or journalists as "religion" versus "science" battles.

By comparing conscientious objections by health care providers to nonmedical exemptions by parents, we gain another way of looking at the emotions and reactions elicited whenever someone says, with respect to a question of science, "It's against my religion." In both cases, the invocation of a word such as *religion, conscience, morality,* or *belief* may stop conversation, rather than promote efforts to understand why someone is saying no to participating in the delivery of health care and the protection of vulnerable community members.

Conscience Clauses in Historical Context

The history of medicine reveals that policies permitting nonmedical exemptions to vaccinations are indeed conscience clauses. This very term, as well as the term *conscientious objector,* comes from late-nineteenth- and early-twentieth-century British debates over compulsory vaccination against smallpox.[28] In 1853, the House of Commons passed the Compulsory Vaccination Act. Beginning in 1867, local vaccination officers were authorized to track down noncompliant parents, who were subject to fines and even prison. A class-driven "anti-vaccinationist" resistance strengthened, buoyed by reports of injuries and deaths from unsterile vaccination procedures and through appeals to germ theory. To placate the resisters, the governing Tories added a conscience clause to the Vaccination Act in mid-1898, to allow parents to apply for penalty-free exemptions.

Medical historian Nadja Durbach points out that while the antivaccinationists called themselves "conscientious objectors," they did not lobby for the conscience clause that rendered them "licensed lawbreakers." However, they made their peace with the new provision: by the end of 1898, local magistrates had granted more than two hundred thousand exemptions. The *New York Times* London correspondent cabled home the prediction that, "in a year or two England will be an unvaccinated country."[29] He was nearly right. As a result of the 1898 conscience clause and a subsequent amendment in 1907 that made it even easier to opt out, the vaccination rate tumbled from 80 percent in 1898 to below 50 percent in 1914.[30]

Conscience-clause policies for health care providers and for parents are alike in another respect. In seeking to protect individual rights—specifically, the right to say no to an act that violates a core moral conviction—they may give insufficient attention to the limits on these rights.

Health Care Refusal as "Conscience" Refusal—Category Mistake?

Should we permit "conscience" refusals at all? What are the ethical consequences of sacralizing certain health care transactions by invoking religion, morality, or conscience in a refusal to participate in this transaction? And with respect to vaccination refusal in particular, what are the consequences of writing into law exemptions to civic mandates that suggest religious or moral convictions that may not, in fact, exist?

The case of resistance to military conscription may help to clarify the distinction between saying no on the basis of religious or moral conviction and doing so for a different reason. Resistance to conscription is a practice long associated with members of Christian churches with a core commitment to nonviolence and with other individuals whose pacifist principles are expressed in nondenominational or nonreligious terms. Before *Roe v. Wade* and the subsequent enactment of conscience-clause statutes, resistance to conscription was what the phrase *conscientious objection* meant for Americans who may have been unaware of its public health origins: the term was appropriated from the British antivaccination movement and applied to government policies involving World War I–era pacifist resisters, who were also known as "COs" or derided as "conchies."[31]

Draftees need conscience clauses more than physicians do, because they have fewer freedoms and privileges. Although a physician may equate a particular health care procedure with the wrongful termination of human life, this physician can avoid the objectionable situation by refusing certain training or by selecting certain specialties. A draftee is faced with the prospect of being trained to kill human beings and then being ordered to use that training.

However, during times of universal conscription, Quaker or Mennonite conscientious objectors (COs) have not been permitted to say no to the draft and simply go home: conscience refusals were understood to have personal and social consequences. So COs served "without weapons"—often as medics in war zones in World War I or in civilian hospitals or on public works projects in World War II. (Resisters have also been imprisoned for civil disobedience if they refused to comply with any alternative service.) Military personnel and pacifists alike can describe conscientious objection to conscription in terms of a system of rights and responsibilities: to invoke and be granted CO status means to incur different obligations, to perform alternative service, to recognize the consequences of refusing to serve. The ongoing wars in Afghanistan and Iraq have raised new questions about whether American or allied military personnel who volunteered for service should be granted CO status if they believe they can *no longer* serve in combat zones because of their emerging beliefs about the morality of killing—in particular, the killing of noncombatants—or the morality of the conflict itself.[32]

This is where health policymakers—typically, at the state level—tend to come up short in their ethical analysis. Drafting a conscience clause becomes a way to grant someone a right to say no to something, without requiring them to justify this action and without sufficient attention to what will happen to other people— always the test of ethics—after one person says no. Given how few religions take any position on vaccination, allowing "religious convictions" (or their secular, "personal" equivalents) to provide moral shelter for vaccination refusal creates at least two perverse incentives. One encourages parents to valorize fears as "beliefs," as articles of religious faith, and suggests that these beliefs trump civic responsibilities toward their own children and the community in general. Another suggests that any "belief" about science is legitimate, even if there is no evidence to support the belief. While it is possible to respect personal values in the context of civic responsibilities, no member of the public is well served when category errors—fear as belief, belief as science—are written into health policy.

Recall the pacifist conscientious objectors. A core commitment to nonviolence is different from a nonbelief in vaccines. There is no argument about what can happen, as a question of fact, when you pick up a loaded gun and aim it at another person. This is how conscientious objection to military service by citizens has historically been construed: my core convictions are organized around nonviolence; I refuse to take up a job that is organized around violence; I will accept the consequences, by performing an alternative job that contributes to society, places no one in danger (except perhaps myself), and does not violate my convictions.

By this historical standard, an immunization refusal should not be awarded the status of "conscience" refusal but should instead be considered a health care refusal. Giving moral shelter to immunization refusal by calling it conscientious objection, freedom of religion, moral conviction, or personal belief—and making this easy to do—is a bad idea. It promotes semantic and ethical slippage by conflating different reasons for refusals and giving equivalent legitimacy to all beliefs, including demonstrably false beliefs about relevant scientific facts that, if acted upon, have consequences for public health.

Immunization Refusal as Informed Health Care Refusal

There is another way for health care professionals to work with parents who are considering using nonmedical exemptions to opt out of mandatory vaccinations. According to well-established principles of bioethics, a refusal of health care that is based on a false belief "is not an *informed* refusal." In the absence of an underlying cognitive or thought disorder that prevents a person from making an informed health care decision, health care professionals may have a responsibility to "impose unwelcome information" if a person with the capacity to make decisions is about to make a decision based on misinformation.[33] The American Academy of Pediatrics's implementation guidelines for vaccination standards include guidelines and other resources for pediatricians preparing to talk with parents who are considering opting out of vaccinations or pediatricians who want to ensure that a parental refusal is an informed refusal rather than a refusal based on an uncorrected false belief.[34] To date, legislators in several states have proposed that parents who refuse the HPV vaccine on behalf of their adolescent daughters must state, in writing, that this is an *informed* refusal. The proposed legislation in Vermont reads as follows: "If a parent or guardian objects to administration of the HPV vaccination, the parent or guardian must also state in writing, using a form such as the American Academy of Pediatrics form titled 'refusal to vaccinate,' that he or she has received information explaining the connection between the human papillomavirus and cervical cancer, and that the parent or guardian has elected for the child not to receive the human papillomavirus vaccine."[35] Since some parents will always refuse, requiring them to make an informed refusal, and to do so in writing, is preferable to missing an opportunity to elicit any false beliefs, try to correct them, and try to help parents acknowledge the consequences of forgoing vaccination. Even if "informed refusal" statutes may have been crafted to ease physicians' concerns about liability should an unvaccinated minor later become infected, it is sometimes possible to do the right thing for not quite the right reason.

By contrast, physicians who encourage parents to work the system to avoid vaccinations are shirking their professional and civic responsibilities. An Associated Press article in late 2007 included an interview with a pediatrician who told the reporter that because her state's law does not allow nonmedical exemptions based on personal belief, she advises parents who wish to be exempt to claim "religion" instead: "It [state law] says you have to state that vaccination conflicts with your religious belief. It doesn't say you have to actually have that religious belief. So just state it."[36] Encouraging parents to lie, while failing to inform them of the consequences of the decision they are about to make on behalf of their children and their community, is not an admirable ethical stance for a physician.

A 2005 survey of pediatricians' attitudes toward families who refuse vaccines focused on families with strong personal beliefs about vaccine safety.[37] Around 30 percent of the pediatricians surveyed said they would dismiss a family who refused vaccination from their practice. Most said they would not dismiss such families, being unwilling to turn children out of the health care system because if their parents' beliefs. One pediatrician and medical ethicist, Benjamin H. Levi, urges pediatricians faced with parents who are considering vaccine refusal to be attentive to their civic responsibilities in their communities as well as their professional responsibilities in their own practices.[38] He encourages pediatricians to think like public health advocates: what are the actual immunization rates in your city or state, not just in your own practice? How do these rates compare with the rates needed for herd immunity to be maintained? How many nonmedical immunization refusals can your community tolerate from any single practice relative to these statistics? And what are you doing, as a community health care provider, to get your city or state's immunization rate higher than the herd immunity minimum? If pediatricians are doing all of these things— are, in effect, becoming fully informed themselves—their communities, their practices, and their own consciences can safely accommodate case-by-case individual refusals.

Levi also believes that alerting parents who refuse or contemplate refusing vaccinations to the rate of immunization needed for herd immunity is both good ethics and good medical practice. It reminds parents who refuse immunizations that they are, indeed, free riders who are relying on herd immunity, and it presents herd immunity in terms of collective responsibility and collective action, rather than as lots of individuals exercising "personal choice." The public's health cannot be left up to personal choice. While any of us may choose to purchase bottled water for our own use, we may not make personal decisions about the safety of the water that comes out of everyone else's faucets.

Parents who may refuse vaccines, or specific vaccines, but are otherwise attentive to the health and medical care of their children believe that they are acting in their children's best interests, that they are keeping their children safe. And they may have difficulty sorting fact from fear, given the number of highly charged Web sites dedicated to this particular issue and the rumors that circulate within communities, including intentional communities, such as private schools. Pediatricians who are already in a caregiving relationship with these families can try to clarify the science, dispel fears, encourage compliance with public health standards, and remind them of the social cost of refusing immunization. Other community members—teachers, clergy, fellow parents, public officials—also have an ethical obligation to promote the common good: to the child with leukemia who cannot be vaccinated and needs the protection of herd immunity and to the undervaccinated children in low-income neighborhoods.

The HPV Vaccine as "Car Seat" Public Health

Given that laws permitting nonmedical refusals present real dangers to the public's health, why do policymakers continue to make these laws? The custom of respecting appeals to "religion" under the First Amendment and in American culture can devolve into reflexively writing "religion" into certain laws as an all-purpose justification for a refusal. Yet laws that claim to respect religious freedom may be profoundly disrespectful to the health of the community. Vaccination directly involves children's bodies, and we rely on parents to protect, and want to protect, their children's bodies. But vaccination is also a public health measure, akin to clean water and mandatory seat belts. Discussing vaccination apart from the public health context makes it easier to refuse and harder to acknowledge the consequences of refusal. Granted, the HPV vaccine is different from other childhood vaccines: its principal aim is not strengthening herd immunity. As a public health measure, it is more like a car seat, protecting the individual from harm. Bearing in mind that the HPV vaccine is a new vaccine whose long-term effects are as yet unknown, policymakers should nevertheless be wary of enacting new, customized, opt-out policies, particularly if they do not clearly describe an informed refusal process. Given the extent of vaccine skepticism as a documented public health problem in the United States and in other countries, parents who oppose vaccines may cite these HPV-vaccine-specific exemptions as precedents in arguing for customized exemptions to future vaccines.

Conclusion

When one person's "right" or "belief" affects another person's health or access to health care, we always have an obligation to ask for an explanation. This can be a very uncomfortable business. However, we must recognize that the sudden invocation of a "right" or a "belief," in the context of health care, may signify fear, anxiety, confusion, coercion, or something other than an informed decision. And semantic slippage—failing to use language clearly, or to explain *why*—can lead to unhealthful public policy. It is not hard to make a bad law, but it can be hard to undo such a law. Parents, physicians, and policymakers should question what or whom they are respecting through nonmedical exemptions, and whose lives they are not treating with respect.

NOTES

1. Centers for Disease Control and Prevention, "National Immunization Program: Guide to contraindications to vaccination," Sept. 2003, www.cdc.gov/nip/recs/contra indications_guide.pdf.

2. The National Conference of State Legislatures counts twenty states as offering "philosophical" exemptions: Arizona, Arkansas, California, Colorado, Idaho, Louisiana, Maine, Michigan, Minnesota, Missouri, New Mexico, North Dakota, Ohio, Oklahoma, Pennsylvania, Texas, Utah, Vermont, Washington, and Wisconsin; see D. Grady, "Measles in U.S. at Highest Level since 2001," *New York Times*, May 2, 2008.

3. D. A. Salmon, J. W. Sapsin, S. Teret, R. F. Jacobs, J. W. Thompson, K. Ryan, and N. A. Halsey, "Public Health and the Politics of School Vaccination Requirements," *American Journal of Public Health* 95 (2005): 778–783.

4. D. A. Salmon and A. W. Siegel, "Religious and Philosophical Exemptions from Vaccination Requirements and Lessons Learned from Conscientious Objectors from Conscription," *Public Health Reports* 116 (2001): 289–296.

5. National Conference of State Legislatures, *HPV Vaccine*, www.ncsl.org/programs/ health/HPVvaccine.htm. According to the Centers for Disease Control and Prevention

(CDC), HPV infects approximately 20 million people in the United States and there are 6.2 million new cases each year. There are more than thirty strains of HPV that affect at least half of sexually active people in their lifetime. Most strains of HPV do not produce any symptoms and disappear on their own. Cervical cancer is the second-leading cancer killer of women worldwide. In the United States, nearly 10,000 women are diagnosed with cervical cancer each year and 3,700 women die. Most cases occur outside of the United States, where the Pap test is not available. If states make the vaccine mandatory, they must also address funding issues, including for Medicaid and State Children's Health Insurance Program coverage and youth who are uninsured, and they must decide whether to require coverage by insurance plans. The CDC makes the HPV vaccine available through the federal Vaccines for Children (VFC) program in all fifty states and in Chicago, New York, Philadelphia, San Antonio, and Washington, DC. VFC provides vaccines for children ages 9 to 18 who are covered by Medicaid or who are Alaskan Native or Native American children and for some other underinsured or uninsured children in this age range.

6. Of the 24 states that drafted legislation during the 2006–7 legislative session requiring middle school girls to be vaccinated against HPV, 10 states and the District of Columbia described specific exemptions to the HPV vaccine. These states were Colorado, Illinois, Kansas, Kentucky, Michigan, Minnesota, New York, Ohio, Vermont, and Virginia. Another 12 states mentioned exemptions in their draft legislation.

7. Centers for Disease Control and Prevention, "Outbreak of Measles among Christian Science Students—Missouri and Illinois," *Morbidity and Morality Weekly Report* 43 (July 1, 1994): 463–465.

8. *Prince v. Commonwealth of Massachusetts*, 321 US 158 (1944).

9. M. D. Hinds, "Judge Orders Measles Shots in Philadelphia," *New York Times*, March 6, 1991.

10. D. G. McNeil, "Worship Optional: Joining a Church to Avoid Vaccines," *New York Times*, Jan. 14, 2003.

11. T. Poling, "Vaccines, Autism, and Our Daughter, Hannah," *New York Times*, April 5, 2008.

12. J. Rovner, "Case Stokes Debate about Autism, Vaccines," *NPR*, March 7, 2008. John Gilmore, executive director of the group Autism United, claimed, "For the first time the court has conceded that vaccines can indeed cause autism."

13. G. Harris, "Deal in an Autism Case Fuels Debate on Vaccine," *New York Times*, March 8, 2008.

14. The Immunization Safety Review Committee of the Institute of Medicine published eight reports over a four-year period. Each report refuted any causal connection between vaccines and harmful aftereffects (such as autism). The reports are available at www.iom.edu/?id=5977.

15. M. Hornig, T. Briese, T. Buie, M. L. Bauman, G. Lauwers, et al., "Lack of Association between Measles Virus Vaccine and Autism with Enteropathy: A Case-Control Study," *Public Library of Science* ONE 3, no. 9 (2008): e3140 doi:10.1371/journal.pone.0003140.

16. J. Steinhauer, "Public Health Risk Seen as Parents Reject Vaccines," *New York Times*, March 21, 2008.

17. The CDC reports that 1 in 10 children with measles develops ear infections, 1 in 20 develops pneumonia, and 1 in 1,000 develops encephalitis. CDC, www.cdc.gov/ncidod/dvrd/revb/measles/measles_general_info.htm#sec_5.

18. J. Barnard, "US Doctors Pay to Hear Ore. Town's Vaccine Views," *Associated Press*, Jan. 9, 2009.

19. See, for example, the Web sites of the National Vaccine Information Center, www.909shot.com/Default.htm, and of *Mothering* magazine, www.mothering.com.

20. Statistic cited by Saad B. Omer, Johns Hopkins Bloomberg School of Public Health, in Steinhauer, *Public Health Risk*, 2008.

21. Editorial, "We Must Vaccinate Kids," *Los Angeles Times*, April 29, 2008.

22. T. May and D. Silverman, "Clustering of Exemptions as a Collective Action Threat to Herd Immunity," *Vaccine* 21 (2003): 1048–1051.

23. Grady, "Measles in U.S. at Highest Level since 2001."

24. Centers for Disease Control and Prevention, "Update: Measles—United States, January 2008-July 2008," *Morbidity and Mortality Weekly Report* 57, no. 33 (2008): 893–896.

25. P. J. Smith, S. Y. Chu, and L. E. Barker, "Children Who Have Received No Vaccines: Who Are They and Where Do They Live?" *Pediatrics* 114 (2004): 187–195.

26. E. T. Luman, K. M. Shaw, and S. K. Stokley, "Compliance with Vaccination Recommendations for US Children," *American Journal of Preventive Medicine* 34, no. 6 (June 2008): 463–470.

27. The National Immunization Survey includes state-by-state comparisons of immunization rates, as well as vaccination rates for specific diseases and other data sets. CDC, www.cdc.gov/nis/.

28. N. Durbach, *Bodily Matters: The Anti-Vaccination Movement in England, 1853–1907* (Durham, NC: Duke University Press, 2005).

29. H. Norman, "The News in London: Unvaccinated England," *New York Times*, Dec. 18, 1898, 19.

30. W. A. R. Thomson, "Rider Haggard and Smallpox," *Journal of the Royal Society of Medicine* 77, no. 6 (1984): 506–512.

31. Durbach, *Bodily Matters*, 196–197.

32. For descriptions and analysis of recent conscientious-objection cases among soldiers serving in Afghanistan and Iraq, see the 2007 documentary *Soldiers of Conscience* and its Web site, POV, www.pbs.org/pov/pov2008/soldiersofconscience/update.html.

33. T. L. Beauchamp and J. F. Childress, *Principles of Biomedical Ethics*, 6th ed. (New York: Oxford University Press, 2009), 131.

34. National Vaccine Advisory Committee, "Standards for Child and Adolescent Immunization Practices," *Pediatrics* 112, no. 4 (Oct. 2003): 958–963; the implementation guidelines (*Maintaining Standards of Excellence: Part 6, A Series in Support of the National Vaccine Advisory Committee Standards for Child and Adolescent Immunization Practices*) and related documents, including the "Refusal to Vaccinate" form, are available on the American Academy of Pediatrics's immunization Web site, www.cispimmunize.org/.

35. V.S.A. sec. 1122, HB 256 (Vermont).

36. S. LeBlanc, "Parents Use Religion to Avoid Vaccines," *Associated Press*, Oct. 17, 2007.

37. E. A. Flanagan-Klygis, L. Sharp, and J. E. Frader, "Dismissing the Family Who Refuses Vaccines: A Study of Pediatrician Attitudes," *Archives of Pediatrics & Adolescent Medicine* 159 (Oct. 2005): 929–934.

38. Benjamin Levi, personal communication to Nancy Berlinger, Hershey, PA, April 2006. See also B. H. Levi, "Addressing Parents' Concerns about Childhood Immunizations: A Tutorial for Primary Care Providers," *Pediatrics* 120 (2007): 18–26.

CHAPTER TWELVE

Sex, Science, and the Politics of Biomedicine

Gardasil in Comparative Perspective

Steven Epstein and April N. Huff

The advent of human papillomavirus vaccines such as Gardasil and Cervarix—vaccines designed to interrupt transmission of a sexually transmitted infection in order to prevent the development of cancer—holds enormous public health significance. But these developments also provide insight into central aspects of political life by demonstrating the complex interplay among biopolitics, biomedicalization, and the often bitterly fought politics of sexuality in the present-day United States.[1] We address this interplay by examining an apparent contradiction suggested by the case of Gardasil, in light of episodes in the politics of science and the politics of sexuality during the presidential administration of George W. Bush.

With regard to sexuality, both domestic and international public health policy became sharply conservative during the Bush administration. On his third day in office, President Bush reinstated the "Mexico City Policy," which prevents international nongovernmental organizations from receiving federal funding if they provide or promote abortions.[2] The president's Emergency Plan for AIDS Relief of 2004 gave a significant amount of funding to HIV prevention programs that emphasized abstinence from sex outside marriage and monogamy inside marriage as the central methods for reducing the spread of HIV abroad.[3] Of particular interest, however, is the relation between such sexual conservatism and the politics of science and medicine.

In repeated cases involving sexuality policy, powerful actors within key federal government agencies privileged a Christian Right moral agenda over the mainstream scientific consensus.[4] More precisely, right-wing activists both inside and outside government sought to influence federal policies by deploying a scientific counterexpertise that aligns with conservative Christian values, using the idiom of science to call into question the conventional scientific wisdom. While "science wars" have played out in diverse arenas and with reference to a range of topics (global climate change being a noteworthy example), it is striking how often these debates have turned to matters of sexuality and sexual health.[5] Right-wing activists have influenced federal health agencies by marshaling data intended to show that condoms have a high failure rate, that abortion increases the risk of breast cancer, and that teenage abstinence is the only sure path to well-adjusted adulthood. They also have sought to block federal funding for research on sexual topics that they find distasteful, nearly succeeding in some instances.[6] This activism at the intersection of science and morality was facilitated by what one analyst has described as an "avalanche of religious right appointments" to various federal agencies, including the advisory committees to federal health agencies such as the National Institutes of Health (NIH), the Food and Drug Administration (FDA), and the Centers for Disease Control and Prevention (CDC).[7]

In relation to such developments, the puzzle is that Gardasil prompted comparatively *little* right-wing opposition at the level of federal policymaking. Despite strong concerns expressed by some commentators about promoting teen sexuality, despite insistence from some quarters that the best defenses against HPV infection were abstinence and monogamy—indeed, despite the fact that a doctor affiliated with the prominent conservative Christian advocacy group Focus on the Family sat on the CDC's Advisory Committee on Immunization Practices—the HPV vaccine received a rapid, ringing endorsement from two federal agencies: the FDA, which approved the vaccine just six months after receiving the manufacturer's application, and the CDC, which recommended it for universal use in girls just weeks later.[8] An openness to Gardasil was manifested even in the White House: "There's nothing new about requiring a vaccine that will protect the health of people in our country," First Lady Laura Bush told CNN's medical reporter, Dr. Sanjay Gupta, when asked about the new vaccine in a televised interview, adding, "It's just like getting the flu shot."[9]

To be sure, sharp controversy did subsequently break out into the open, but it was largely confined to the question of whether, at the level of state government, vaccination with Gardasil ought to be made mandatory for school attendance or should simply be recommended. Opposition to mandatory vaccination policies was voiced not only by social conservatives but by actors from across the political spectrum who made diverse arguments against this policy approach. However, at the level of federal health policy, the tone around Gardasil was rather different from what one might have predicted on the basis of recent past experience with sexuality policy. Our goal in this chapter, therefore, is to move the discussion beyond the basic and now familiar story line that "politics" trumped "science" during the Bush years. The case of HPV vaccines suggests that such claims are insufficiently nuanced to provide a thorough or convincing analysis.

What, then, accounts for the support for Gardasil? What kinds of arguments have been made on its behalf, and what gives these arguments their force? We develop our analysis in two steps. First, we summarize the turning points in the case of Gardasil, based on interviews conducted with key actors along with archival research. Then, we explain how the case of Gardasil compares with other recent cases of policy clashes involving morality and science.

The Cancer Frame and the Politics of "Disinhibition"

The story of Gardasil clearly demonstrates the power of framing[10]—specifically, the effect of depicting the vaccine in a thoroughly biomedicalized way and in a more or less completely desexualized way (on this point, see also chapters 3 and 7). Even though Gardasil's most immediate function is to prevent the sexual transmission of the human papillomavirus (and even though the vaccine's efficacy has been demonstrated in relation to precancerous cervical lesions and not cervical cancer itself),[11] the vaccine has been discursively constructed—ubiquitously—as a "vaccine against cancer." Jon Abramson, the chair of the CDC's Advisory Committee on Immunization Practices, said, "[Gardasil] was the first . . . vaccine truly created to prevent cancer, more than preventing the specific infection."[12] It is noteworthy that several proponents of Gardasil have been quoted in the mass media explicitly characterizing Gardasil as *not* having to do with prevention of a sexually transmitted infection. For example, one Republican state representative from Virginia who sponsored a mandatory immunization bill flatly told the *Washington Post:* "This is not a prevention for a sexually transmitted disease. This is a prevention for cancer."[13]

The desexualizing of Gardasil and its framing as a vaccine against cancer are no accident. They appear to have been the strategy of its corporate sponsor, Merck, which, before the vaccine was licensed, sought the advice of women's health advocates about how to avoid portraying cervical cancer prevention from the standpoint of sexually transmitted disease.[14] The primary emphasis on cancer is central to Merck's extensive public advertising for Gardasil, which consistently emphasizes the goal of "One Less" case of cervical cancer. Television and magazine ads for Gardasil feature a multiethnic assortment of well-informed, self-assertive girls engaged in healthy physical activities and conversing with their mothers about the vaccine (see chapter 7; for a discussion of alternative representations of the vaccine, see chapters 8 and 4). The specter of sexuality is kept entirely at bay: no potential heterosexual partners—no males, period—invade the frame of the ads, and the only social dyad portrayed as relevant is that binding mother to daughter.[15]

As Janet Gilsdorf, the chair of the CDC's working group on HPV vaccines, observed, the general public is just not terribly familiar with the HPV–cervical cancer link. As she put it, until fairly recently, "no one knew that cervical cancer was caused by a virus except for a few virologists."[16] This lack of clarity in public awareness makes it easier to frame Gardasil as a vaccine against cancer without having to dwell on its actual target, HPV, or the associated issues of sexual conduct. And by an interesting route, the lack of public knowledge of the potential health consequences of HPV infection also caused the most conservative member of the CDC's Advisory Committee on Immunization Practices to set aside his concerns about the effect of vaccination on sexual morality. Previously affiliated with Focus on the Family, Reginald Finger would seem to be the classic case of a far-right appointee to a Bush-era federal health advisory committee. A physician trained in public health, Finger notes, on his personal web page: "In everything I do, I seek to put Jesus Christ and His kingdom first."[17] However, Finger concluded that the availability of an HPV vaccine would be *unlikely* to promote sexual promiscuity on the part of young people. Finger couched his argument in relation to the theory of "disinhibition"—that is, the idea that new medical technologies or interventions can have a disinhibiting or liberalizing effect on sexuality if participants no longer fear that sexual activity may bring negative consequences. In this case, however, Finger eventually reasoned it made no sense to fear that a vaccine would make teens sexually reckless if they didn't know enough about HPV to have feared it in the first place. In his view, "for disinhibition to be a factor inhibition has to be a factor.

And nobody really has evidence to show that fear of HPV is an inhibition factor for teenagers. Certainly, HPV is not HIV and neither is it pregnancy."[18]

Right-Wing Pragmatism and Corporate Persuasion

Finger's eventual realization that he had scant reason to oppose Gardasil is suggestive of the larger story here: far-right conservative organizations either moderated their views about Gardasil over time or else found that their unmoderated opposition simply offered them little political traction; however, the specter of state-government-imposed mandatory vaccination policies eventually provided them with a powerful way to frame an oppositional message.[19] On one hand, the idea that Gardasil promoted promiscuity was hard to maintain or put across because of the lack of discussion of HPV as a sexually transmitted infection; on the other hand, it was difficult to stand in the way of the bandwagon that formed behind a "vaccine against cancer." In the assessment of one women's health advocate who supported the licensing of Gardasil, conservatives "clearly made a strategic [decision] that coming out against a vaccine that could save women from getting cancer was not a place that they wanted to be. So they pinned their position on this idea that it shouldn't be forced on anyone and that parents should always have the decision making authority."[20]

To be sure, stances and arguments have varied among conservative advocacy groups. Toward the hard-line end is the Traditional Values Coalition (TVC), the self-described "grassroots church lobby" based in southern California. On the organization's Web site, the TVC tries to paint cervical cancer as the "U.S. epidemic that isn't," suggesting that while deaths from the disease are a tragedy, "this is not a national health crisis." Not going quite so far as to argue against the vaccine altogether, TVC does claim: "HPV is contracted through sexual contact and is not contagious. Therefore, almost all cases of HPV could be prevented through responsible sexual behavior, including fidelity in marriage and abstinence outside of marriage."[21]

However, other conservative groups have taken a more provaccine position, including Focus on the Family. Although its position statement on HPV vaccines also endorses "abstinence until marriage and faithfulness after marriage as the best and primary practice in preventing HPV and other STIs," the statement goes on to say: "Focus on the Family supports and encourages the development of safe, effective and ethical vaccines against HPV, as well as other viruses. The use of these vaccines may prevent many cases of cervical cancer, thus saving the lives of millions of women across the globe."[22] Interestingly, Linda Klepacki, a sexual-health analyst for Focus on the Family, was vehement in her criticism of media coverage of the organization's position on Gardasil, arguing that the media reflexively cast the organization as antivaccine: "We were consistently mischaracterized in the media. We were consistently said [*sic*] that we were against the vaccine. . . . There was even one report that we were trying to kill all of these teenagers from cervical cancer because we won't allow them to get this vaccine. The treatment in the press was amazingly inaccurate."[23]

Even allowing for potential media misrepresentation, it does seem that some conservative organizations, such as Concerned Women of America and the Family Research Council, initially approached HPV vaccines from a hostile or very skeptical standpoint but softened their views over time.[24] For example, in 2005 a spokesperson from the Family Research Council was quoted as claiming that "abstinence is the best way to prevent HPV" and that "giving the HPV vaccine to young women could be potentially harmful, because they may see it as a licence to engage in premarital sex."[25] But by 2007 the group's vice president for policy stated on its Web site: "There is no 'culture war' over the existence of the HPV vaccine—it's a boon to medicine."[26] Such softening may reflect legitimate interest in preventing cervical cancer, or it may simply indicate a pragmatic assessment that any innovation framed as a "vaccine against cancer" is just too hard to fight head-on. Additionally, conservative religious groups faced criticism that unvaccinated girls who maintained their sexual purity until marriage might still be put at risk for infection by future spouses who did not. A recognition of this practical consideration may have played a role in the diminution of opposition to the vaccine. Indeed, in her prepared statement presented to the Advisory Committee on Immunization Practices, the Family Research Council's policy analyst made a pitch for abstinence but then noted: "However, we also recognize that HPV infection can result from sexual abuse or assault, and that a person may marry someone still carrying the virus. These provide strong reasons why even someone practicing abstinence and fidelity may benefit from HPV vaccines."[27]

Vaccine manufacturers were by no means passive as advocacy groups sorted out their positions with regard to HPV vaccines. To the contrary, Merck and GlaxoSmithKline (GSK)—both of them pharmaceutical titans with substantial resources and considerable experience in promoting their products—worked proactively to build support for their products and to defuse potential opposition among particular stakeholders. Well before Gardasil received FDA ap-

proval, Merck representatives reached out to the leading conservative organizations, sat down to meet with their representatives, and sought to persuade
them that the vaccine posed no threat to their moral views. GlaxoSmithKline
has done the same in relation to its product Cervarix.[28] Similarly, the companies
sought allies among health-related, community-based organizations on the liberal end of the political spectrum. In 2005, a GSK spokesperson told community groups in an e-mail: "Because cervical cancer takes many years to develop
and remains asymptomatic for many years, it is easier to turn the question of
introduction in girls into a hypothetical or political debate about adolescent
sexuality. We will need your support in staying focused on . . . the real problem
of cervical cancer."[29]

Federal Consensus and State-Level Conflict

These developments help to explain why, at the FDA and CDC hearings, so little
public opposition to Gardasil was voiced. "I actually thought that there was going
to be more resistance on all fronts," commented Gilsdorf, the head of the HPV
working group within the CDC's advisory committee.[30] Women's health advocates who attended the FDA hearing at which Gardasil was approved for sale also
noted the substantial difference in tone compared to meetings where more controversial technologies, such as Plan B emergency contraception, were debated.[31]
According to Amy Allina, the program and policy director at the National
Women's Health Network: "The level of tension was just much lower, there
wasn't a feeling that there was something very controversial or contentious being
discussed. It was much more what you'd see at another advisory committee meeting where you're just talking about a major medical breakthrough. Everyone was
aware that there was this undercurrent of discomfort with talking about anything that has to do with sex, which we have in this country, but by focusing
on the cancer prevention instead of on the HPV as a sexually transmitted infection, the company managed to shift the conversation a little bit."[32]

The tide changed rapidly, however, once politicians in various states began
seeking to make vaccination with Gardasil mandatory for school attendance by
girls. In Texas, for example, Governor Rick Perry established a mandatory
vaccination policy, though it was ultimately overturned by legislative action.
Around the country, state legislators affiliated with the organization Women
in Government also introduced legislation to mandate the vaccine. The specter
of mandatory policies gave conservative groups a powerful way of framing an

oppositional message. Linda Klepacki of Focus on the Family stated, for example, "We are an organization that supports parental rights. . . . We believe that parents are the medical decision-makers for their children."[33] Although mandatory policies generally contained accommodating parental opt-out provisions, conservative groups have been highly successful in opposing mandates in nearly every state where they have been proposed.

There are several reasons why mandatory policies for HPV vaccination fared poorly overall. First, mandates could be framed as "big government" trampling the rights of the individual—a frame that appeals to a broad swath of people of a more or less libertarian persuasion. Second, the prospect of mandates opened the door for opposition from the generic movement of vaccine skeptics that includes, for example, those who claim that vaccination has caused a spike in autism rates,[34] along with those who simply mistrust public health injunctions (see chapter 9). Third, although support for mandates came from diverse quarters, skepticism about mandates also proved to be diverse (see chapter 1) and was voiced by prominent scientists who had no ties to conservative groups. For example, both Jon Abramson and Janet Gilsdorf from the Advisory Committee on Immunization Practices were against mandates, despite being very far from the religious Right's position on this issue, simply because the vaccine was so new and mandatory policies seemed premature.[35] In addition, women's health advocates worried about the potential cost of mandates and the effect they might have in draining resources away from valuable public health measures such as cervical cancer screening.[36]

Finally, popular support for mandates substantially evaporated as a backlash developed against Merck, which was aggressively promoting mandatory vaccination policies for its expensive product in an attempt to earn profits and to do so quickly, before the competitor vaccine, Cervarix, came onto the market in the United States. Soon news reports surfaced noting that Women in Government had received funding from Merck, and Governor Perry of Texas was also accused of having ties to the company. The resulting backlash drew support from across the political spectrum, from the Web site of the Traditional Values Coalition to the pages of *Nation*. The comment of Allina of the National Women's Health Network was that just as "we didn't want to see [Gardasil] held back for political reasons," neither "[did we want it to be] pushed forward for economic reasons."[37]

Case Comparisons: Condoms, Plan B, and HPV Vaccines

To sum up the argument so far: The initial acceptability of HPV vaccines was heightened both by direct efforts on the part of manufacturers and by the intensive medicalization (and desexualization) of Gardasil via the promotion of the "vaccine against cancer" frame. Conservative Christian organizations, along with their allies within the government, have taken a range of positions toward HPV vaccines, but most of them have not been opposed outright to the vaccines' use (or at least, have not found it strategic to say they are against it), and those that did stand opposed gained few converts to their cause. These developments help to explain why Gardasil did not become embroiled in a moral panic in relation to sexuality. Only subsequently, with the push to mandate vaccination at the level of the states, did right-wing groups find a public oppositional voice that resonated widely—not against the vaccines per se, but rather against the idea of mandatory policies.

But we want to take the argument further. Additional insight into the trajectory of Gardasil comes from identifying a set of factors that distinguish the case of HPV vaccines from other recent cases involving medical science, sexuality, morality, and politics. Here we return our attention to the broader political environment concerning sexuality and science in the administration of George W. Bush, including the conservative positions taken with regard to abortion policy, sexuality research, sex education, and so forth. However, to simplify matters, we juxtapose the story of Gardasil with abbreviated accounts of two other recent cases that, like Gardasil, involve medical technologies that have implications for sexual morality and that seek to preempt the risk of an unwanted outcome:[38] debates about condom efficacy and debates about over-the-counter sales of the emergency contraceptive known as Plan B. All three of these debates have involved the FDA, the CDC, or both agencies.

The "condom wars" began in 2001, when a fact sheet on condoms mysteriously disappeared from a CDC Web site, replaced by a notice indicating that it was "being revised."[39] Around the same time, Tom Coburn, an extreme conservative U.S. representative from Oklahoma at the time, began pressuring the FDA to change the packaging for condoms to include a statement warning that condoms are not effective in preventing sexually transmitted diseases.[40] (HPV infection served as a wedge issue here, as Coburn and other condom critics argued that the data on whether condoms prevented transmission of HPV were inconclusive.)[41] Under pressure, the NIH convened a cross-agency panel to review all the data on condom efficacy and condom failure rates.

Condom defenders found themselves in a difficult position, because it was easy for opponents to exploit uncertainty in condom efficacy data, whereas conclusive proof that condoms are effective would have required randomized trials that would simply have been unethical to conduct.[42] Although the NIH review largely vindicated condoms, particularly for preventing HIV, claims about their limitations continued to inform public policy. For example, federal funds have supported abstinence-only education programs in the United States that, as a rule, mention condoms only with reference to their failure rates.[43]

A second comparison case is that of Plan B. In 1997, reproductive rights activists created the Women's Capital Corporation (WCC) with the exclusive purpose of developing and marketing emergency contraception. Their product, Plan B, was approved by the FDA for prescription sale the following year.[44] In 2003 WCC sought permission to sell Plan B over the counter. However, in May 2004 the acting director of the FDA's Center for Drug Evaluation Research rejected not only the advice of the agency's expert advisory committee but also that of his own staff when he refused to permit over-the-counter sales of Plan B, which by then had been acquired by Barr Laboratories. Former FDA employees said it was "unheard of" for a high FDA official to overrule the recommendations of both an advisory committee and agency staff.[45]

Within the advisory committee, a Christian Right ally named David Hager was one of three Bush appointees who advanced a minority position against approval.[46] Hager argued that the availability of Plan B would only encourage women to be more sexually promiscuous or even to dispense with traditional forms of contraception.[47] Although Plan B's primary mechanism of action is to prevent ovulation, opponents claimed that there was uncertainty about how the drug actually worked, and via this claim they sought to make scientific arguments that Plan B was actually functioning as an abortifacient and not as a contraceptive at all.[48]

In 2005 the agency again delayed approval of the medication, despite heated objections from Democrats in Congress, who threatened to hold up the nomination of a new appointee to head the FDA. The delayed approval also prompted the angry resignation of Susan Wood, the FDA's assistant commissioner for women's health and director of the FDA's Office of Women's Health, in a widely publicized move.[49] Approval for over-the-counter sales eventually was granted, but only for women over the age of 18.

Comparing the cases pertaining to condoms, Plan B, and Gardasil, we can identify four factors that seem to vary across them. The first is the theme of biomedicalization, which has already been discussed. Gardasil, a vaccine obtained only in a doctor's office, and one marketed as a vaccine against a dread disease, is highly medicalized. Plan B, a pharmaceutical drug, though one that is self-administered, had been a highly medicalized approach to birth control, but the point of seeking over-the-counter use was precisely to de-medicalize it somewhat, in order to facilitate access to the product.[50] And condoms, which are sold in drugstores but also in many other locales including restroom vending machines, used in the absence of any direct interaction with health professionals, and often viewed more as a kind of sexual device than as a medical technology, are far less medicalized.

The second factor is sexualization. As we have argued, Gardasil was promoted in ways that seemed deliberately to downplay its link to a sexually transmitted infection—a framing that was facilitated by the lesser degree of knowledge among the general public about HPV and its etiological role in the development of cancer. Moreover, the fact that vaccination with Gardasil is temporally separated from any act of sex (perhaps preceding sexual debut by some number of years) makes it very different from condoms and Plan B. By contrast, condoms are a technology whose use occurs precisely at the moment of sexual behavior, and Plan B must be used within seventy-two hours of unprotected sexual intercourse. As a result, the associations with themes of sexual morality are much harder to disrupt in the latter two cases.

The greater degree of sexualization of Plan B was especially evident in the sharp debate over the age at which over-the-counter use could be countenanced, with the specter of teen sexuality serving as the greatest stumbling block to approval. By contrast, Gardasil mostly escaped such scrutiny despite its administration to girls nearing their sexual debut. Susan Wood, the former director of the FDA's Office of Women's Health, summarized the difference in perceptions: "People get really nervous about teenage promiscuity if you're talking about contraception because it's clear contraception is about sexual behavior. [But] it's not at all clear to most people that cervical cancer vaccine has anything to do with sexual behavior."[51]

The sexualization of these technologies is additionally affected by the social controversies that surround issues of sexual diversity. Adding to the sexualized character of condoms, perhaps, is their association in popular culture with both gay and straight sex, as a consequence of the HIV/AIDS epidemic. By contrast, Plan B is completely heterosexualized, and so is Gardasil in nearly all popular discourse (see chapter 4).

The third factor concerns biopolitics and reproduction. A striking difference here is that both Plan B and condoms are technologies that interrupt reproduction and are thought of more generally as forms of birth control. By contrast, not only does Gardasil have no impact on reproductivity, but in fact its use is intended to preserve future health and reproductive potential. To support Gardasil, therefore, is to support the biopolitics of healthy heterosexuality.

The fourth factor relates to the political economy of medical technologies. As we described, Gardasil was aggressively promoted by a powerful manufacturer, Merck, one of the leading global drug companies, which is concerned with maximum earnings while its product remains under patent protection and which also sought to solidify its market before the rival product, GlaxoSmithKline's Cervarix, was approved by the FDA for sales in the United States.[52] Both companies reached out to advocacy groups in proactive attempts to smooth the waters and ensure the uptake of their vaccines. Barr Laboratories (a much smaller pharmaceutical company than Merck or GlaxoSmithKline) pushed for over-the-counter status of Plan B, and both Barr and Women's Capital Corporation worked closely with women's health groups and the FDA to develop their research goals before seeking over-the-counter approval.[53] But these companies lack the financial resources of Merck. Meanwhile, few condom manufacturers rose to the defense of their products, and such companies lack the kind of close relationship with a federal agency that pharmaceutical companies have with the FDA.

This approach suggests a multifactorial argument about what happens, in a hostile political climate, to medical technologies that have implications for sexuality policy. Everything else being equal, it would seem that a medical technology that in public and policy discourse is medicalized and not very sexualized (or homosexualized); does not interfere with reproduction; and is promoted aggressively by a powerful corporate sponsor is more likely to "slide through" than a technology that is less medicalized and more sexualized, is perceived as interfering with reproduction, and is not promoted aggressively by a powerful manufacturer. Of course, a highly schematic claim such as this one can take the analysis only so far; as always, "the devil is in the details." However, even this comparison demonstrates that the basic story line according to which the Bush administration uniformly repressed sexual science in the name of a right-wing moral agenda is insufficient to characterize the complex relationship of morality, politics, economics, and science in the United States. Future comparative work may help reveal the often circuitous pathways by which medical technologies such as Gardasil wind their way through the political and cultural environment in ways that may variously affect their uptake and their capacity to preserve health.

ACKNOWLEDGMENTS

We are grateful to the participants in the "Cancer Vaccines for Girls?" conferences held at Rutgers University, and especially to Keith Wailoo, Robert Aronowitz, Julie Livingston, and Betsy Armstrong, for helpful suggestions on an earlier draft. Steven Epstein also wishes to thank audiences at the UCLA Center for Society and Genetics and Stanford's Center for Advanced Study in the Behavioral Sciences for their comments. The authors' work was supported by the Department of Sociology, the Science Studies Program, and the Academic Senate Committee on Research, all at the University of California, San Diego. Steven Epstein completed this work while a resident Fellow at the Center for Advanced Study in the Behavioral Sciences.

NOTES

1. By *biopolitics* we mean the state's promotion and management of population health (Michel Foucault, *The History of Sexuality*, vol. 1, *An Introduction*, trans. Robert Hurley [New York: Vintage Books, 1980]), and by *biomedicalization* we mean the use of biomedical frames of understanding and technologies to transform human life (Adele E. Clarke, Janet K. Shim, Laura Mamo, Jennifer Ruth Fosket, and Jennifer R. Fishman, "Biomedicalization: Technoscientific Transformations of Health, Illness, and U.S. Biomedicine," *American Sociological Review* 68 [2003]: 161–194; Peter Conrad, *The Medicalization of Society* [Baltimore: Johns Hopkins University Press, 2006]). On the politics of sexuality, see, for example, Gayle Rubin, "Thinking Sex: Notes for a Radical Theory of the Politics of Sexuality," in *Pleasure and Danger: Exploring Female Sexuality*, ed. Carole S. Vance (New York: Routledge, 1984), 267–318; Elizabeth Bernstein and Laurie Schaffner, *Regulating Sex: The Politics of Intimacy and Identity* (New York: Routledge, 2005); Kristin Luker, *When Sex Goes to School: Warring Views on Sex—and Sex Education—since the Sixties* (New York: Norton, 2006); Diane Di Mauro and Carole Joffe, "The Religious Right and the Reshaping of Sexual Policy: An Examination of Reproductive Rights and Sexuality Education," *Sexuality Research and Social Policy* 4, no. 1 (2007): 67–92; Jessica Fields, *Risky Lessons: Sex Education and Social Inequality* (New Brunswick, NJ: Rutgers University Press, 2008).

2. "Bush Moves to Outflank Democrats on Abortion Finance Limits," *New York Times*, March 25, 2001.

3. Donald G. McNeil, "Bush's Global AIDS Effort Limited by Restrictions," *New York Times*, March 31, 2007.

4. Esther Kaplan, *With God on Their Side: How Christian Fundamentalists Trampled Science, Policy, and Democracy in George W. Bush's White House* (New York: New Press, 2004); Chris Mooney, *The Republican War on Science* (New York: Basic Books, 2005).

5. Steven Epstein, "The New Attack on Sexuality Research: Morality and the Politics of Knowledge Production," *Sexuality Research and Social Policy* 3, no. 1 (2006): 1–12.

6. Ibid.; Joanna Kempner, "The Chilling Effect: How Do Researchers React to Controversy?" *PLoS Medicine* 5, no. 11 (2008): 1571–1578.

7. Kaplan, *With God on Their Side*, 84.

8. Center for Biologics Evaluation and Research, *Product Approval Information—Licensing Action: Gardasil Questions and Answers*, Food and Drug Administration, www.fda .gov/cber/products/hpvmer060806qa.htm, June 8, 2006 (accessed Nov. 4, 2007); Advisory Committee on Immunization Practices, *Vaccines for Children Program Resolution on HPV Vaccine—Resolution Number 6/06-2 on June 29, 2006*, Centers for Disease Control and Prevention, www.cdc.gov/vaccines/programs/vfc/downloads/resolutions/0606hpv.pdf, June 29, 2006 (accessed Aug. 1, 2007). See also Elizabeth Rosenthal, "The Evidence Gap: Drug Makers' Push Leads to Cancer Vaccines' Rise," *New York Times*, Aug. 20, 2008.

9. *House Call with Dr. Sanjay Gupta: Laura Bush: Changing the World One Heart at a Time*, Web transcript, CNN, http://transcripts.cnn.com/TRANSCRIPTS/0712/29/ hcsg.01.html, Dec. 29, 2007 (accessed Sept. 20, 2008).

10. Erving Goffman, *Frame Analysis: An Essay on the Organization of Experience* (New York: Harper and Row, 1974); Charles E. Rosenberg and Janet Lynne Golden, *Framing Disease: Studies in Cultural History* (New Brunswick, NJ: Rutgers University Press, 1992).

11. Charlotte J. Haug, "Human Papillomavirus Vaccination—Reasons for Caution," *New England Journal of Medicine* 359, no. 8 (2008): 861–862.

12. Jon S. Abramson, M.D., Department of Pediatrics, Wake Forest University School of Medicine; chair, CDC Advisory Committee on Immunization Practices, telephone interview, Sept. 27, 2007.

13. Susan Levine and Hamil R. Harris, "Wave of Support for HPV Vaccination of Girls," *Washington Post*, Jan. 12, 2007, B1.

14. Susan Wood, Ph.D., former director, Office of Women's Health, FDA, telephone interview, July 18, 2008.

15. Merck has also been very quiet about its recruitment of young men who have sex with men as a subset in its trial of the vaccine in males (chapter 4), no doubt because any public focus on this group would immediately sexualize the discussion.

16. Janet Gilsdorf, M.D., Department of Pediatrics, University of Michigan Medical School, and Department of Epidemiology, University of Michigan School of Public Health; member, CDC's Advisory Committee on Immunization Practice and former chair (until June 2007), ACIP's Working Group on HPV Vaccines, telephone interview, Sept. 27, 2007.

17. Reginald Finger, "Welcome to Reg Finger.com," www.regfinger.com/ (accessed Dec. 25, 2009).

18. Reginald Finger, M.D., medical analyst for Focus on the Family, Colorado Springs, CO (2001–5); former member, CDC Advisory Committee on Immunization Practices, telephone interview, Sept. 14, 2007.

19. An argument similar to ours about conservative groups' moderation of their opposition to Gardasil has recently been made by Monica J. Casper and Laura M. Carpenter, "Sex, Drugs, and Politics: The HPV Vaccine for Cervical Cancer," *Sociology of Health and Illness* 30, no. 6 (2008): 894.

20. Amy Allina, program and policy director, National Women's Health Network, Washington, DC, telephone interview, Aug. 7, 2008.

21. Andrea Lafferty, *No Mandatory HPV Vaccine for Girls!* Traditional Values Coalition, www.traditionalvalues.org/modules.php?sid=3015, Feb. 21, 2007 (accessed Aug. 26, 2007).

22. Focus on the Family, *Focus on the Family Position Statement: Human Papillomavirus Vaccines*, Focus on the Family, www.family.org/sharedassets/correspondence/pdfs/Pub licPolicy/Position_Statement-Human_Papillomavirus_Vaccine.pdf, Feb. 21, 2007 (accessed July 22, 2007).

23. Linda, Klepacki, R.N., sexual health analyst, Focus on the Family, Colorado Springs, CO, telephone interview, Sept. 27, 2007.

24. Rita Rubin, "Injected into a Controversy," *USA Today*, Oct. 20, 2005, 8D.

25. Debora MacKenzie, "Will Cancer Vaccine Get to All Women?" *New Scientist*, April 18, 2005.

26. Peter Sprigg, *Don't Mandate HPV Vaccine—Trust Parents*, Family Research Council, www.frc.org/get.cfm?i=PV07D03, 2007 (accessed July 30, 2007).

27. Moira Gaul, *Family Research Council Statement Regarding HPV Vaccines*, Family Research Council, www.frc.org/get.cfm?i=LH06B03, Feb. 21, 2006 (accessed July 30, 2007).

28. Klepacki interview; Rob Stein, "Cervical Cancer Vaccine Gets Injected with a Social Issue," *Washington Post*, Oct. 31, 2005, A3; Gaul, *Family Research Council Statement*.

29. David Gilden, "Protecting against HPV: The Next Battleground?" *Treatment Issues: Newsletter of Current Issues in HIV/AIDS* 19, nos. 5–6 (2005), cited from www.thebody.com/content/art13356.html.

30. Gilsdorf interview.

31. Kirsten Moore, M.A., president and CEO, Reproductive Health Technologies Project, Washington, DC, telephone interview, Aug. 18, 2008.; Allina interview.

32. Allina interview.

33. Klepacki interview.

34. James Colgrove, "The Ethics and Politics of Compulsory HPV Vaccination," *New England Journal of Medicine* 355, no. 23 (2006): 2389–2391; L. Udesky, "Push to Mandate HPV Vaccine Triggers Backlash in USA," *Lancet* 369, no. 9566 (2007): 979–980.

35. Abramson interview; Gilsdorf interview.

36. Allina interview.

37. Ibid.

38. On Gardasil as a "vaccine against risk," see chapter 2.

39. Marie Cocco, "White House Wages Stealth War on Condoms (Op-Ed)," *Newsday*, Nov. 14, 2002.

40. Judith Auerbach, Ph.D., director of public policy, San Francisco AIDS Foundation; formerly with the Office of AIDS Research, National Institutes of Health, interview, San Francisco, Sept. 20, 2007.; Kaplan, *With God on Their Side*, 170–171; Mooney, *Republican War on Science*, 215.

41. Auerbach interview.

42. Ibid.

43. Special Investigations Division, Committee on Government Reform—Minority Staff, *The Content of Federally Funded Abstinence-Only Education Programs*, U.S. House of Representatives, http://oversight.house.gov/documents/20041201102153-50247.pdf, report, Dec. 2004 (accessed July 23, 2005).

44. Moore interview.

45. Gardiner Harris, "Morning-After-Pill Ruling Defies Norm," *New York Times*, May 8, 2004, 13.

46. Kaplan, *With God on Their Side*, 115–118.

47. Mooney, *Republican War on Science*, 219.

48. Wood interview; Frank Davidoff and James Trussell, "Plan B and the Politics of Doubt," *Journal of the American Medical Association* 296, no. 14 (2006): 1775–1778; L. L. Wynn and James Trussell, "The Social Life of Emergency Contraception in the United States: Disciplining Pharmaceutical Use, Disciplining Sexuality, and Constructing Zygotic Bodies," *Medical Anthropology Quarterly* 20, no. 3 (2006): 297–320.

49. Gardiner Harris, "Official Quits on Pill Delay at the F.D.A.," *New York Times*, Sept. 1, 2005, A12.

50. Allina interview.

51. Wood interview.

52. After unanticipated delays, Cervarix was approved by the FDA in October 2009.

53. Moore interview; Allina interview; Nonprescription Drugs Advisory Committee in Joint Session with the Advisory Committee for Reproductive Health Drugs Meeting, Dec. 16, 2003, Center for Drug Evaluation and Research, Food and Drug Administration, www.fda.gov/ohrms/dockets/ac/03/transcripts/4015T1.htm (accessed Aug. 15, 2008).

Part IV / In Search of Good Government

Europe, Africa, and America at the Crossroads of Cancer Prevention

CHAPTER THIRTEEN

Vaccination as Governance

HPV Skepticism in the United States and Africa, and the
North-South Divide

Julie Livingston, Keith Wailoo, and
Barbara M. Cooper

Perhaps more than other medical technologies, vaccines—invasive processes
whose benefits emerge through counterfactual reflection on the *absence* of
disease—condense and highlight relationships of trust or skepticism between
the state and its citizens and subjects. So it is not surprising that across the globe,
wherever the prospect of human papillomavirus vaccination has been raised as
a possibility, it has brought to the surface nascent debates about governance and
control—about the troublesome relationship between government, big capital,
adolescent girls, and the family, and about issues of sexuality and social control.
These debates over HPV and the cervical cancer vaccines frame a challenge
with multiple ironies. As one American policymaker noted, the vaccine "encapsulates so many issues that are at the core of politics and health policy right
now."[1] In what follows we consider the possible introduction of HPV vaccines
to compare issues of skepticism and pharmaceutical governance in the United
States and various sites in Africa. In contrasting the political complexities surrounding the potential introduction of a single technology in one region of the
largely privileged global North and one region of the largely resource-poor
global South, political economy and epidemiological realities of contemporary

geopolitics mean that such discussions at their broadest iteration take on starkly divergent form. Yet, there is also much overlap in the skepticism and suspicion that have surrounded the HPV and other vaccines. And so we also complicate this North-South divide, noting both diversity within these regions and occasions when the situation of poor patients in wealthy contexts comes close to resembling the status of relatively well-off patients in impoverished settings.

Considering the role of the vaccine in governance, we must contend with one important paradox. The burden of cervical cancer mortality is far greater in the global South, and so the vaccine story unearths a perverse risk calculus. More than 80 percent of the burden of cervical cancer is in the developing world, yet much of the most spirited debate occurs in relatively wealthy northern nations, such as the United States, where the vaccine is marketed, where the infrastructure for vaccine delivery is robust, and where (despite disparities and obstacles far less pervasive than in the global South) the vaccine debate unfolds amid a resource-rich culture of comparative entitlement. Thus, the rise of the HPV vaccine allows us to explore the global ironies of risk, the nature of vaccine skepticism, and government's role in shaping and mitigating risk.

This essay illuminates the paradoxes of this global discussion about the state's oversight and management of population and health, comparing the United States (from Washington, D.C., to Houston, Texas) and African contexts (where we consider two contexts in particular, Niger and Botswana). On the one hand, our research confirms the well-known dictum that all politics is local. We argue that the diversity of responses to the possibility of an HPV vaccine reveals diverse political calculations not merely over the management of health risk, but over questions of governance. In all these locales, the question of vaccination unearths familiar forms of skepticism not only over the technique, but over ways in which it relates to government's role in people's lives. The HPV vaccine also provokes skepticism, to be sure, about capital (and in particular, the pharmaceutical industry). But, like James Colgrove (chapter 1), we focus primarily on its place in guiding the role and capacity of the state in managing the health of the people. As skepticism about the vaccine grew, many debates revolved around the vulnerabilities, protection, and control of cancer in teenage girls. Here, questions of governance (the role of the state in the management of sexuality and in family relations) came clearly into view. On the other hand, we find common themes structuring local and national vaccine politics: the special status of adolescent girls as a particularly heightened proxy for popular attitudes toward the state; a reminder that while poverty may increase the biological vulnerability of a given population, it also has the potential to fuel mistrust in the motives of the state; and the routine questioning of expert claims through the use of scientific reasoning and evidence by culturally and politically diverse peoples. The HPV debate, then, is about much more than a medical technology. In each context, the vaccine is also a political tool, opening new possibilities for governance and control, but also carrying risk for government—unleashing the possibility of protest and backlash.

Our work is mainly based on a reading of the literature and (looking ahead to the future developments and deployments of the HPV vaccines) our informed speculation about the ways they might be incorporated or resisted in diverse settings.[2] Comparing debates over vaccination in a range of African contexts with debates among selected U.S. states where the vaccine debate was particularly heated (Michigan, Texas, and Virginia), we encounter people who see themselves as both privileged and entitled and people who do not. Without stereotyping these perspectives in the global North and South, we find diversity in both arenas, as well as a common set of concerns about governance that animate the HPV debate across the map.

Landscapes of Risk and Entitlement

While approximately 4,000 deaths in America could be attributed to cervical cancer in 2009, this was a tiny portion of the estimated 260,000 deaths globally—the majority of such deaths will occur among women in the developing world.[3] Not only are an estimated 80–85 percent of all invasive cervical cancer cases occurring in Africa, Asia, and Latin America, but the implications of the disease are far more dire for women in the global South.[4] In Africa, patients are more likely to be diagnosed at later stages of the disease and to lack access to effective treatment options, thus increasing the likelihood of sterility and death from such cancers.[5] In addition, the synergy between HIV and viral cancers means that we will see rates of cervical cancer growing in places where robust programs of antiretroviral therapy (ART) are implemented to ameliorate the effects of serious HIV/AIDS epidemics (many of which are concentrated in Africa), much as our colleague Doreen Ramogola-Masire (chapter 5) describes for Botswana. African women, while highly vulnerable to cervical cancer, are, for a variety of structural reasons, not equally positioned to debate cervical cancer care; nor are they, compared to mothers in the United States, well positioned to debate the desirability of a vaccine aimed at their daughters.

Moreover, vaccinations, injections, and disease prevention in these contexts reflect starkly different politics. The consequences of these differences in risk and in the landscape of opportunity are profound.

Other differences in health care infrastructure and affordability are glaring. In the United States, Pap smears have been part of the common landscape of cancer prevention for decades. Throughout much of Africa, by contrast, Pap smears and other screening options are lacking, and effective and available oncological services are also extremely scarce. In this context, a vaccine to *prevent* cervical cancer might well be a welcome development. In fact, some see the vaccine as a sort of magic bullet, which can provide protection that is needed especially because of the limits of Pap smear prevention services. Yet, there is no uniformity on this point. Questions remain about the match between the viral strains that are pervasive in Africa and those targeted by the extant vaccines. At nearly $350 for the series of three injections, the HPV vaccines are currently priced well out of the reach of middle-income countries on that continent and the low-income countries that have seen their health budgets drastically shrink through nearly two decades of neoliberal structural adjustment policies. From Burundi to Guinea to Mozambique to Congo, very little is left of public medicine and primary care. Many clinics and hospitals are understood and experienced as facades, or debris of an earlier era of a more expansive social contract, and so patients are often expected to pay clinic and hospital fees and to purchase their own medicines and any supplies necessary for surgery or other procedures; often they go out into the public marketplace to do so.[6] Even in middle-income contexts such as Botswana or South Africa, where health budgets may be larger, the demands of other urgent health needs, most notably HIV/AIDS and diabetes, consume many of these resources. In such contexts, the vaccine is currently priced out of reach as a feasible public health strategy, though some parents with private insurance might be able to obtain a vaccine for their daughters in the private health sector. HPV vaccination campaigns in Africa, as a result, will have to follow the complex politics of pharmaceutical-company drug donation that increasingly structure the flow of patented drugs into Africa.[7] Thus, the emergence of the vaccine itself—even if taken up and widely used—could be only a first tentative step toward a potentially effective intervention to protect African women from the perils of cervical and other genital cancers.

As in the global landscape of risk, HPV mortality within the United States also traces its own economic divisions. Across the nation, the epidemiology of cervical cancer also follows variations in wealth, resources, medical infrastructure, and social structures—state by state, and across regions within states. The cervical cancer burden is higher in poor and minority communities. Sexually transmitted diseases (STDs), it has been widely noted, are "concentrated in poor, segregated neighborhoods."[8] The disparities have shown this consistent pattern from state to state. How governments deal with deficiencies of infrastructure shapes their reception of the HPV vaccine. In the United States, the infrastructure for monitoring the disease is well developed, so that disparities can be measured in terms of both mortality and incidence. A 2008 report notes that in Michigan, for example, cervical cancer incidence for whites was 6.6 per 100,000, for African Americans 11.6, and for Latinas 11.7. With screening rates for whites and blacks around 90 percent, the mortality rate for whites (1.9 per 100,000) was half the rate for African Americans, 4.2 per 100,000.[9] In Texas also, the African American–white mortality ratio (5.7/3.2) was larger than the incidence ratio (11.7/10.1). But among the Latina population, a high incidence (15.1) was paired with a relatively low mortality (4.2).[10]

In the United States, a well-developed yet contested logic of governance and public health shaped the early reception of the HPV vaccines. Governments had long been in the business of amassing statistics on diseases and vaccinations, and state governments had also long viewed schools as the primary site for addressing vaccine-related health concerns. With the emergence of HPV vaccines, attention thus quickly turned to whether to *mandate* vaccination for school-age girls. According to the National Conference of State Legislatures, the fate of vaccine legislation to date has been determined by several factors.

> The debate in states has centered—in part—around school vaccine requirements, which are determined by individual states. Some states grant regulatory bodies, like the Board of Health, the power to require vaccines, but the legislature must still provide funding. Some people who support availability of the vaccine do not support a school mandate, citing concerns about the drug's cost, safety, and parents' rights to refuse. Still others may have moral objections related to a vaccine mandate for a sexually transmitted disease. Financing is another concern: if states make the vaccine mandatory, they must also address funding issues, including for Medicaid and SCHIP coverage and youth who are uninsured, and whether to require coverage by insurance plans. This has caused some to push for further discussion and debate about whether or not to require the vaccine.[11]

The rapid development of HPV state policies thus built upon a vast preexisting infrastructure, while also reflecting American comfort with vaccination and governance. In Colorado, for example, the legislature focused on creating a cervical cancer immunization program by adding the HPV vaccine to the list of Medicaid benefits, while also requiring certain health insurance providers to cover the cost of the vaccine.[12] A handful of other states also passed legislation requiring private insurance companies to cover the cost of HPV vaccines (New Mexico, Nevada, Illinois, and Rhode Island).[13] But many other states (twenty-seven in all) attempted to tackle the HPV issue comprehensively, by aiming for the goal of school-based immunization of all sixth-grade girls. This far-reaching measure succeeded in Virginia and in the District of Columbia and became a topic of heated debate about the limits of public health governance in 2006 and 2007.[14]

Even though these American debates about the HPV vaccines started from fundamentally different positions than the positions in Africa, it would be incorrect to characterize the debates in the global North and South as reflecting a world of privilege versus one of poverty, of entitlement versus limited deprivation. The global South is not without a politics of entitlement, and the North has many (if hidden) intransigent pockets of intense need. It is by looking closely at the skepticism surrounding the vaccine that many of these convergences come into view.

Varieties of Skepticism

When the District of Columbia took up mandatory HPV vaccination legislation in January 2007, for example, one columnist saw the move as "tinged with ugly assumptions." Accusing the City Council of racist paternalism, Courtland Milloy wrote sarcastically that "only the most progressive and caring elected city official—in this case, two nice white people—would propose a program to vaccinate against sexually transmitted disease girls under 13 in a predominantly black school system. After all, if the girls' parents can't protect them—and, God knows, they can't protect themselves—then somebody's got to do it."[15] The concerns unearthed by the HPV policy bore a striking resemblance to anxieties elsewhere. Barbara Loe Fisher, president of the National Vaccine Information Center, sought to link Milloy's objection to a broader skepticism about vaccines' dangers—noting that "his warning also challenges us to think about an important moral question: should citizens be injected against their will with biological agents that can injure and kill for what the state has defined as the common good?"[16] But another letter-writer disagreed with Milloy's claims, espousing instead a familiar faith in the vaccine as beneficial. "Implying that people who would submit to or advocate compulsory vaccination 'are content to live and die as slaves?' Outrageous . . . I simply don't believe that trying to prevent girls from developing cervical cancer is a racist plot."[17]

As the state of Michigan considered mandatory vaccination, one editorial portrayed the move as an encroachment on the rights and prerogatives of parents. The parents "as guardians of specific children living individual lives, wanted proof beyond the quick acceptance of public officials looking to do good by a large swath of their constituents."[18] Across the United States, competing claims about the vaccine, its risks, and its purported efficacy intersected with questions of parents versus state governance in the name of preventing invasive cervical cancer. Skeptical parents raised many questions about the new technology—about whether it worked, what its unintended effect might be, and so on. Pharmaceutical companies, academic scientists, and public health specialists in turn sought to manage this discussion, putting forward claims for the efficacy of the vaccines. And patients, community activists, women's health advocates, health planners, politicians, and individual clinicians were compelled to evaluate these claims in relation to their own understandings of risk and benefit. Discussing the failure of the legislation, the Michigan editorial noted, "What all those well-meaning crusaders forgot to take into account . . . was the ferocity with which parents protect their young. Michigan moms and dads wanted further evidence that the prevention was worth the cure. As a new vaccine, it might have unknown side effects."[19]

Throughout Africa, as in the United States, skepticism over scientific HPV claims, the true relationship of HPV to invasive cervical cancer, and the vaccines' efficacy reflect a complicated epistemological problem. In both contexts, gaps of knowledge are evident; although many suffer from, or are at risk for, cervical cancer, few have heard of cervical cancer or of HPV with its sexual mode of transmission and its long latency period. Its acronym, which sounds similar to HIV (in Anglophone contexts), means that any claims about the vaccine are bound to be caught up, at least in part, with current understandings and ideas about HIV and STDs. In Botswana, for example, it has become difficult to communicate with patients about viruses *other* than HIV, now that HIV has been rendered as *The* Virus (*mogare*) in the Setswana language.

In addition, anthropologists Melissa Leach and James Fairhead found that throughout West Africa, many people did not distinguish between vaccines and other putatively curative or therapeutic injections; they used the same term for both categories.[20] This is not to invoke tired stereotypes of incredulous or unscientific Africans, but rather to note that the meaning of the HPV vaccine in these different contexts will engender different kinds of skepticism and different notions of risk and benefit.

Claims about vaccination and its efficacy, viruses, and cervical cancer have encountered skepticism worldwide. In places like Botswana or like Khayelitsha in South Africa, where a technological optimism has emerged (perhaps only temporarily) out of widespread, and in some cases seemingly miraculous, experiences with ART, a vaccine against HPV is likely to find a more receptive audience. But claims about efficacy are shaped in relation to skepticism about governance. Until recently, women in Khayelitsha received their antiretrovirals from the international nongovernmental organization (NGO) Doctors without Borders, not from their national government, which was historically reluctant to provide these drugs. Though there has since been a change of government in South Africa, in such a context the sudden appearance of a new government-promoted vaccine might raise suspicion, despite potential receptivity for new drugs that the ART experience has engendered. In other African contexts, where the burden of cervical cancer is great but where questions of governance and trust differ dramatically, one might expect a different skeptical logic to shape the uptake of an HPV vaccine.

Beyond questions of trust, the HPV vaccine debate reveals vast differences in national health priorities. In Niger, as in West Africa more generally, prevention and treatment of cervical cancer is not a high priority given other urgent health needs. HIV infection rates in Niger are relatively low compared with those in countries of eastern and southern Africa, reducing the coinfection of HIV and HPV. As a general rule, the countries that have expressed the greatest interest in the vaccine have relatively high HIV infection rates. Furthermore, oral contraceptive use, another cofactor in the development of cervical cancers, is extremely low in Niger, and very few women smoke. In other words, some of the conditions present elsewhere on the continent that might contribute to high rates of cervical cancer are not relevant in Niger, setting aside the question of HPV infection rates.[21] With roughly 532 cervical cancer deaths per year in Niger, it is easy to see why, despite a relatively high mortality rate (15.7) compared with the world figure (9), health professionals might not advocate for the widespread use of the HPV vaccine when other health concerns appear far more pressing.[22] A woman is far more likely to die in childbirth or from a respiratory infection, diarrhea, or malaria than of cancer of any kind.[23] The overwhelming majority of efforts directed at women's health focus on reducing perinatal mortality.

However, in addition to the likelihood that health professionals in the Nigerien government will prefer to deploy scarce resources (Niger is consistently among the poorest countries on the globe) toward more urgent health concerns, the particular shape of debates about vaccination and the public good in this region would make the introduction of a vaccine specifically targeted at peripubescent girls extraordinarily politically fraught. In July of 2003, just as the World Health Organization (WHO) was poised to enter a key phase of the polio eradication campaign in Nigeria (Niger's powerful southern neighbor), the Supreme Council for Islamic Affairs in Nigeria claimed that the oral polio vaccine to be employed had been contaminated with carcinogenic, antifertility, and AIDS-inducing agents. Niger has since then struggled consistently to allay parental fears of the polio vaccine in a resistance movement that spread from Nigeria into Niger, a crisis that links distrust of government intentions; skepticism about the benevolence of intergovernmental agencies like the WHO, particularly in the context of the "war on terror"; an awareness of the profusion of counterfeit drugs in the West African market; and long experience with governments and agencies promoting fertility reduction through injections and implants. Parents and community leaders have tended to ask why, when their own health priority would probably be to reduce the ravages of malaria, governments, nongovernmental organizations (NGOs), and intergovernmental organizations have returned again and again to polio eradication campaigns—intrusive, repetitive, and evidently not altogether effective. Is the antipolio campaign merely a pretext, they wonder, for a population-reduction agenda developed elsewhere that is to be implemented by intrusive injections into the bodies of vulnerable girls?[24]

Given this atmosphere of intense mistrust, the 2005 vaccination coverage rates for the diphtheria-pertussis-tetanus vaccine (third dose completed) at 89 percent and the polio vaccine (third dose completed) at 83 percent must be read as something of a public relations victory. While obviously these figures could bear improvement, they reflect the tireless efforts of the WHO and major political figures in Niger to repair the damage created by the crisis in confidence in the medical system. In such a context, to propose vaccinating nubile girls (and only girls) in order to prevent a sexually transmitted virus that is a necessary but not sufficient cause for a cancer that few Nigeriens have even heard of would appear to be politically imprudent, to say the least. There would be much to lose and little to gain in such an intervention. There will probably be quarters of the globe in which vigorous debates about HPV will not emerge because in the existing political climate such a debate would make no sense.

Thus, different forms of skepticism about vaccination are shaped by different realities across the globe, particularly relating to the social contract and what citizens expect from government. In the United States, ambivalence toward government is deep-seated. American attitudes often reflect a robust suspicion of government, but also they reflect a strong sense of middle-class entitlement to the resources of government. In places like Nigeria or Guinea, by contrast, where patients and their relatives must procure medicines prescribed at biomedical sites in an open marketplace of injectionists and drug sellers, the social contract is little more than a sham. The abundance of counterfeit and expired drugs on the market engenders skepticism about biotechnical efficacy.[25] In such settings, where one must pay a "fee" for a vaccine, many people will question the motives behind its recommendation even if an HPV vaccine is heavily subsidized. In such contexts, "fees" provide flimsy cover for bribes for health workers who may go months without receiving their (already meager) salaries. Such fees, it is commonly assumed, are critical components of the economy for nominally free public services. In such an environment, the motives and interests attached to vaccination differ from those in the U.S. context; patients may strongly suspect the motives of the health worker recommending vaccination and, by extension, the efficacy of the vaccine itself. But African societies and economies are not monolithic in this regard. Elite Nigerians or Senegalese, many of whom have insurance or private funds and are tapped into media sources that heighten awareness of cancers and their risks, may decide to purchase vaccination at private clinics, but for the majority of poorer citizens (and many who are better off) claims about the benefits of such a vaccine may prove at best dubious, and at worst spurious.

Despite the large differences in context, the skepticisms over vaccination in the United States and African contexts share some similarities—particularly regarding the ways in which interests guide expert views about vaccination. The logic in some African settings is similar to the skepticism found in various U.S. contexts in 2007 when states took up HPV legislation. Even (or perhaps especially) among better-educated people, in Africa as in the United States, it rarely escapes notice that scientific claims are interested claims. Efficacy, then, is judged in part through evaluation of the interests that are assumed to underlie scientific knowledge.

At the height of the HPV vaccine debate in the United States, patients and politicians alike saw the question of efficacy against the perceived predations of Euro-American capital and scientific knowledge production, but their skepticism also took many other forms. In the United States, it was possible to find both unambiguously positive claims about the vaccine's efficacy, like those of Cosette Wheeler, a professor of genetics and obstetrics-gynecology in New Mexico, who greeted Food and Drug Administration approval of the vaccine with these words: "Now we can say that the vaccine is 100 percent effective against HPV Types 16 and 18, which are responsible for the majority of the cancers."[26] Other experts made more nuanced recommendations, arguing that "the debate is not about whether the vaccine is effective, but about who should be vaccinated, and when."[27] But it was also common to hear rejoinders like those in the *Grand Rapids (Michigan) Press*, which noted in May 2007 that "new data on the controversial HPV vaccine . . . have raised serious questions about its efficacy." One of the concerns was that "blocking the targeted strains may have opened an ecological niche that allows the flourishing of HPV strains previously considered to be minor players, partially offsetting the vaccine's protection."[28] Other forms of skepticism, as Jennifer Reich (chapter 9) explains, had to do with the lack of knowledge about how long the vaccine might be effective or what its "unknown side-effects" (whether on the body or on the behavior of the vaccinated individual) might be.[29] These varieties of skepticism suggested, at the very least, the unfolding of a complex sociological understanding of the HPV vaccine—moving the discussion of efficacy well beyond the narrow immunological one.

South Africa provides perhaps the best-known comparative example of skepticism. There well-educated politicians, and some local scientists, challenged claims about the efficacy and risks of antiretroviral therapy for several years in the late 1990s and early 2000s. These public figures suspected intertwined racist and capitalist motives by American and European scientists, and their suspicion was buttressed by the punitive language of public health and informed by a long history of medicine that was complicit with an institutionally racist state. They established their skepticism, however, not by denying the truth claims of science or biomedicine as a domain of knowledge and a system of practices; rather, they challenged scientific claims by marshaling counterclaims and debate from within the scientific literature itself.[30]

In northern Nigeria, the international polio eradication campaign ran into a similar situation, in which political, medical, and religious leaders countered claims of efficacy and risk with their own reading of the scientific literature.[31] Resistance to the polio vaccine, manifested in widespread fears that the vaccine would render girls sterile, merged a complex array of factors ranging from fears of globalized religious persecution in the wake of September 11 to political and demographic anxieties over ethnic and regional politics within the Nigerian state. But such fears also resonated with widely publicized abuses that occurred in 1996 in Kano, where families accused Pfizer pharmaceutical of experimenting on Nigerian children without their parents' knowledge or consent, during a meningitis vaccine program.[32] Across Africa, for a century or more, drugs have been tested, bodily fluids and materials have been appropriated and sampled, and medicine has at times been coercive in ways that are neither easily nor safely ignored by parents of daughters who might be vaccinated with a new drug from Merck or any other major pharmaceutical company, nor by community leaders themselves.

In the global North as in the global South, then, efficacy claims—whether or not taken at face value—became the flashpoint for historical debates over citizenship and rights and the limits of government. With this backdrop, it becomes easier to see how (across contexts and even in the same setting) the HPV vaccine could symbolize both the realization of citizenship rights to life and self-determination for some, while embodying a truly malignant force and an incursion into the sacred realm of family for others. One of the crucial differences between U.S. and African settings, however, is the fact that in the United States administrative mechanisms are well developed to battle these forms of skepticism—and to override them if necessary. When Kentucky's legislature took up an HPV mandate (including an opt-out clause for parents), some vaccine advocates saw the opt-out clause as even more problematic: "If parents feel empowered to opt out of HPV vaccinations, they may feel empowered to opt out of other vaccinations, for reasons that may not be medically sound," noted one clinician.[33] Although Kentucky's bill was not passed (despite Milloy's objections), the District of Columbia did pass a law mandating vaccination for sixth-grade girls. Because the District of Columbia is governed by Congress, the law requires congressional approval before it can become effective.[34]

Different Administrative Realities

The problem of governance operates at the level of the state and at the level of administrative bureaucracies—and different administrative realities do inform how the HPV vaccine fits into different social settings. In the United States and across Africa, fundamentally different administrative realities shape the possibilities for acceptance of the HPV vaccine and the trajectory that acceptance might take. The vaccine is, after all, administered in three parts. So it requires a certain amount of administrative infrastructure for its use. In the U.S. context, the schools, long established, offer themselves as an ideal administrative vehicle. The same cannot be said of African contexts. Then there is the availability of a public health infrastructure in the United States (student health centers, public health services), compared to the hollowness or even absence of the administration of health services in many African settings. These different administrative realities produce their own economic calculus.

In the global North, the public health imperative has long been closely tied to the logic of state police powers over its citizens. Vaccination, and public health more broadly, is fundamentally part of a social contract between the state and its citizens. In the United States, the ease and rapidity with which some states took up mandatory school vaccination bills was made possible by the state budget and the public health infrastructure, which stood ready to carry out the mandate. In Michigan, for example, as the legislature considered the mandate in March 2007, the *Detroit Free Press* noted, "Last week, the [Macomb County] department [of health] had 682 of the 1,600 doses it ordered last fall, all of which are earmarked for females 18 and younger through the Medicaid insurance program for poor children. The department administered 256 of the doses and sent 662 to area physicians who requested them." In the United States, from state to state, the availability (and the implied administrative efficiencies of a public health system well situated to deliver vaccination) was one of the factors prompting the question "Why Aren't Girls Getting the HPV Vaccine?"[35]

The contrast between the speed and readiness of public health administration in the global North, versus the realities in the global South, is stark. In many settings where the government is unable or unwilling to prioritize services, NGOs have stepped in to fill the breach, and here too we may see some organizations that decide to take on the issue of vaccination. In some other African contexts, the social contract is already very much in doubt, distorted by decades of privatization, corruption, and in some cases outright state collapse. In other contexts, such as in Gambia or Botswana, robust claims of medical citizenship, already animated by and enacted through childhood vaccination, ART, and other programs, may provide a setting in which patient demands drive state provision of HPV vaccines. But in all cases, concerns about the suitability of the available vaccines and pragmatic challenges of delivering a three-part vaccine may make the state reticent about adding an HPV initiative to already over-loaded health planning. These are concerns particularly in the many places, such as Tanzania, where the social contract is politically meaningful and yet oncology services are private or extremely limited. A different economic calculus is in effect there than in the United States, where the state and the insurance companies must weigh the cost of the vaccine now against the expense of cancer care later.[36] By contrast, in Botswana, where oncology services are provided free to all citizens, an argument can be made about the cost-effective nature of the vaccine.

Capital, the State, and Sexual Governance

State power is one thing. The state power combined with the powers of industry is quite another. And yet again, state power tied to industry in the service of issues relating to sexuality, morality, and health control adds political and cultural complexity to the HPV vaccine debate. Indeed, in the United States (and across Africa), one of the factors that spawned particularly heated commentary and skepticism about the vaccine was the prominent (and some would argue inappropriate) role of the HPV vaccines' makers in a complex public health debate where matters of sex were so closely tied to the vaccine.

One of the most potent accusations against those promoting the HPV vaccine mandate in the United States was, and remains, that Merck's economic interest was controlling the public health debate in America. In the midst of the Texas debate, commentators noted that organizations like Women in Government, which had become quite prominent in "holding a series of luncheons and conferences nationwide to discuss its fight against cervical cancer," were partners with Merck and had received "funding from Glaxo, as well as Digene, a company that makes a test to detect the presence of HPV." Critics of Texas governor Rick Perry's executive order mandating vaccination for sixth-grade girls complained that "the Governor's former chief of staff was now a lobbyist for Merck."[37] "This is 'follow the money' if I've ever seen it," pronounced the president of the ultraconservative Texas Eagle forum, which had also staked out a position against the vaccine on the grounds that it promoted promiscuity.[38]

From early on in the HPV debate, experts saw the new vaccines as an economic product—a commodity with an as-yet-undefined market. As early as April 2005, one author in the *Muskegon Chronicle* of Muskegon, Michigan, an-

ticipated that the expected approval and release of such a vaccine would provoke a "culture clash." It combined sex, money, and parental authority. "Sex is
a scary thing in this culture, and the age of girls to be vaccinated will really be
an issue," a sociology professor predicted at the time. Although these concerns
might have worried the vaccine makers, the global market potential was impressive. Unlike other pharmaceutical products, vaccines had not been long associated with capitalistic motivations, but this vaccine—a formidable commodity
particularly if tied to state mandates—could be different. Even without public
mandates, it became widely characterized as a financial boon to industry. Even
before the Texas controversy in 2007, the financial prospects for this vaccine
had been stellar. It "could hit $4.3 billion in revenue by 2010," noted the news-
paper; it was a projection based on the assumption that adolescent girls and
boys, as well as women in their twenties and thirties, would be vaccinated. As
one observer noted, however, "for the companies to realize those billions of
dollars in annual revenue, they are expected to advertise widely and charge a
lot. . . . Merck will likely charge $100 for each of three needed doses . . . while
Glaxo may charge $80 per dose." These amounts were considerably higher
than the costs of other common childhood vaccines (measles, mumps, and rubella, around $35; chickenpox, $58).[39]

As a new economic and public health product, the vaccine generates anxieties
in many settings about how the profit motive shapes the push to vaccinate. In
the African context, these same concerns exist alongside even more complex
economic realities, including vast markets of counterfeit drugs, crumbling or
overstretched infrastructures, and the emergence of a robust clinical-trials industry in the absence of other health services, that define the health care system.
It is hard to imagine a parallel in U.S. contexts to the common African practice
of getting a vitamin or antibiotic shot from the local injectionist or receiving
such an injection while standing in a marketplace. So along with skepticism
about the motives of large pharmaceutical entities, there is in Africa also a growing dependency on partnerships between private pharmaceutical companies
and national governments and NGOs to provide badly needed health care. In
places like Botswana, skepticism about Merck is tempered by the realization that
the national ART program is built on an innovative public-private partnership
that includes Merck.

For those who supported the vaccine mandate, it promoted a vision of government as a superparent. One Michigan obstetrician-gynecologist argued in
a March 2007 article, "Your daughter may be a virgin when she goes on her

honeymoon, but unless her husband was pristine, he may bring the virus to the marriage bed. . . . [Without the vaccine] One has no ability to protect themselves from the virus," he concluded.[40] But the HPV vaccine debates, as they played out in state after state, revealed the limits of this vision. Demonstrated instead was powerful skepticism about the government's role in "protecting" families and girls as they navigated an apparently dangerous sexual terrain. The most prominent disputes, and the ones best covered in the American media and public health literature, have been the controversies over whether such a vaccine would actually encourage adolescent girls to engage in "risk-free" sex. Social conservatives prominently complained "that vaccination could encourage adolescents to be more promiscuous"; others alleged that "parents' authority over their daughter's health care would be usurped."[41] The promiscuity debate became so heated that it overshadowed many other features of how HPV intersects with sex debates and attitudes. Such sexuality arguments pervaded the language of vaccine advocates, who could suggest that the vaccine allowed a father or a mother to protect even a virginal daughter from men—long after she left home.

The polio vaccination crisis discussed above would seem to suggest that parents elsewhere also fear usurpation of their authority over their daughters' sexuality and reproductive careers by government, but even more by supragovernmental institutions such as the WHO that are not elected and may have tremendous sway over how states determine their medical priorities. Because debate about HPV vaccines has not been heard yet in places like Niger, it is hard to know whether the same kinds of fears that a higher safety threshold would prompt girls to be more sexually active would emerge. It seems more likely, however, that the debates would focus instead on the specter of infertility that so frequently shapes individual and community choices in West Africa. What parents and daughters fear in the context of vaccination is not pregnancy out of wedlock but a reduction in the capacity to bear children. Because injections are so fully associated with contraception, and because their effects seem to be irreversible, anxiety tends to focus upon the implications of vaccination not for individual sexuality per se, but for reproduction for the social group as a whole.

In the United States, the combination of suspicion over economic interest and moral anxiety over the "sex effects" of the vaccine shadowed the debate in state after state, linking the issue of cervical cancer closely to the sexuality of adolescent girls and the state's oversight of good health. In state after state, legislators sought to act out their roles as surrogate parents of at-risk girls. In Georgia, as in Texas, the combination of issues proved devastating to vaccine advocates. "A powerful state Republican lawmaker proposed making the vaccine mandatory for girls entering sixth grade," began a Michigan article reflecting on the fate of Georgia's legislation, "and the governor included $4.3 million in his budget to make it available to about 13,000 girls whose family's insurance wouldn't cover it." But the goal of the state acting as parent ran up against another set of fears. As the article continues, "state lawmakers nixed the plans after aggressive lobbying by religious conservatives, who argued that vaccinating young girls could promote promiscuity."[42] This argument—that the vaccine fostered promiscuity—became a new linkage for those resisting the expanded reach of government (see chapter 12 on the limits of this discourse).

The fetishized focus on girls' sexuality in the American media and political culture stood in contrast to more structural interpretations of HPV and cervical cancer, and to mortality differences, across the map.[43] As a news report in Minnesota commented, rates of cervical cancer and HPV were high in poor and minority communities and pointed to different problems of sexuality, control, and governance. Why were rates higher in such communities? It was not that black teens engaged in more sexual activity than their white counterparts; what nurtured the disparity was, rather, the segregated nature of that sexual activity, combined with the silent nature of HPV and the deficiency of screening services. One factor, the report insisted, was that "chlamydia and HPV are silent diseases. Men and women are often unaware that they are infected. But left untreated, chlamydia can cause pelvic inflammatory disease (PID), which leads to potentially fatal ectopic pregnancies and infertility. HPV causes most cervical cancer." Another factor was the structure and dynamics of communities: "Experts point to poverty and less access to health care as reasons STD rates are higher among blacks. But that does not explain everything, they say. Young blacks are no more likely to be sexually active than their white counterparts and about half of each group use condoms regularly. However, STD rates have been higher in the black community for many years, and that could be causing what experts call a cohort effect." An adolescent-health expert insisted that sex behavior could not be disentangled from the organization of society: "Most people tend to choose sexual partners from their social network—the people they go to school with, who live in the neighborhood, friends of friends. So these higher rates are perpetuated, especially in more segregated areas."[44] This more complex social governance of sex and sexual realities has been, to some extent, submerged in the American political debate over cervical cancer, girls' sexuality, and the vaccine.

In the U.S. and the African contexts, the sexual conundrums surrounding vaccination take on vastly different forms, yet they find common roots in fears about state governance. In both contexts we find that debates over the age of sexual debut are common, but in Africa questions and fears about sterility and population control are also present, coupled to preexisting tensions about sex and (in this case) prepubescent girls. The discussion of who is responsible for these girls is worth looking at; we aim simply to draw attention to the different assumptions and possibilities surrounding this conversation in diverse American and African contexts. Girls, not boys—why? At play is a paternalistic ethos to protect our girls, not a logic of herd immunity; nor is it about personal liberty and the public self. Why don't boys have to give up a little bit of their liberty for the broader public? If scientific links between HPV and a widening spectrum of cancers—from cervical, to anal, to oral—continue to be discovered, as appears likely, then the ethics of vaccines is likely to shift with each new finding, as Steven Epstein explains (chapter 4).

To those who harbor fears related to fertility, the state does not seem beneficent but rather seems to be aligned with predatory bioscientific forces seeking to appropriate or disrupt fertility. This view arises out of widespread late-twentieth-century experiences with family planning and population control policies, which were central to bilateral and international health efforts in the 1970s and 1980s in Africa. There was what scholars often call a social emphasis on "wealth in people" in figuring value in social, political, and productive networks on the continent. While some women certainly appreciated the possibilities for reproductive control that these programs brought, for many others, *infertility*, not overfertility seemed to be the pressing problem of their lives. At its most extreme, family planning (often in the form of *injections* of Depo Provera) could be put to coercive and eugenic uses. In late-twentieth-century colonial Namibia (the 1980s) and Zimbabwe (the 1970s), for example, family planning was instituted by demographically anxious minority white states, and this context shaped how birth control technologies were put to use.[45] When condom campaigns to prevent HIV followed closely on the heels of such programs, initially they were feared as thinly narrated justifications for continued population control in diverse contexts from Tanzania to Botswana to Zambia.

In both the U.S. and the African contexts, when a vaccine is specifically targeted at girls, fears about state motives are understandably magnified. Such anxieties cannot be ignored, nor should one assume that a little coercion, while unpleasant, is necessary to attain a greater good. The potential consequences are significant. In Cameroon in 1990, schoolgirls leaped from windows to evade a vaccination team attempting to prevent neonatal tetanus through a national campaign. The girls themselves and many of the adults in their midst, including teachers, parents, and religious leaders, feared that the vaccine would cause sterility and interpreted it as a politically motivated policy on the part of the central state to punish the grasslands region during a period of political unrest. Such an interpretation was facilitated by poor communication and organization by the vaccinating team and the public health program from which it stemmed. But one of the most troubling outcomes of this crisis should give us pause when thinking through the significance of vaccine-sparked sterility fears. Researchers found that many of the girls who were vaccinated felt compelled to "test the vaccine's alleged sterilizing effect, either to prove their fears or to reassure themselves of their reproductive potential . . . 20 percent of the girls interviewed in 1995 claimed to have explicitly 'tested' their ability to conceive following the rumor." Of those who became pregnant, anecdotal evidence, including some first-person testimonies from vaccinated girls and from local health workers, suggests that many in turn induced abortion, a risky prospect given the illegality of abortion in Cameroon.[46]

Conclusion

We have tried, in this very preliminary sketch, to illuminate some of the ironies and tensions in how vaccination in general, and the HPV vaccine in particular, "works" as a political tool and a mode of health management in these various contexts. In exploring HPV in this way, we have attempted to move beyond the easy contrast between a resource-rich global North and a resource-poor global South. What emerges is a sketch of complicated landscapes where varieties of skepticism are evident, in which the administrative and social disparities are stark, in which entitlement and skepticism are rooted in local, historical experiences. We have illustrated some differences as well in the administrative structures of public health in the global North and South. These differences, of course, will be crucial to the "success" of any vaccine effort, in any context. Regardless of the context, making the HPV vaccine the right tool for the job will continue to be a complex challenge for government (allied with private industry in multiple ways) and a complex challenge of administrative, social, and sexual governance.

NOTES

1. "Cancer Vaccine Bills Stall: Sex, Parenting, Politics: 'Perfect Storm' of Controversy Slows Acceptance," *Grand Rapids Press*, May 27, 2007; the same article noted, "Over the past months, a vaccine that once was hailed as a breakthrough to prevent cancer deaths has become embroiled in some of the nation's most politically charged issues: teen sex, parental control, state mandates, a backlash against vaccines and a suspicion of drug companies."

2. We are not alone in pursuing such a comparative framework. Melissa Leach and James Fairhead have recently published work, *Vaccine Anxieties: Global Science, Child Health, and Society* (London: Earthscan, 2007), comparing the politics and challenges of childhood vaccination in the United Kingdom and various sites in west Africa.

3. American Cancer Society, "What Are the Key Statistics About Cervical Cancer?" www.cancer.org/docroot/CRI/content/CRI_2_4_1X_What_are_the_key_statistics_for_cervical_cancer_8.asp?rnav=cri; "Summary of Key Points: WHO Position Paper on Vaccines against Human Papillomavirus (HPV)," www.who.int/immunization/documents/HPV_PP_summary_LE_03-04-09.pdf, April 9, 2009.

4. Cynthia Dailard, "The Public Health Promise and Potential Pitfalls of the World's First Cervical Cancer Vaccine," *Guttmacher Policy Review* 9, no. 1 (Winter 2006): 6.

5. Epidemiological returns for cancer in Africa need to be taken with a grain of salt, given the lack of cytology facilities and the sparseness of national cancer registries on the continent, but it seems safe to assume that the figures represent an underreporting, rather than an inflated modeling of the burden of disease. The most recent World Health Organization (WHO) report (2007) appears to use statistics from 2002. That report indicates a crude incidence rate across Africa (including North Africa) of 19 per 100,000 women per year and an age-standardized rate of 29.3. This represents 78,897 cases. The reported burden varies across the continent, with the bulk of reported cases (more than two-thirds) in West and East Africa. We might expect these statistics to shift as southern Africa expands its ART coverage. World Health Organization, *Human Papilloma Virus, and Cervical Cancer, Summary Report for Africa* (updated 2007), www.who.int/hpvcentre/statistics/dynamic/ico/country_pdf/XFX.pdf?CFID=808801&CFTOKEN=23785438 (accessed May 5, 2008).

6. The literature on the privatization of public health in Africa is vast. For an excellent overview, see Meredeth Turshen, *Privatizing Health Services in Africa* (New Brunswick, NJ: Rutgers University Press, 1999). For detailed local or national examples, see Adeline Masquelier, "Beyond the Dispensary's Prosperous Façade: Imagining the State in Rural Niger," *Public Culture* 13, no. 2 (2001): 267–292; Ogoh Alubo, "The Promise and Limits of Private Medicine: Health Policy Dilemmas in Nigeria," *Health Policy and Planning* 16, no. 3 (2001): 313–321.

7. This process has begun, with an announcement by Merck at the 2009 Clinton Global Initiative Annual Meeting. There, Merck pledged to donate up to 5 million doses of Gardasil to women in low-income but high-HPV-burden contexts. "Merck and Qiagen Collaborate to Accelerate Access to Cervical Cancer Vaccination and Screening in Developing Countries," www.merck.com/newsroom/news-release-archive/corporate-responsibility/2009_0923.html (accessed Dec. 29, 2009).

8. "Black Teen STD Rate Needs Our Attention," *USA Today*, April 4, 2008, 12A. In the United States (as a recent Centers for Disease Control and Prevention report noted, and as quoted in the *USA Today* article), the HPV gap and the STD gap followed similar patterns of ethnicity, age, and economic status, with "black teens [having] far greater rates of STDs, including human papillomavirus (HPV), chlamydia, herpes simplex virus, and trichomoniasis."

9. "The 'State' of Cervical Cancer Prevention in America—2008: Turning Challenges into Opportunities," Women in Government Report, 43, www.womeningovern ment.org/prevention/state_report/2008.

10. Ibid., 64.

11. National Conference of State Legislatures, "HPV Vaccine Legislation: 2007–2008," updated May 2008, www.ncsl.org/programs/health/HPVvaccine.htm.

12. Ibid.

13. "The 'State' of Cervical Cancer Prevention in America," 12.

14. Other states, including Iowa, also took up HPV vaccine legislation as an insurance coverage requirement, ensuring that "unlike some laws being considered by other states, money to pay for the vaccination would be coming out of insurer's pockets, not taxpayers'." Thus the issue was framed as a consumer's challenge to insurance companies. Alyssa Cashman, "Iowa Senate OKs HPV Move," *Iowa City Daily Iowan*, April 4, 2008.

15. Courtland Milloy, "District's HPV Proposal Tinged with Ugly Assumptions," *Washington Post*, Jan. 10, 2007, B1.

16. "Risk, Racism, and the HPV Vaccine," letter to editor from Barbara Loe Fisher, *Washington Post*, Feb. 1, 2007, A14.

17. "Risk, Racism, and the HPV Vaccine," letter to editor from Bob Dardano, *Washington Post*, Feb. 1, 2007, A14.

18. "Lawmakers Rethinking How to Cure Cancer," *Grand Rapid Press*, June 9, 2007, A11.

19. Ibid.

20. This also appears to be the case in Botswana (and quite possibly throughout southern Africa), though this sense may be changing as biomedical ideas about "the immune system" are crystallizing in the popular imagination through extensive public heath campaigns and widespread experiences of ART therapy. Melissa Leach and James Fairhead, *Vaccine Anxieties: Global Science, Child Health and Society* (London: Earthscan, 2007), 110.

21. HPV infection rates for Niger are not available; most reports cite the West Africa regional estimate, which is 16.5. But that figure is probably much higher than would be found in Niger, a landlocked country with quite strict sexual mores, reflected in the relatively low HIV infection rate of 1.1.

22. WHO/ICO Information Centre on HPV and Cervical Cancer (summary report on HPV and cervical cancer statistics in Niger, 2007), World Health Organization, www.who.int/hpvcentre (accessed Sept. 20, 2008).

23. WHO Mortality Country Fact Sheet 2006: Niger, www.who.int/whosis/mort/ profiles/mort_afro_ner_niger.pdf (accessed Sept. 20, 2008).

24. For a fuller discussion of the polio vaccination crisis in Niger and Nigeria, see Adeline Masquelier, *Women and Islamic Revival in a West African Town* (Bloomington: Indiana University Press 2009); and Elisha Renne, *The Politics of Polio in Northern Nigeria* (Bloomington: Indiana University Press, forthcoming).

25. Susan Reynolds Whyte, Sjaak Van Der Geest, and Anita Hardon, *Social Lives of Medicines* (Cambridge: Cambridge University Press, 2003); Kristin Peterson and Olatubosun Obileye, *Access to Drugs for HIV/AIDS and Related Opportunistic Infections in Nigeria: A Status Report on the Sociopolitical, Economic, and Policy Climate on Drug Availability for People Living with HIV/AIDS (PLWHA) and Recommendations for Future Access* (Washington, DC: Policy Project/Futures Group International, 2002.

26. Ben Dietz, "Vaccine Prevents Cervical Cancer: So, What's the Down Side?" *New York Times*, May 23, 2006, F5.

27. Richard Ma, "Editorial: Vaccinating Girls against HPV," *Practitioner*, July 31, 2007, 7.

28. "Vaccine Faces Scrutiny: Effectiveness against HPV, Cervical Cancer Questioned," *Grand Rapid Press*, May 10, 2007, A3. This news report was based on a study published in the *New England Journal of Medicine*.

29. "Lawmakers Rethinking How to Cure Cancer."

30. For an extensive analysis of these debates, see Didier Fassin, *When Bodies Remember: Experiences and Politics of AIDS in South Africa* (Berkeley: University of California Press, 2006).

31. Maryam Yahya, "Polio Vaccines—"No Thank You!" Barriers to Polio Eradication in Northern Nigeria," *African Affairs* 106, no. 423 (2007): 185–204.

32. Ibid.; see also P. W. Geissler and R. Pool, "Editorial: Popular Concerns about Medical Research Projects in Sub-Saharan Africa—A Critical Voice in Debates about Medical Research Ethics," *Tropical Medicine and International Health* 11, no. 7 (2006): 975–982; Luise White, *Speaking with Vampires: Rumor and History in Colonial Africa* (Berkeley: University of California Press, 2000).

33. Sarah Vos and Ryan Alessi, "HPV Vaccine Bill Blocked for Now: House GOP Leader Says It Needs Cost Analysis," *Lexington (KY) Herald-Leader*, Feb. 20, 2008.

34. "The 'State' of Cervical Cancer Prevention in America," 29.

35. Ruby L. Bailey, "Why Aren't Girls Getting the HPV Vaccine?" *Detroit Free Press*, March 20, 2007.

36. It appears, however, that some doctors in Tanzania have recently been advocating the incorporation of an HPV vaccine into the standard childhood immunization package. Stacey Langwick, personal communication, Oct. 2008.

37. Saul and Pollack, "Furor on Rush to Require Cervical Cancer Vaccine."

38. Janet Elliott and Todd Ackerman, "Perry Orders Vaccine for Young Girls," *Houston Chronicle*, Feb. 3, 2007.

39. Ed Silverman, "Parents Balk at Idea of STD Vaccine for Kids," *Muskegon (MI) Chronicle*, April 12, 2005, D2.

40. Bailey, "Why Aren't More Girls Getting the HPV Vaccine?"

41. Susan Levine, "Parents Question HPV Vaccine, Push to Mandate Shots Rapidly Creates Backlash," *Washington Post*, March 2, 2007, C1.

42. "Cancer Vaccine Bills Stall," A10.

43. David Tuller, "New Vaccine for Cervical Cancer Could Prove Useful in Men, Too," *New York Times*, Jan. 30, 2007, 5. "The cervix is similar biologically to the anus, so there's plenty of hope that it will work there also," noted a professor of medicine at the University of California, San Francisco. "The anal cancer rate for gay men is similar to cervical cancer rates before the advent of Pap smears," added Tuller.

44. "STD Rate Keeps Rising in State," *Minneapolis Star Tribune*, March 31, 2008, 1A. The expert on adolescent health was Marla Eisenberg, an assistant professor at the University of Minnesota.

45. Jenny Lindsay, "The Politics of Population Control in Namibia," in *Women and Health in Africa*, ed. Meredeth Turshen (Trenton, NJ: Africa World Press: 1991), 143–168; Amy Kaler, *Running after Pills: Gender, Politics, and Contraception in Colonial Rhodesia* (Portsmouth, NH: Heinemann, 2003).

46. Pamela Feldman-Savelsberg, Flavien Ndonko, and Bergis Schmidt-Ehry, "Sterilizing Vaccines or the Politics of the Womb: Retrospective Study of a Rumor in Cameroon," *Medical Anthropology Quarterly* 14, no. 2 (2000): 159–179, quotation on 166.

Public Discourses and Policymaking

The HPV Vaccination from the European Perspective

Andrea Stöckl

By 2007 most "old" member states of the European Union (those states that either had been founding members or had joined the European Union by 1995) had introduced the human papillomavirus vaccine into their national public health programs or had recommended that the general public undergo the series of vaccinations. The United Kingdom, Germany, France, Italy, Denmark, the Netherlands, and Spain provided the vaccine through their respective health organizations, whereas in Austria the vaccine could be purchased in pharmacies even though its cost would not be reimbursed by health insurance companies. These countries have been the first in the European Union to legalize the HPV vaccines, Gardasil (developed and produced by Sanofi Pasteur MSD, the European branch of Merck, in 2006) and Cervarix (developed and produced by GlaxoSmithKline in 2007). Following legalization, the United Kingdom and Germany acted on the recommendations of their public health advisory bodies to provide the public with free or subsidized HPV vaccination. These nations, along with Italy, then took the lead in using public funds for the implementation of vaccination programs for their young female population against the HPV infection.[1]

Despite these common public health policies, a comparison between the British and the Continental European context shows that there are wide-ranging differences in public responses and debates that can only be attributed to culturally and socially diverse knowledge practices that in turn determine how information and innovation are distributed, socially reorganized, and restructured. In this chapter I compare health policies in the United Kingdom, Germany, Austria,[2] and Italy. Further, I analyze lay and expert debates surrounding the introduction of HPV vaccines in these countries (for the debates in France, see chapter 15). My hypothesis rests on the assumption that medicines have social lives. By comparing several countries, with different cultures of preventive medicine, one can see that despite the arguments for an evidence-based internationalized medicine, cultural background influences the prevention of infectious diseases. The way societies organize and institutionalize public health plays a huge role in how new medical technologies are discussed and socially coproduced.

Vaccines are a special kind of materia medica or "material things of therapy," to use the language of Susan Reynolds Whyte and colleagues (2002, 3). They are not therapies but hold the promise that therapy will never be needed. They are directed toward the future, toward possibilities, and thus encompass uncertainties, probabilities, and contingencies. The prospective materia medica, in this case the HPV vaccine, has a different meaning and impact depending on the context in which it is used. For instance, on the one hand, in English-speaking countries the HPV vaccine as materia medica was associated not only with a new way of preventing cancer, but also with female promiscuity and the moral dilemmas that parents would have to face if they decided to vaccinate their daughters. This debate has largely been absent from Continental Europe. On the other hand, while there has been a notable absence of debate in the United Kingdom on the issue of whether boys should be included in the vaccination program, this debate occurred in both Germany and Austria. Only Germany and Austria have included boys in their recommendations for the vaccine. There are clearly differences in the way new vaccines are discussed and introduced. However, the biggest difference in comparison to the United States is the absence of advertisements for the prospective Gardasil or Cervarix user: the European Union still forbids direct advertisements for prescription drugs to the public.[3] Thus, the European cases reveal different strategies of convincing the public about the necessity for the HPV vaccine.

Methods and Assumptions

In order to understand the diversity of the debates on the introduction of the HPV vaccine in Europe, I performed a content analysis (Krippendorff 2004) of the following material: selected newspaper reports from all countries from 2006 up to November 2008 (when this chapter was written) and official information on Web sites sponsored by governments, such as the National Health Service (NHS) in the United Kingdom and the Ministero della Salute in Italy; charity-based research organizations, such as Cancer Research UK, Deutsches Krebsforschungszentrum (DKFZ), and Österreichische Krebshilfe; and other national organizations dedicated to health matters, for instance, italiasalute.it. This list of sources is by no means exhaustive, but within this selected group, I have tried to balance left-leaning newspapers such as the *Guardian* and center-right-wing newspapers such as the *Times*. For instance, sexual conduct is discussed much more in the *Guardian* than it is in the *Times*, the former catering to a younger readership whose lifestyle includes being open about sexuality, and the latter catering to a readership that considers sexual conduct a private matter. Similarly, the Austrian newspaper *Der Standard* is aimed at a left-leaning, younger readership whose lifestyle supposedly revolves around enjoying life, eating well, traveling, and being open about sexuality, while also entertaining concerns about the environment and about ethical conduct more generally. The Austrian *Die Presse* and the German *Die Zeit*, in contrast, are mostly read by what was once referred to as the upper bourgeoisie, those who would be called the upper middle class in English-speaking countries. A similar situation presents itself in Italy, with *La Repubblica* aimed at the urban young and the middle-aged and *Il Corriere Della Sera* at the richer, northern bourgeoisie.

My sampling strategy was to search the online archives of the newspapers that I cite here. The search terms were "HPV vaccine," "cervical cancer" (which I translated into Italian and German), and "Harald zur Hausen & DKFZ." Several questions led my inquiry. First, I wanted to know who argued for the implementation of the vaccination and how the issue of providing the budget for the vaccination was resolved. I was curious about this issue because not every country has the same health care system: in some countries, such as the United Kingdom, patients are dependent on primary care provided by General Practitioners (GPs), whereas in other countries patients can go directly to a specialized clinical practitioner of choice, such as a pediatrician or a gynecologist.

My second set of questions had to do with notions of knowledge distribution and awareness: What do stakeholders know, and what do parents know? Who was given the task of informing parents and their daughters of a relevant age? This issue is linked to the question of what kind of "public" is created (see Irwin and Michael 2003); who recommends what to whom, and how is the public supposed to react? Because direct consumer advertising for prescription drugs is not allowed, awareness that there is a new drug available that would prevent cancer has to be achieved differently in Europe. However, as has been discussed widely by science and technology scholars (e.g., Wynne 2005), no scientific endeavor is undertaken in a social vacuum, and no product of research is value-free. On the contrary, scientific products, and especially pharmacological products, are endeavors that compress labor, hope, and potential into tiny pills or liquid serums. If we compare the current situation to times when scientific advances were less discussed by the public, one could argue that there now exists a heightened awareness of risk and uncertainty that is inherent in any scientific innovation. Policymakers' awareness of this context has led to the utilization of the public in all attempts at introducing scientific novelties, and especially new materia medica. We can thus observe a shift in which the so-called public is not only informed of new medical technologies but also involved in making them socially acceptable.

My third line of inquiry was focused on questions of gender and lifestyle that had long dominated the Anglo-American discussion on the vaccine's introduction: would Europe replicate the U.S. debate on HPV vaccination as 'another passport' to more promiscuity among young girls and women? If not, what debates would substitute for this one that ultimately had to do with the usage and construction of new gender stereotypes? Why were there provisions and recommendations for boys' vaccination in some countries but not in others?

In subsequent sections, I briefly describe the historical and social context of each country and then focus on one discourse that epitomizes the debate in that context. Vaccines as materia medica per se elicit controversies and debates because they reverse one of the pillars of the relationship between medicine, the state, and society (see Colgrove 2006). Instead of offering a cure, they force members of a given society to take action for themselves before the possible onset or outbreak of a disease can even be imagined on a personal level. Persuading the public to participate in a vaccination program is thus never a straightforward public health issue. This situation has been exacerbated in societies of late modernity, which define themselves as knowledge societies: members of the public have to know why they participate in vaccination programs, and they have to be informed of possible risks.

The National Health Service and the HPV Vaccine Debate in the United Kingdom

Transparency is one of the mainstays of modern public health policy in the United Kingdom, but it has nonetheless created much distrust (see O'Neill 2002b). The introduction of the HPV vaccine was overshadowed by prior British debates surrounding the measles, mumps, and rubella (MMR) vaccine—a late-modern debate that united several elements in an explosive assemblage (see Casiday 2007 on MMR; Brownlie and Howson 2006 on trust and immunization work; Poltorak et al. 2005 on MMR and vaccination choices; and Hobson-West 2007 on resistance to childhood vaccination).[4] This assemblage was impacted by an already developing distrust vis-à-vis public health measures (such as vaccination programs) as well as by the repercussions of the 1990s scandals in Liverpool's Alder Hey Hospital and at the Bristol Royal Infirmary, in which children's organs and tissue were used for research after their deaths without asking for parental consent. The relationship between medical authorities and parents was thus already strained when the MMR vaccine controversy broke out in 1998 (see Salter 2004).

The MMR controversy was sparked by a paper published by Andrew Wakefield in the *Lancet* in 1998. Wakefield claimed, based on a study of twelve children, that the rising incidence of autism diagnoses in children was linked to the introduction of the MMR vaccine. Wakefield's study raised the question whether the vaccine against measles, mumps, and rubella should be given all in one dose or as single doses, as recommended by the *Lancet* report that Wakefield had written. The MMR controversy, which had its heyday in 2001 and 2002, has since been resolved from the perspective of medical research. Casiday (2007) argues that the MMR controversy in the United Kingdom was about not only the management of children's and parents' decision-making processes, but also about the tension between private and public good. The triple dose of the MMR is covered by the NHS, the United Kingdom's publicly funded health service, whereas the single dose has to be paid for by parents themselves. Despite the debunking of Andrew Wakefield's theory, levels of immunization never got back to the same level as before the controversy. And there have been numerous outbreaks of measles reported in the United Kingdom ever since the MMR crisis.

In hindsight, the MMR debate epitomized the public's distrust of the New Labour government and the invasion of Iraq. The MMR debate was, one could argue, a metonym for the general mood in the millennial United Kingdom. Tony Blair, then prime minister, never disclosed whether he had vaccinated his own son, who was an infant at the time, and the prime minister was thus heavily criticized for being a hypocrite and for playing a double game. On top of this, his wife, Cherie Blair, was also criticized for consulting "health gurus" who took a radical anti-MMR-vaccine stance.[5] Consulting health gurus rather than taking the advice of the national health system was something that only well-to-do middle-class people would consider and, more importantly, could afford. The MMR vaccine controversy thus said a lot about the changing relationship between the state and its citizens in the early years of the new millennium: parents who had to make a decision about the MMR vaccine felt either that the state was letting them down by not taking care of them or that it would be wrong to do what the state in its role of caretaker had asked them to do. The MMR debate was thus not just about classical themes of late modernity such as risk perception and management of uncertainty. One could also read it against the backdrop of a changing idea of what the state, and with it public health, should provide for its citizens and how much it should protect them from disease.

The MMR controversy greatly influenced the introduction of the HPV vaccine in the United Kingdom. By 2008, there was some initial epidemiological evidence of a trial introduction in some schools in Manchester in 2007, before the vaccination program was rolled out nationwide (Brabin et al. 2008). This research, published in the *British Medical Journal* (*BMJ*), suggested that the MMR controversy had influenced the public's trust in vaccination programs; however, parental opposition to the HPV vaccine was based on concerns about the vaccine's long-term safety and questions about whether or not one could trust public health officials. The *BMJ* article also indicated that in the United Kingdom a national vaccination program was to start in September 2008. Schoolgirls aged 12 and 13 were to be offered the HPV vaccine by their respective primary care trusts, delivered through their schools. Trials had already been carried out in the Greater Manchester area. Then, starting in autumn 2009, a two-year catch-up campaign would vaccinate girls up to 18 years of age. This campaign would offer to vaccinate girls between ages 16 and 18 beginning in autumn 2009, and girls between 15 and 17 beginning in autumn 2010. According to NHS Direct, an information Web site set up to inform the general public on medical matters and prevention of disease, there was to be no vaccination for women over the age of 18 in 2008 "because it would not be cost effective in preventing cervical cancer."[6]

The NHS Direct Web site,[7] which has been specially designed to impart information on HPV vaccination, is divided into eleven sections: the first section is a general introduction, the second gives reasons "why it should be done," the third is about "facts" and "who can use it," and following sections explain how vaccination is performed, contain information on side effects, and offer cautions. The style of address changes from an informational style to directly addressing the audience: several sections address the reader as "you," especially when dealing with sexual health: for instance, "You should not use this vaccination if you have a weakened immune system." A leaflet distributed by the NHS to inform primary care doctors makes subtle statements about women's sexual behavior and warns that "it is common for women to state that their partner is their only ever sexual partner and that their partner states that the woman is his only sexual partner. . . . Yet we know that women in some such relationships test HPV positive."

Up until June 2008, it seemed that Sanofi Pasteur's Gardasil would be used as the vaccine of choice by the Department of Health. The Joint Committee on Vaccination and Immunisation had not made any recommendations regarding which vaccine brand should be used. However, in June 2008 news emerged that the contract would go to GlaxoSmithKline's Cervarix, even though this vaccine offers protection against only two types of HPV and Gardasil protects against four different strains. This news prompted the *BMJ* to report in its news section that "an opportunity was missed in choice of cervical cancer vaccine" and that the government would prefer to save money over lives.[8]

Further, there is no provision for boys and men to be vaccinated with either Gardasil or Cervarix. The possibility is not seriously discussed. The most recent reason for this policy decision is given on the updated NHS information Web site, which states that "the vaccine's effectiveness in adult men aged 16 and above has not been evaluated. Infection with high risk HPV does not cause symptoms in men. HPV infection has fewer implications for cancer developing in men. Cancer of the penis is rare, compared to other forms of male cancer."[9]

So what factors have shaped the uptake of the HPV vaccine in the United Kingdom? Consider studies that examine public knowledge of the link between the HPV infection and cervical cancer in the United Kingdom: in August 2007, the BBC reported a study published in the *British Journal of Cancer* that claimed only 2.5 percent of women in a survey of sixteen hundred women were aware of the link. The BBC reporter concluded, "Experts said the public needed to be better informed before widespread vaccination was introduced."[10] However, the 2.5 percent awareness was already a major improvement over that found by a study carried out in 2002, in which only 0.9 percent of women made a link between HPV and cervical cancer. The authors suggest that noninformation causes a major problem because women who are not fully aware of the link cannot give informed consent. Furthermore, most people who were interviewed were not aware that cervical cancer is actually a sexually transmitted disease (Marlow, Waller, and Wardle 2007).

The 2008 study I mentioned before, published in the *BMJ* about uptake of the trial vaccination in Manchester, showed that 20 percent of parents refused the vaccination of their daughters. Of the 20 percent, 54 percent gave "too many unknowns" as a reason for refusal, 47 percent were worried about the safety of the vaccine, especially the long-term safety, and only 4 percent argued that the vaccine "condones sexual activity" (Brabin et al. 2008). Even though this study explicitly states that uptake rates were not so bad and that parents' biggest worries were about long-term safety, the *Guardian* reports that "some may have concerns that allowing vaccination may promote promiscuity, because the cancer-causing virus which the vaccination targets is passed on in sexual intercourse."[11] At least in the press, if not more generally, the theme of sexual promiscuity seems to linger on, despite contradictory evidence that parents are more concerned about safety in the wake of the MMR debacle of recent years.

The HPV Vaccine in Germany and Austria

The situation in Germany and Austria is slightly different from that in the United Kingdom, because even though there is opposition to vaccination programs, mostly on religious grounds, there have been no big vaccination controversies over the past few years. The only minor point of controversy in Germany was on an issue raised at the introduction of the HPV vaccination program, and even though it does not concern the issue of sexuality, it says a lot about the globalization of materia medica. This issue concerned Harald zur Hausen, who discovered the link between cervical cancer and the human papillomavirus in the early 1970s and who published his research in 1976, leading to the development of the vaccine. The German newspaper *Die Zeit* repeatedly reported on the fact that the German research success that first linked a type of cancer to a virus had not been acknowledged. This distress was immediately put to rest by the international recognition of Harald zur Hausen as one of the recipients of the Nobel Prize for medicine in 2008.[12]

Research suggests that in Germany, sixty-five hundred women are diagnosed with cervical cancer each year. According to the Robert Koch Institute, 70 percent of sexually active women test HPV positive and 70 to 90 percent of these are no longer infected with the virus after two years. Of all infections, 74 percent occur in women from 15 to 24 years old. Gardasil and Cervarix were both admitted to the German and Austrian pharmaceutical market in 2007. The Ständige Impfkommission (standing committee on vaccination; STIKO) of the Robert Koch Institute strongly recommended the implementation of a coordinated immunization program of HPV vaccination involving physicians, pediatricians, gynecologists, and public health practitioners, to assure the immunization of 12-to-17-year-old girls before their first sexual intercourse. In Germany, 72 percent of girls of age 16 have already seen a gynecologist. The STIKO recommended that the three vaccinations be given before the first sexual activity of girls, but, in contrast to the authorities in the United Kingdom, they also admitted that it is not yet known for how long immunity is granted.[13] Additionally, the institute conceded that there are no precise epidemiological data on the occurrence of HPV infection in Germany; thus, recommendations for the vaccination are based on epidemiological data from the United States. Because Germany's political system is federal (i.e., the various federal states have a certain autonomy when it comes to decision making on public health matters), there is no nationwide immunization program.[14]

In Austria, as in Germany, both HPV vaccines are recommended, but the Austrian public health officials have decided not to finance the vaccination through the national insurance system. Girls and young women have to pay around 600 euros for all three vaccines, should they want to undergo the tripartite vaccination program. Yet, despite the high costs for the vaccine, it was reported that by February 2008, already thirty thousand young women had acted on the recommendation.[15] Yet, some newspapers argue that the Austrian population is considered to be one of the least "vaccine-friendly" publics of the European Union. Only 17 percent of Austrians are vaccinated against the influenza virus, and Austria is one of the countries of the European Union, together with Portugal, that has the least-vaccinated population. No vaccination in Austria is mandatory, but the Oberste Sanitätsrat, the Austrian advisory committee to the Parliament on public health issues, has argued that being vaccinated is a "moral obligation" and that people who work with the public should be vaccinated against Hepatitis A, for instance.[16]

Despite this criticism, the then Austrian health secretary, Andrea Kdolsky, who was in Parliament when the HPV vaccine was introduced,[17] argued that the long-term effects were not yet well established and that Pap smear testing was a more cost-effective and less invasive technology.[18] Austrian health economists had suggested that it would be necessary to vaccinate all girls and all boys to eliminate the virus completely. The health economists argued that even if an entire cohort of boys and girls was vaccinated, the benefits would not be felt until twenty years later; furthermore, the practice would have to be continued for more than fifty years to gain a reduction of 70 percent in newly diagnosed cervical cancer.[19] The Austrian health minster therefore decided not to incorporate the HPV vaccine into the national vaccination program, and even though the vaccination is recommended for both girls and boys, it is not covered by the national insurance companies.[20] As a result, Austria is one of the few countries in the European Union in which the HPV vaccine is not considered to be beneficial for the "greater good" and where the costs are not borne by the general public.

In contrast to public discourse in the United Kingdom, German and Austrian information leaflets (for instance, the one designed by the German center for cancer research [DKFZ] and by the Austrian Krebshilfe) point out that boys and men can transmit the infection and should ideally also be vaccinated, especially because this way they could be simultaneously protected from anal carcinomas and warts.[21] Going a step further, an Austrian Web site dedicated to women's health argues that there should be an explanation as to why boys are not included in the vaccination programs, given that HPV can cause genital warts in boys and men and that herd immunity can be attained only by including boys in the programs. The authors of the Web site argue that when the focus is on women and girls only, the female population is held responsible for the reproductive health of both sexes.[22] Even Harald zur Hausen has repeatedly advocated for the vaccination of boys. He goes so far as to claim that the vaccination program makes sense only if boys are included.[23]

There were some well-known cases that influenced policymaking and public discussion in Austria and had repercussions in Germany as well. One case was of a girl who suffered serious complications after receiving the HPV vaccine; she was affected by acute disseminated encephalomyelitis and had to be

hospitalized.[24] Another girl died three weeks after she received the first vaccination in January 2008.[25] GlaxoSmithKline, the manufacturer of Cervarix,
with which the girls had been vaccinated, denied any link between their vaccine and the adverse effects and the death. Eventually, the European Medicines
Agency declared that the death had no relation to the girl's vaccination and that
the vaccine was safe.[26] Yet, by April 2008, the debate was far from concluded,
especially since the publication of an open letter on the Internet by the parents
of the girl who died. In their letter, the parents repeated their claim that their
daughter's death was directly linked to the vaccine.[27] They also repeated the
claim in a letter to the newspaper *Salzburger Nachrichten* and argued that the
death of their daughter had never been fully explained and that the links between the "vaccine lobby," the department of health, and the "vaccine experts"
were "harrowing."[28] Even though the link between the girl's death and the previous vaccination had never been established,[29] Austrian newspapers reported
that Austrian and German parents felt "completely left alone" by the public
health authorities when it came to deciding whether to allow their daughters to
participate in the vaccination program.[30]

Italy's National Vaccination Program

Italy's health system, like most health systems of Continental Europe, is insurance based; that is, employers have to pay for health insurance for their employees. Provision of health care is notoriously controversial in Italy, and that
fact is perhaps why in 2006 the then health secretary, Livia Turco, took pride
that Italy would be the first state within the European Union where the HPV
vaccine would be provided for all girls free of charge, beginning in February of
that year. According to her, there were about two hundred eighty thousand
12-year-old girls in Italy in 2006, and the cost of providing initial and follow-up
vaccine was estimated at 75 million euros per year. Because the vaccination
program was seen as a public health issue, it was felt that its cost should be paid
for by the state and its citizens.

As in Austria and Germany, gynecologists or even pediatricians would provide the vaccine. Even though Italy had had its own small "MMR" scandal (in
2003, parents were afraid that children allergic to egg protein would have an
allergic reaction to the morbillo-parotite e rosolia [MPR] vaccine,[31] and a boy-
cott of the then mandatory MPR vaccination program ensued), the problem was
not that the parents were opposed to the vaccine but rather that large numbers

were uninformed about it and about why it was needed. *Corriere della Sera*, a center-right-wing Italian newspaper, reported that at the start of 2008, 47 percent of mothers who had a teenage daughter had heard of the vaccine and were informed about the link between HPV infection and cervical cancer.[32] However, 63 percent of women who did not have children in that age group were unaware that HPV could also be the cause for cervical cancer, and among women under age 25, 82 percent were reported not to know that there was a link. Expert knowledge was also limited in Italy: a research paper in the journal *Vaccine* (Esposito et al. 2007) reports that in 2006 there was very scant knowledge of HPV among Italian medical practitioners. Only 45 percent of surveyed pediatricians knew that HPV was a DNA virus and knew about its transmission and preventing transmission. Despite this scant knowledge of the link between HPV and cervical cancer, the president of the Societá Italiana di Medicina Generale argued that the introduction of the HPV vaccine should not lead anybody to make a statement about women's lifestyles. He explicitly pointed out that women should be persuaded to participate in the vaccination program and should be informed that this vaccine would also protect them from other consequences of having acquired the HPV virus.[33] The debate intensified when *Corriere della Sera* reported in May 2008 that it would be advisable to also include boys in the vaccination program, because not only would they be protected from genital-urinary cancers, including anal cancer, but also such vaccination would in turn possibly protect a future female partner.[34]

Conclusion

As we can see, the debates and controversies about introducing the HPV vaccine take on very different forms in each analyzed cultural field. In Germany and the United Kingdom, public health officials recommend vaccination at age 12; in Austria the recommended age is 9. Further, while in the United Kingdom primary care doctors and schools provide the main information as well as vaccination, in Germany and Austria it is gynecologists and in Italy it is pediatricians and gynecologists who do so. In the UK system, anyone who seeks medical treatment has to go through a general practitioner first. The GP will make suggestions to the patient, and, ideally, the GP and the patient will come to a shared decision. In Continental Europe, people can choose a medical professional specialized in a certain area of medicine without having to go through the GP route. As a result of this greater autonomy, in Germany and Austria most women have seen a gynecologist by the age of 16 and are thus familiar with "sexually laden" medical consultations. In Continental Europe, going to see a gynecologist with an adolescent daughter is part of the growing-up ritual, whereas in the United Kingdom sexual health and education are more a question of schools, GPs, and charities. Yet, there is also an element that these countries share in common: in all these contexts, parents are concerned about long-term safety and about the effects that vaccination might have on their children's future lives.

As mentioned in the introduction to this chapter, the question of lifestyle and the sexuality of young girls is a major issue in the Anglo-American debate. However, in the German-speaking world and in Italy there is virtually no debate on lifestyle at all. The reason might be that the knowledge that HPV is linked to cervical cancer, and thus a sexually transmitted disease, is not yet as common in the population as it is in Anglo-American societies. The analyzed materials from Germany and Austria do not even mention a relationship between lifestyle and HPV infection. On the other hand, in the United Kingdom there is no debate about whether boys should be vaccinated, and indeed vaccination is not provided for boys, not even on a private level.

European societies define themselves as knowledge societies and pride themselves on having an informed and educated public. Not only is knowledge seen as a major commodity that keeps the economy afloat (Felt and Wynne 2007), but most member states of the European Union would also define the relationship between their expert advisory committees and their publics as based on the free flow of information. For instance, it has become standard practice in the advisory bodies to the government of the United Kingdom to include a lay member of the public among its "experts." And yet, the Joint Committee on Vaccination and Immunisation is a purely expert panel whose membership is announced but whose debates are secret until the time its members make a decision and give a recommendation. Such a practice is striking in times when transparency is a major topic of public concern.

I suggest that further research and debates are needed. There are four points that require further inquiry: First, we need to understand why there is no debate on lifestyle, early maturity, and sexuality in German-speaking countries and Italy. Is it because girls are expected to seek a consultation with a gynecologist anyway once they enter puberty? Or is it because a gynecologist is seen as a trustworthy expert, whereas a GP is "part of the family" and thus discussions on sexuality are actually more difficult to conduct with the GP? Next, we need to analyze why there is such a big debate in Anglo-American countries about sexuality, lifestyle, and HPV vaccination as a gateway to promiscuity, while this debate is nearly absent in German-speaking countries and Italy. A possible cause might be the scant knowledge about the link between HPV infection and cervical cancer as a sexually transmitted disease. But some emerging research from the United Kingdom suggests that the concern with having to introduce young girls prematurely to the thought of sexual intercourse might have been media hype. As already mentioned, parents everywhere seem far more concerned about safety and long-term implications. This leads us to a third topic of research, namely how much is known about the link between HPV infection and cervical cancer among the population in general. The knowledge of the link between HPV infection and cervical cancer will definitely spread within the months and years to come.

Lastly, research ought to be concerned with understanding who will likely oppose the vaccination and on what grounds. As we have seen, in all four countries the economic cost-effectiveness issue is already an argument, expressed both by public health officials and by the general public. As with other health movements, the Internet plays a major role in the way that opposition is formed. The Internet is used as a forum to raise awareness and concerns, which then usually lead to a warning against the vaccine based on safety.[35]

In conclusion, it might be argued that the debates surrounding the introduction of the HPV vaccine in Europe are largely focused on the relationship between the state and its citizens and on questions of transparency. In this the European debate differs fundamentally from the U.S. debate, in which personal responsibility for a moral lifestyle is at the core of public criticism.

NOTES

1. www.medicalnewstoday.com/articles/85624.php.

2. I look at Germany's and Austria's policy together because even though they have different policies, they share similar public debates. What happens in one country is taken up by the other country.

3. www.haiweb.org/campaign/DTCA/BMintzes_en.pdf.

4. I use the term *assemblage* in the same sense as Manual de Landa (2006); that is, assemblages are historically contingent actual entities whose elements are in constant fluctuation and have to be analyzed in relation to each other and to the new assemblages they form.

5. www.guardian.co.uk/society/2008/aug/30/mmr.health.media.

6. www.nhsdirect.nhs.uk/articles/article.aspx?articleId=2336&sectionId=161.

7. www.nhsdirect.nhs.uk/articles/article.aspx?articleId=2336&sectionId=159.

8. *British Medical Journal* 336:1456–1457.

9. http://cks.library.nhs.uk/patient_information_leaflet/hpv_vaccination.

10. http://news.bbc.co.uk/2/low/health/6940478.stm.

11. www.guardian.co.uk/society/2008/apr/25/health.cancer.

12. He shares the Nobel prize with Luc Montaigner and Françoise Barré-Sinoussi, both virologists who discovered the AIDS virus. www.zeit.de/2008/42/M-Nobel-Viren.

13. www.rki.de/cln_048/nn_205760/DE/Content/Infekt/EpidBull/Archiv/2007/12__07,templateId=raw,property=publicationFile.pdf/12_07.pdf. The Robert Koch Institute is the official governmental institution of Germany concerned with infectious and noninfectious diseases.

14. www.eurosurveillance.org/ViewArticle.aspx?ArticleId=3169.

15. www.aekwien.or.at/media/impfen_PKImpftag.pdf.

16. http://diepresse.com/home/panorama/oesterreich/357628/index.do?from=si marchiv.

17. As of summer 2008, there is a new minster of health in the Austrian government.

18. http://diestandard.at/?url=/?id=1200563120711.

19. www.derstandard.at, accessed May 13, 2008.

20. www.krebshilfe.net/information/vorsorge/frauen/hpv.shtm#01.

21. www.krebsinformationsdienst.de/themen/vorbeugung/hpv-impfung.php.

22. www.kfunigraz.ac.at/dafbwww/HPVfakten.pdf.

23. www.medizin-welt.info/aktuell/aktuell.asp?newsID=122.

24. See www.derstandard.at, from Feb. 1, 2008.

25. http://ooe.orf.at/stories/249675/.

26. www.kleinezeitung.at/nachrichten/chronik/470562/index.do.

27. www.aekwien.or.at/news_pdf/5818_1.pdf.

28. *Salzburger Nachrichten*, April 14, 2008.

29. www.frauenaerzte-im-netz.de/de_news_652_1_373.html.

30. See www.derstandard.at, from Jan. 18, 2008.

31. www.epicentro.iss.it/focus/morbillo/Vaccino_Morbillo.pdf.

32. www.corriere.it/sportello-cancro/articoli/2007/12_Dicembre/06/hpv_vaccino.shtml.

33. Claudio Cricelli, interview on www.italiasalute.leonardo.it/news2pag.asp?ID=7912.

34. www.corriere.it/sportello-cancro/articoli/2008/05_Maggio/26/hpv_maschi.shtml.

35. The Internet-based patient advocacy group Zentrum der Gesundheit argues that the HPV vaccine is only introduced into societies by politicians who are corrupted by pharmaceutical companies.

REFERENCES

Brabin, L., S. A. Roberts, R. Stretch, D. Baxter, G. Chambers, H. Kitchener, and R. McCann. 2008. "Uptake of First Two Doses of Human Papillomavirus Vaccine by Adolescent Schoolgirls in Manchester: Prospective Cohort Study." *British Medical Journal* 336, no. 7652: 1056–1058.

Brownlie, J., and A. Howson. 2006. "'Between the Demands of Truth and Government': Health Practitioners, Trust, and Immunisation Work." *Social Science and Medicine* 62:433–443.

Casiday, R. E. 2007. "Children's Health and the Social Theory of Risk: Insights from the British Measles, Mumps, and Rubella (MMR) Controversy." *Social Science and Medicine* 65:1059–1070.

Colgrove, J. 2006. *State of Immunity: The Politics of Vaccination in Twentieth-Century America*. Berkeley: University of California Press.

De Landa, M. 2006. *A New Philosophy of Science: Assemblage Theory and Social Complexity*. New York: Continuum International.

Esposito, S., S. Bosis, C. Pelucchi, E. Begliatti, A. Rognoni, M. Bellasio, F. Tel, S. Consolo, and N. Principi. 2007. "Pediatrician Knowledge and Attitudes regarding Human Papillomavirus Disease and Its Prevention. *Vaccine* 25:6437–6446.

Felt, U., and B. Wynne. 2007. "Taking European Knowledge Society Seriously: Report of the Expert Group on Science and Governance." Science, Economy and Society Directorate, European Commission, European Union, Brussels.

Hobson-West, P. 2007. "'Trusting Blindly Can Be the Biggest Risk of All': Organised Resistance to Childhood Vaccination in the UK." *Sociology of Health and Illness* 29, no. 2: 198–215.

Irwin, A., and M. Michael. 2003. *Science, Social Theory, and Public Knowledge*. Maidenhead, PA: Open University Press.

Krippendorff, K. 2004. *Content Analysis. An Introduction to Its Methodology*. Thousand Oaks, CA: Sage.

Marlow, L. A. V., J. Waller, and J. Wardle. 2007. "Public Awareness That HPV Is a Risk Factor for Cervical Cancer." *British Journal of Cancer* 97:691–694.

O'Neill, O. 2002a. *Autonomy and Trust in Bioethics*. Cambridge: Cambridge University Press.

———. 2002b. *A Question of Trust*. Cambridge: Cambridge University Press.

Poltorak, M., M. Leach, J. Fairhead, and J. Cassell. 2005. "'MMR Talk' and Vaccination Choices: An Ethnographic Study in Brighton." *Social Science and Medicine* 61:709–719.

Reynolds Whyte, S., S. van der Geest, and A. Hardon. 2002. *Social Lives of Medicines*. Cambridge: Cambridge University Press.

Salter, B. 2004. *The New Politics of Medicine*. Basingstoke, England: Palgrave MacMillan.

Wynne, B. 2005. "Reflexing Complexity: Post-Genomic Knowledge and Reductionist Returns in Public Science." *Theory, Culture, and Society* 22, no. 5: 67–94.

HPV Vaccination in Context

A View from France

Ilana Löwy

Medical innovations need to make a place for themselves among prevailing concepts, practices, professional traditions, material constraints, institutional variables, beliefs, attitudes, and power relations. They are also, to a large extent, local developments. The globalization and homogenization of biomedical knowledge was not accompanied by a parallel homogenization of medical practices. Even countries with comparable levels of income and medical infrastructures frequently develop divergent approaches to disease prevention and treatment.

Debates about human papillomavirus vaccination typically have failed to take note of the disparate patterns of dissemination of the HPV vaccine in industrialized countries—an omission that Andrea Stöckl successfully challenges in chapter 14. Stöckl reveals important differences in the reception of the HPV vaccine in the United Kingdom, Germany, Austria, and Italy. This current chapter identifies possible reasons for such differences, based on an in-depth analysis of a single case: the reactions to the vaccine in France. To understand the way professionals and the lay public in France reacted to the promise to control cervical cancer through vaccination, I temporarily put aside the global debates on broad themes such as sexuality and morality, individual versus collective responsibility, ethical aspects of mandatory health measures, adolescent sexuality and parental rights, and the risk-to-benefit ratio of medical innovations with unknown long-term consequences. Instead, I focus on specific topics such as local medical and public health traditions as well as on national variables and contingent developments. I propose that the introduction of HPV vaccines in France was significantly shaped by two earlier histories: the difficulties encountered in attempting to implement a national plan to screen for cervical cancer and the turbulent history of the introduction of hepatitis B vaccination in the 1990s. The failure to ensure that all French women undergo regular cervical smears stimulated a search for alternative ways to reduce the risk of cervical cancer—thus making the HPV a source of much interest within the medical community. However, the recent attempts to vaccinate all French school children against hepatitis B dampened enthusiasm for another nationwide campaign of vaccination of children against a sexually transmitted disease.

Exfoliate Cytology in France: A Patchy Landscape

One of the arguments of opponents of Gardasil is that vaccination against cervical cancer risk is not necessary, at least in Western countries.[1] Critics argue that nowadays, cervical cancer is a rare disease, which can be nearly completely prevented by appropriate public health measures. The majority of women diagnosed with cervical malignancies were not screened properly for precancerous lesions of the cervix. Countries that successfully introduced screening for such lesions, such as the Scandinavian countries, drastically reduced the prevalence of cervical cancer and the mortality from this disease. Moreover, as all the experts emphasize, vaccination will not abolish the need for regular vaginal smears. Why should one introduce an expensive and potentially risky public health measure instead of investing in better means of disseminating an existing and proven approach and reinforcing other reproductive health programs that, in all probability, will provide better returns?

The Pap smear's cost-to-effectiveness ratio has already been widely acclaimed, thus making it an important investment in the prevention of cervical cancer. In the United States, exfoliate cytology was introduced in the late 1940s. In 1941, scientists found a correlation between the presence of abnormal cells in cervical smears and the diagnosis of cervical tumors; this finding was rapidly followed gynecologists' adoption of the Pap smear.[2] In 1947, the Boston City Hospital opened a clinic for early detection of cervical cancer, and all the departments of the hospital were requested to refer as many women over age 35 as possible to the Gynecological Clinics for screening.[3] This was probably the first use of the term *screening* in this context. Women with suspicious vaginal smears were subjected to a biopsy by curettage, a cervical biopsy, or both. In many cases these procedures led to a diagnosis of previously unsuspected malignancies. At first, the sole aim of exfoliate cytology was the detection of invasive carcinoma,[4] but it rapidly became a test to detect preinvasive, in situ lesions.[5]

In the United States, the use of Pap smear technology was strongly supported by the American Cancer Society (ACS).[6] The ACS sponsored the First National Cytology Conference (Boston, 1948) and funded the training of pathologists by George Papanicolaou, the physician who developed the vaginal smear. The ACS initiative was later supported by the National Cancer Institute and the U.S. Public Health Service. In the United Kingdom also, one of the first initiatives to widen application of the Pap smear came from a nongovernmental organization, the Medical Women Federation (MWF). In 1964, this organization set up a committee for cervical cancer screening. A Labor parliament member, Joyce Butler, joined the MWF's committee and helped to put the Pap smear issue on Parliament's agenda.[7] In 1966, screening for cervical cancer was proclaimed a national service by the British government, and the National Health Service (NHS) established local co-coordinating committees to implement such screening. Simultaneously, the NHS decided to promote the establishment of regional screening centers to centralize the collection of vaginal smears and their readings.[8]

The NHS first established a regional program of screening and then, in 1988, a national one grounded in a network mobilizing general practitioners, public health doctors, cytology laboratories, gynecologists, and oncologists.[9] The majority of UK experts are persuaded today that the Pap smear is a highly efficient public health measure and that it reflects the current understanding of the natural history of cervical cancer.[10] However some specialists continue to maintain that the efficacy of the Pap smear was never proved. There were no randomized clinical trials of the effects of screening for cervical cancer on mortality from this disease, and while there is undoubtedly a sharp decline in the incidence of invasive cervical cancer in all industrialized countries, the decline started in the 1930s, well before the introduction of mass screening. It may be related to changes in lifestyle, biological properties of HPV that affect host-parasite relationships, or both. The incidence of another malignancy induced by an infectious agent, stomach cancer (associated with infection by the bacterium *Helicobacter pylori*) declined precipitously in the majority of Western countries, without any medical intervention.[11]

In France, the introduction of the Pap smear was not sustained by initiatives originated outside the medical profession, as it was in the United States and the United Kingdom. Professional organizations of gynecologists recommended systematic Pap smears, but the issue was discussed only among professionals. Neither cancer activists nor women's health activists supported their initiative. In the 1970s, debates on a national screening program for cervical cancer began at the instigation of the French Health Ministry. A 1975 report of the ministry's Cancer Commission advocated a nationwide campaign for the screening of cervical cancer destined to cover 50 percent of the target population in its first year. In order to keep costs in check, the commission proposed that women with a first normal smear be screened only every five years. The report modeled the costs of a nationwide screening campaign under these conditions, compared it with the burden of cancer, and concluded that the screening was cost-effective.[12]

An additional report discussed the training of cytology technicians. It stated that such training was insufficiently regulated and proposed opening a state-sponsored school that would provide a year-long training program for cytology technicians, with a possibility of studying for another year for a diploma of higher rank. Cytology technicians were expected to provide a true "screening"; that is, they would have to screen out all the obviously normal slides and send all the suspicious ones to a qualified pathologist, who would examine them and provide a diagnosis.[13] The recommendations of these two reports were never implemented, however. Although France has an efficient national health system, the introduction of the Pap test followed the U.S. rather than the British pattern. A national program of screening for cervical cancer in France did not started until 2003, with the introduction of the National Cancer Plan.

Screening for cervical cancer in France was nevertheless strongly encouraged by gynecologists' associations and by the Health Ministry, and its costs were fully reimbursed by the National Health Insurance (Sécurité Sociale). The ministry promoted several pilot projects in selected regions—projects that proposed a thirty-month interval between screenings and that played an important but uneven role in the propagation of cervical smears outside major cities.[14] Intervals between screenings were left to the discretion of individual doctors, and practices varied greatly.[15] In 2003, the National Cancer Plan adopted the principle of screening women every three years if they had had two consecutive normal annual smears; this recommendation reflected the opinion of many European experts.[16]

The uptake of the Pap smear in France is uneven. According to data published by the French Institut National du Cancer, 35 percent of French women have never had a Pap smear, while others have had too many tests.[17] Another particularity of the French case is the unique role of medical gynecology. Gynecologists in France, unlike those in the majority of Western countries, are not all trained in obstetrics. Some (from the 1950s on, mainly women) specialize exclusively in medical management of female disorders.[18] In the 1960s and 1970s, medical gynecologists were at the forefront of struggles for the liberalization of contraception and abortion. Simultaneously they became interested in the hormonal treatment of menstrual and menopausal disorders and in women's sexual problems.

The French configuration was thus quite different from the one in the United States. French medical gynecologists were not seen by feminists as enemies of women but rather as professionals who had elected to be on their side. French gynecologists were seldom suspected of aggressively "pushing drugs" or of trying to medicalize normal events in women's lives, and they were usually trusted by their patients. The feminization of medical gynecology helped to project the image of a profession of "women helping women." The alliance between medical gynecologists and women—including many feminists—is probably one of the reasons for the absence of a French women's health movement and the paucity of feminist criticism of medical practices. The main exception has been feminist criticism of the overmedicalization of childbirth and pressures to provide a more woman-friendly environment in maternity wards, given that gynecology-obstetrics has traditionally been a male-dominated specialty that is seen as conservative. From the late nineteenth century onward, French obstetricians were leaders in the effort to stop the decline in French population; consequently, they celebrated women's traditional role as mothers and opposed birth control and abortion.

Today, French medical gynecologists assert their authority over the supervision of all routine aspects of women's reproductive functions, with the exception of childbirth. They also promote the early detection of female cancers of the breast and uterus, and thus the performance of Pap smears. Medical gynecologists mostly practice in cities, however, and have mostly middle-class, educated clients. Women who do not regularly visit a gynecologist—those who live outside big cites and those of more modest socioeconomic strata—are expected to have their vaginal smears made by their general practitioners (GPs). But many GPs feel uneasy doing cervical smears. Moreover, in the absence of a national plan, French GPs (unlike their British colleagues) are not systematically encouraged to propose this test to their patients. Interviews conducted with GPs revealed that a significant proportion of them (especially men) are reluctant to provide Pap smears and that many find taking the smears to be difficult.[19] Women who visit medical gynecologists often have a yearly Pap smear (and thus overuse the technique, according to European standards); those who consult a GP, however, may have an insufficient number of smears, or none at all. Women who consulted a gynecologist were found to be four times more likely to be prescribed regular Pap smears than those who consulted a GP.[20]

The incidence of cervical cancer in France decreased by more than 40 percent between 1978 and 2000 (from 22 per 100,000 women in 1975 to 8 per 100,000 women in 2000).[21] The rate has stagnated since: no change of incidence was recorded between 2000 and 2005.[22] In 2005 there were 3,400 new cases of invasive cervical cancer in France, and approximately 1,000 women died from the disease. In addition, 40,000 women were treated for preinvasive cervical lesions. In its statement of December 12, 2007, the National Organization of French Gynecologists and Obstetricians deplored the insufficient implementation of the Pap smear in France. Among women diagnosed with cervical cancer in 2006, 24 percent had never had a cervical smear, and 43 percent had not had a smear in the previous three years. However, 27 percent of those who developed invasive malignancies had had a normal smear less than three years before, a finding that points to the intrinsic limits of this method (inaccurate readings of smears or the presence of "interval tumors" with a rapid growth rate). Moreover, 3 percent of women treated for cervical dysplasia developed invasive tumors in spite of the treatment. The National Organization of French Gynecologists and Obstetricians also deplored the fact that Pap tests continued to be diffused in an "opportunistic" way. That is, the test was proposed to women only when they saw their gynecologist. Such a pattern failed to reach all the women, especially those from lower socioeconomic strata. In spite of efforts to implement a national plan, in 2007 an estimated 40–45 percent of French women were not adequately screened.[23]

To sum up, the Pap smear continues to be seen as a problematic technique in France. Its uptake continues to be highly uneven; many GPs refuse to perform cervical smears; and experts estimate that the percentage of false positive and false negative results is unacceptably high. Difficulties with the implementation of the Pap test heightened the experts' interest in other solutions, including testing for the presence of carcinogenic HPV as a "first screen" (a method that could reduce the need for expert sampling) and a vaccination against HPV (an approach that could limit the need for frequent smears). Cost-benefit calculations of HPV vaccinations in France have also taken into account the lessening of the burden of the disease, the considerable costs of treating cervical intraepithelial neoplasia (CIN), and the surveillance of women diagnosed with CIN and atypical squamous cells of undetermined significance.[24] New biomedical technologies, some specialists believe, will allow France to overcome its persisting failure to organize an efficient screening for cervical cancer.[25]

The First STD Vaccine: The Tumultuous History of Hepatitis B Vaccination

In the United States, Gardasil is a product of Merck. In France, it is a product of Sanofi Pasteur. Pasteur Institute is a famous "trade name," expected to convey the aura of cutting-edge but also safe biomedical products. Historians of science may provide a more guarded view of the Pasteur tradition, however. The rabies vaccine, a medical innovation from the early days of the institute, was not always a model for the careful implementation of a new scientific approach. Nevertheless, the transformation of Louis Pasteur into a "lay saint" in the French Third Republic, together with real achievements of the institute in the domain of public health (the institute was associated with the development of an antidiphtheria serum; Bacillus Calmette-Guérin [BCG], an antituberculosis vaccine; sulfa drugs; and pioneering studies of HIV), increased the scientific credibility of the Pasteur Institute and thus of its commercial products. However, the founding of the Agence de Médicament and of the Agence Française de Sécurité Sanitaire des Produits de Santé (AFSSAPS)—organizations regulating drugs and sanitary products—coincided with a series of public "health scandals" in which the Pasteur Institute was shown to have played a direct or indirect role. For instance, there was the "contaminated blood scandal," in which the delay in introducing tests for the presence of HIV virus in the blood supply was linked to the French government's wish to promote tests developed by the Pasteur Institute; and then there was the "growth hormone scandal," in which a growth hormone produced by the Pasteur Institute in the 1980s induced Creutzfeldt-Jakob disease in its recipients, and France was especially slow in replacing a hormone originating in cadaver pituitary glands by a recombinant hormone—again, mainly to favor products of the local industry.[26]

In addition to potential skepticism about the Pasteur Institute, the problematic vaccination against hepatitis B in the 1990s is especially relevant to the debates on the introduction of the HPV vaccine in France, or rather to the absence of such debates. Vaccination against hepatitis B, energetically promoted by the French government, was later linked with the development of neurological diseases in vaccinated people.[27] The unhappy French experiences with this first attempt at mass vaccination against a sexually transmitted disease (STD) may explain the absence of pressures to make HPV vaccination obligatory or even to highly recommend it. From the eighteenth century onward, the treatment of healthy people to prevent disease was seen as a problematic endeavor. The d'Alambert-Bernoulli debate in the French Academy of Science of the 1760s already outlined some of the main dilemmas of vaccination: the choice between individual and collective good, the definition of an "acceptable risk," and the distinction between short-term dangers and long-term benefits. The history of vaccination is, to a large extent, the history of balancing contradictory aspirations.[28]

Furthermore, sexually transmitted diseases are seen as different from childhood diseases, the usual targets of vaccination campaigns. Decisions about vaccination for highly contagious childhood diseases take into account positive effects of herd immunity and negative effects of the presence of unvaccinated individuals in a given population. Thus, specialists evaluated that a 1 percent increase in the refusal of pertussis vaccine would increase by 12 percent the risk of a pertussis outbreak in the school system, a good reason for making this vaccine mandatory.[29] The transmission rate of STDs is usually slower, however, and the danger of immediate epidemic outbreak is lower. In addition, the modification of individual behavior affects to an important extent the rate of transmission of STDs (while it has a much lower effect on, e.g., airborne diseases). STDs are therefore less obvious targets for mandatory vaccination campaigns.[30]

Until recently, the question of mandatory vaccination against STDs was a theoretical one. There were no efficient vaccines against the most prominent STDs of the past, such as syphilis and gonorrhea, or of the present, HIV being the most notable one. The development of hepatitis B vaccine transformed the theoretical question into a practical one. Hepatitis B is transmitted through contaminated blood and blood products like syringes; occasionally, but rarely, through a casual contact with secretions such as saliva or tears; and above all through sexual relations that involve an exchange of body fluids. The disease is widespread in developing countries, where it is frequently vertically transmitted from mother to fetus.[31] The main groups at risk of hepatitis B are health professionals, users of injected drugs, and people who have multiple sexual encounters. By extension, all sexually active people can be seen as being at risk. The latter group, which includes nearly everybody, coupled with a confidence in the vaccine's efficacy and safety, led to the French proposal of widespread vaccination for this disease.

The first hepatitis B vaccine was developed in 1981. In 1982, the Direction Générale de la Santé (DGS), roughly, the French equivalent of the American Public Health Service, recommended the vaccination of all health professionals, but also other people at risk—including those who travel to endemic areas, those who undergo frequent blood transfusions, intravenous drug users, people with multiple sexual partners, and families of chronic carriers of hepatitis B virus. In 1992 the French Health Ministry made obligatory a test for the detection of the HB antigen (which reveals contamination with hepatitis B virus) in pregnant women, and in 1994 it promoted a large-scale vaccination campaign against hepatitis B in junior high schools. The vaccine was fully reimbursed by the French National Health Insurance.

In the 1990s, the Health Ministry energetically promoted vaccination of sixth-graders, arguing that hepatitis B is a major public health threat and a highly contagious disease that can be transmitted through casual contact, especially with saliva. The latter argument was probably intended to deflect negative associations of this disease with sexual promiscuity and the use of intravenous drugs. Simultaneously, the DGS promoted the vaccination of all newborns, and the vaccination campaigns recommended the use of the hepatitis B vaccine produced by Pasteur-Aventis. These campaigns were initially deemed a success in that, between 1994 and 1996, more than 20 million people were vaccinated in France, and the majority of the vaccinated people had no known risk factors. The campaign continued in the following years. Thanks to efficient promotion of the hepatitis B vaccine in schools, 77 percent of the sixth-graders were vaccinated in 1994–95, 73 percent in 1995–96, and 6 percent in 1997–97.[32]

Between 1994 and 1997, however, the Agence de Médicament, responsible for monitoring the effects of new pharmaceuticals, began to receive reports of acute demyelinating pathologies that were observed shortly after hepatitis B vaccination. In 1997, the Agence assigned a special committee to study possible risks of this vaccine. The systematic vaccination of preteens was suspended in 1998. Yet, the Ministry of Health recommended the continuation of voluntary vaccination of sixth-graders by family GPs and stressed that the only group that would undergo obligatory vaccination against hepatitis B were health care workers.[33]

In 2002, following reports that links had not been found between hepatitis B vaccination and multiple sclerosis, the DGS named a new working group on hepatitis B vaccination.[34] The group noted that countries in the European Union did not have uniform policy concerning vaccination against hepatitis B. Some countries, such as Germany, had a very high level of vaccine coverage of newborns (more than 90%) and low coverage (less than 10%) of teenagers; others, like France, had an inverse situation, with 25 percent vaccine coverage of newborns and 85 percent coverage of teenagers. Italy had a high level of both (90% for newborns and 80% for teenagers), and Belgium a low level (10%) for both groups. The DGS working group considered a low level of coverage of young children problematic, since contamination by hepatitis B at a young age often becomes chronic.[35]

The DGS working group identified 771 cases of acute demyelinating episodes (mainly multiple sclerosis) among an estimated 22–29 million vaccinated people. In the great majority of cases, symptoms first appeared shortly (two months or less) after vaccination against hepatitis B. On the one hand, the working group concluded that accidents linked with vaccinations were rare events and that hepatitis B vaccine had a favorable risk-to-benefit ratio. On the other hand, the group's report also noted that the risk was equally distributed among all the vaccinated people, while the benefit was especially elevated for a small group of high-risk individuals. It concluded that because the vaccinated children faced a small but nevertheless real short-term risk, it was important to inform their parents about such a risk. Accordingly, it recommended vaccination by family doctors, who could provide a detailed explanation of the advantages and dangers of a vaccine. Finally, the DGS group upheld the 1998 interruption of vaccination campaigns in junior high schools and recommended giving priority to the vaccination of newborns, who have a negligible risk of developing a demyelinating disease.[36]

The DGS's 2002 report may be seen as a quasi-official recognition of the link between vaccination against hepatitis B and neurological diseases.[37] Such recognition was consolidated through an agreement to compensate all the health workers who had developed demyelinating disease after hepatitis B vaccination.[38]

In spite of the absence of new scientific data, suspicions about undesirable effects of the hepatitis B vaccine have seemed to increase with time.[39] Marc Girard, a former director of Fondation Mérieux (one of the main French producers of serums and vaccines) and a staunch adversary of the hepatitis B vaccine, might have contributed to the increasingly negative view of this vaccine in France. Girard, a recognized specialist on viral vaccines and a media-savvy person, became an expert witness at trials against vaccine manufacturers. He also proposed a biological hypothesis, based on "molecular mimicry" to explain demyelinating disorders following a vaccination.[40]

A public hearing on hepatitis B vaccination organized by AFSSAPS in 2004 reiterated the conclusion of the DGS 2002 working group about a positive risk-to-benefit ratio of the hepatitis B vaccine. Yet, the AFSSAPS document accented increasing evidence of demyelinating episodes following hepatitis B vaccination, especially in adults. Accordingly, it recommended switching to the exclusive vaccination of newborns. The vaccination of preadolescents, it proposed, should be gradually phased out, while the vaccination of adults should be limited to at-risk groups.

The AFSSAPS working group also addressed the complex issue of compensation for vaccination accidents in France. Since hepatitis B vaccination is obligatory for health professionals, postvaccination disease in this group was recognized as a "work accident." Such a qualification freed vaccinated health workers who contracted a demyelinating disease from the obligation to prove a causal link between the vaccine and their neurological problems. They were thus able to obtain financial compensation from the French state. By contrast, people who considered themselves victims of nonobligatory hepatitis B vaccination and who tried to sue vaccine producers were legally obliged to prove the existence of a causal link between their disease and vaccination. Because this was a very difficult task, they failed to obtain compensation. Responding to this confusing situation, the AFSSAPS working group recommended severing the tie between the scientific and the juridical aspects of determining links between vaccination and disease and, if possible, finding a way to compensate other (presumed) victims of hepatitis B vaccination.[41]

In 2007 an epidemiological study found that hepatitis B vaccination is underemployed in France. The vaccination coverage of newborns remained stable—and very partial. Only approximately 20 percent of newborn infants are vaccinated each year (as compared with 90% in Germany). The study stated that between 2004 and 2006 there were 469 cases of acute hepatitis in France. These cases, the study's authors argued, could have been prevented with appropriate vaccine coverage. The absence of such coverage was regrettably attributed to excessive suspicions linked with the hepatitis B vaccination.[42] In addition to the generalized lack of trust in this vaccine, there was also suspicion of collusion between vaccine producers and the authorities—a situation that recalls mistrust of vaccines in some African countries, for instance, suspicions that hampered Niger's polio vaccination campaign in 2003 (see chapter 13).

People who believed that they were harmed by the hepatitis B vaccine attempted to obtain damages from vaccine producers. Trials against manufactures of hepatitis B vaccine continued in 2007 and 2008, even though the French courts systematically refused to recognize the manufacturers' responsibility in accidents attributed to the use of the vaccine. In January 2008, GlaxoSmith-Kline (previously SmithKline Beecham) and Sanofi Pasteur (previously Pasteur Mérieux MSD–Aventis Pasteur) were accused by a group of victims (29 people, including 5 who represented people who died presumably as a result of hepatitis B vaccination) of "involuntary killing" and "marketing of dangerous products." The suit estimated that approximately thirteen hundred French people suffered from secondary effects of hepatitis B vaccine.[43] The accused companies, it so happens, are also the producers of the HPV vaccines, Gardasil and Cervarix. It is to these vaccines that I now return.

HPV Vaccination in France: A Low Profile

The French have a complicated relationship with their pharmaceutical industry. While the government is eager to promote national producers, people tend to be suspicious of the alliance between industrialists and the state. Press articles published at the start of the trial against producers of the hepatitis B vaccine pointed to the "juicy profits" made by the French producer of Gardasil, Sanofi Pasteur, between 1994 and 1998, during the mass campaign promoting hepatitis B vaccination in schools. The French Health Ministry, these articles stressed, energetically promoted hepatitis vaccination by circulating exaggerated claims and by favoring the simultaneous use of a Sanofi-Aventis vaccine.[44] Collusion between a national industry and the French government is not new. From the 1890s on, the Pasteur Institute received important governmental subsidies. The fact that the institute maintained its status as a "charitable institution" in spite of its handsome profits on sales of vaccines and serums led in 1907 to accusations of excessive privilege and of an unfair advantage over competitors. These accusations were rejected by the institute's director, Emile Roux, who used his extensive political contacts to prevent the loss of the institute's special financial status.[45]

Gardasil obtained a marketing permit in the European Union in September 2006.[46] In December 2006, the Infectious Diseases Section of the French High Council for Public Health (Conseil Supérieur de l'Hygiène Publique), affiliated with the Health Ministry, declared that it could not yet recommend a nationwide HPV vaccination campaign. The council also stressed the need to improve the surveillance of cervical lesions in France and to extend the use of the Pap smear to 80 percent of the target population.[47] Three months later, the High Council for Public Health determined that it had sufficient data to advise Gardasil vaccination. The council, together with the Committee on Vaccination of the French Health Ministry, recommended vaccination of all 14-year-old girls, girls and women aged 15–23 who had not yet had sexual relationships, and those who had started such relationships only recently. It emphasized the need to explain that the vaccine provided partial protection only and did not abolish the need for regular Pap smears. This information, it recommended, should be provided by both doctors and vaccine producers. It advised the reimbursement by the National Health Insurance of 65 percent of the price of HPV vaccination.[48] Finally, it recommended the creation of a National Reference Center dedicated to papillomaviruses, which would conduct studies on the long-term effects of HPV vaccination.[49]

In August 2007, the French High Authority on Health (Haute Authorité de la Santé, an independent consulting body created in 2004 to supervise health-related issues) also recommended HPV vaccination of all 14-year-old girls and of selected young women aged 15–23. The authority's statement emphasized, however, the absence of evidence of Gardasil's clinical efficacy (reducing the incidence of invasive cancers) and its unknown duration of protection. It also pointed to the lack of information on effects of HPV vaccination in pregnant women (those who are vaccinated before they realize they are pregnant). Gardasil, like every other new pharmaceutical product, the authority declared, should be submitted to careful scrutiny, and its undesirable effects should be immediately reported to AFSSAPS.[50] A statement made by the French Association of Gynecologists and Obstetricians in December 2007 also associated its recommendation to vaccinate adolescent girls with questions about the duration of the protection provided by the HPV vaccine and about the vaccine's long-term effects. The Association of Gynecologists and Obstetricians raised two additional issues: the possibility of also vaccinating boys in order to limit circulation of carcinogenic strains of HPV in the population (thus establishing herd immunity) and the risk that vaccination against known "carcinogenic" HPV strains would lead to the replacement of these strains by other strains as the main cause of cervical cancer.[51] The media reported these developments, usually without any comment.

The Vaccination Committee of the French National Academy of Medicine also endorsed Gardasil. This endorsement was cosigned by Marc Girard, the main scientific opponent of hepatitis B vaccination in France. The academy followed, in the main, the Health Ministry's recommendation, with one important difference: it proposed lowering the age of vaccination to 11–13 years. The age of 14 was chosen in France because, according to recent studies, only 3 percent of French girls have had sexual relations before the age of 15, and because of uncertainty about the duration of protection provided by the vaccine. The academy committee proposed, however, to lower the age of vaccination against HPV in order to harmonize French recommendations with those of other European countries. The academy report, like other statements on this topic, emphasized that these recommendations should be revised when more information became available about the clinical efficacy of vaccination, its undesirable consequences, its effects on infection with potentially carcinogenic HPV strains, the length of the protection period, and the cost-to-benefit ratio of the HPV vaccine.[52] Although French expert bodies endorsed the HPV vaccine, their cautious tone was unusual for a country where vaccines ("Pasteur's science") were usually presented in a very positive light, where there had never been an important antivaccination movement, and where official bodies nearly always enthusiastically promoted medical innovations. In the case of the HPV vaccine, all the documents produced by the French experts agreed that vaccination should be voluntary and should not be conducted in schools, a possible echo of bitter memories of the 1994–97 hepatitis B vaccination campaign.

Even today, French health professionals typically have either a cautiously positive or a cautiously negative reaction to vaccination against HPV, and these attitudes have not invited strong responses by the lay public. The vaccine nevertheless has been a moderate success. During the first year after the introduction of Gardasil, about a half million young women were vaccinated, at a cost to the state health system of approximately 81 million euros, probably a testimony to a relatively efficient distribution of the vaccine manufacturer's message among gynecologists and general practitioners.[53] More passionate reactions against the

HPV vaccine were rare and were found mainly at the margins of the medical profession. The physician and writer Martin Winckler protested on his blog against a YouTube spot in French that praised Gardasil.[54] The YouTube spot in French was very short (twenty-three seconds) and showed adolescent girls and women making the following statement: "Cervical cancer is produced by a virus, the papillomavirus; the virus can infect the majority of us, from adolescence on. We should react; all women must react." At the end, two women, one young, one older (possibly a mother-daughter pair, widely used in Gardasil's advertisements), declare, "A vaccine exists; talk about it with your doctor: this is important."[55] Winckler criticized the spot's blunt message; a gynecologist, Véronique T., strongly disagreed; and the two discussed the HPV vaccine on Winckler's blog.[56]

Another, somewhat puzzling exception to the rule of the low profile of French debates on the HPV vaccine was the reaction of the union of general practitioners of Réunion. Réunion, an island in the Indian Ocean not far from Madagascar (with a population of about eight hundred thousand) used to be a French colony and was a strategic outpost of trade with India. From 1946 on, Réunion has had the status of "région française d'outre mer." Its citizens, of mixed racial and ethnic origins, have French nationality and many have migrated to France. (In spite of important differences, Réunion's status may be likened to that of the U.S. territory Puerto Rico.) The Regional Union of Liberal Doctors of Réunion (Union régionale des médecins libéraux de la Réunion [URML]) started a series of campaigns against the overuse of tests and medical interventions.[57] The first target of this campaign, named DDI (Dé Dés Information [against misinformation]) was prostate-specific antigen screening for prostate cancer; the second target was the HPV vaccine. URML placed paid advertisements in several Réunion newspapers and magazines and in *Le Monde* (one of the main French daily newspapers) to promote its message. The advertisement on HPV vaccine, published in *Le Monde* of September 8, 2008, affirmed that the efficacy of this vaccine was not proved, that it did not protect against all HPV strains, and that its long-term effects were uncertain. It is not clear why URML took this initiative (contested by some of its members), and the possibility cannot be ruled out that it represents a personal initiative of the union's (white) president, Philippe de Chazournes, who gained important visibility in Réunion media thanks to the DDI campaign. And yet, this energetic activism by some Réunion doctors against the HPV vaccine contrasts with the absence of such activism among their colleagues in France.

Are the French Different?

In view of two and half centuries of French discussions on the acceptability of vaccination, the nearly total absence of public debate about the introduction of the HPV vaccine may seem surprising, especially when contrasted with the heated discussions on this subject in the United States. Although the National Academy of Medicine disagreed with the Ministry of Health recommendations and proposed to lower the vaccination age, the debate on this topic was mostly academic.[58] In an attempt to trace recent articles on HPV vaccination in major French newspapers, I did not find anything on this subject on the health or society pages. This subject was discussed exclusively in economic newspapers and magazines. There I was able to find information on Sanofi Aventis (previously Sanofi Pasteur) and GlaxoSmithKline marketing strategies and their competition for the UK vaccination market. The economic press also reported that in 2007 Sanofi Aventis announced sales of 3 million courses (3 doses each) of Gardasil in Europe.[59]

Vaccination with Gardasil in France now follows the pattern of other voluntary preventive measures reimbursed by the National Health Insurance. It reflects the family's wishes, is probably strongly influenced by views of the family's doctor, and, even though the vaccination is partly reimbursed by the National Health Insurance, it is more frequently adopted by middle-class families than by those from lower socioeconomic strata (the former more frequently also have complementary health insurance to cover the remaining costs of vaccination). Such a pattern of use results in partial and uneven vaccine coverage. Moreover, probably because of the recent "health scandals," French authorities have attempted to monitor carefully the consequences of vaccination against HPV. The Institut de Veillé Sanitaire started a registry of vaccination accidents, another registry of outcomes of pregnancies in women vaccinated when in early stages of pregnancy, and a study that compares the incidence of cervical cancers in vaccinated and nonvaccinated women. The results of the latter studies will probably be available in ten years at the earliest. In the meantime, France's Institut du Cancer continues its campaign to improve the uptake of the Pap smear in France. The scheduled increase in the uptake of cervical smears may make it difficult to interpret the efficacy of the HPV vaccine in lowering the incidence of cervical cancer in France.

The story of the introduction of HPV vaccine in France is one of absence—absence of public debate, of professional controversies, of real engagement with a public health issue. This medical innovation silently "crept in" and was slowly adopted by selected segments of the population without much discussion.[60] Some structural elements may account for the absence of discussion on the HPV vaccine in France: the lack of a French women's health movement (the lack of feminist critique of the medicalization of women's bodies), a low level of interest by the political and religious Right in sexual mores (the lack of propaganda in favor of abstinence from extramarital sex), and greater acceptance of the sexual activity of young people and thus of the need to inform teenagers about the dangers of such activity (exemplified by the lack of parental objections to sexual education in schools and information about STDs and by the distribution of the "day after pill" by school nurses).[61]

In addition to the absence of these factors in the French context, I have argued that the fate of vaccination against HPV in France has been inseparably linked with that of other health measures. It was shaped by the prolonged failure to promote a national program of screening for cervical cancer and was strongly affected by the ill-fated attempt to introduce mass vaccination against hepatitis B in junior high school and by the consequent presentation of this initiative as one of the French "health scandals" of the 1980s and 1990s. These developments, I propose, led to the "privatization" of the HPV vaccine in France and to its presentation (to date) as a purely voluntary measure.[62] Strong influence of local and contingent factors can be found in other European countries too (see chapter 14). A comparison between patterns of implementation of preventive health measures in countries with similar levels of economic development and health care coverage may advance our understanding of mechanisms that shape medical practices and affect the fate of medical innovations. It is very important, as Ludwik Fleck explained in 1946, to pay close attention to the *specific* features of scientific cognition, the historical *singularity* of their development, the structure of the relevant scientific communities, and the unique characteristics of each scientific thought style.[63]

NOTES

1. Abby Lippman, Madeline Boscoe, and Carol Scurfield, "Rebuttal: Do You Approve Spending $300 Million on HPV Vaccination?" *Canadian Family Physician* 54 (2008): 343.

2. George N. Papanicolaou and Herbert F. Traut, "The Diagnostic Value of Vaginal Smears in Carcinoma of the Uterus," *American Journal of Obstetrics and Gynecology*

42 (1941): 193–206; George N. Papanicolaou and Herbert F. Traut, *Diagnosis of Uterine Cancer by Vaginal Smear* (New York: Commonwealth Fund, 1943).

3. L. H. Lombard, M. Middletown, S. Warren, and O. Gates, "The Use of Vaginal Smears as a Screening Test," *New England Journal of Medicine* 238 (1948): 867–871.

4. Charlotte A. Jones, Theodore Neustaaedeter, and Locke L. MacKenzie, "The Value of Vaginal Smears in the Diagnosis of Early Malignancy," *American Journal of Obstetrics and Gynecology* 49 (1945): 159–168.

5. On screening for risk, see Robert Aronowitz, "Situating Health Risks," in *American Health Care History and Policy: Putting the Past Back In*, ed. R. Burns, R. Stevens, and C. Rosenberg (New Brunswick, NJ: Rutgers University Press, 2006), 153–165; and see chapter 2 of this volume.

6. E.g., Edward Rimley to Charles Cameron, medical and scientific director of ACS, July 6, 1955, on the importance of endorsing PAP smears by the ACS, Mary Lasker Papers, box 98; Eugene Pendegrass, chairman of the committee on professional and public education of ACS, to Mary Lasker, Oct. 10, 1956, on the importance of including Pap smears in doctors' annual checkups, Mary Lasker Papers, box 99, both in Columbia University Archives.

7. Eftychia Vayena, "Cancer Detectors: An International History of the Pap Test and Cervical Cancer Screening, 1928–1970" (Ph.D. diss., University of Minnesota, 1999). Vayena's dissertation is an excellent source of information on ACS's role in the implementation of the Pap smear in the United States. Documents of the Medical Women Federation, series SA/MWF, file F.13/3, Archives and Manuscripts Department, Wellcome Library, London.

8. On the history of the Pap smear in the United Kingdom, see Elisabeth Toon, "Demand and Supply, Success and Failure: Managing the Provision of Cervical Cancer Screening in the 1960s UK," paper presented at the conference "How Cancer Changed," Paris, April 2–4, 2009. Press conference of the minister of health, Kenneth Robinson, on cervical cytology, Oct. 21, 1966. Documents of the Medical Women Federation, file F.13/4, Wellcome Library.

9. Vicky Singleton, "Actor Networks and Ambivalence: General Practitioners in the UK Cervical Screening Program," *Social Studies of Science* 23, no. 2 (1993): 227–264; Vicky Singleton, "Stabilizing Instabilities: The Role of the Laboratory in the United Kingdom Cervical Screening Program," in *Differences in Medicine: Unraveling Practices, Techniques, and Bodies*, ed. Marc Berg and Anne Marie Mol (Durham, NC: Duke University Press, 1998), 86–103.

10. Julian Peto, Clare Gilham, Olivia Fletcher, and Fiona E. Matthews, "The Cervical Cancer Epidemic That Screening Has Prevented in the UK," *Lancet* 364, no. 9430 (2004): 249–256.

11. M. F. G. Murphy, M. J. Campbell, and P. O. Goldblatt, "Twenty Years' Screening for Cancer of the Uterine Cervix in Great Britain, 1964–84: Further Evidence of Its Ineffectiveness," *Journal of Epidemiology and Community Health* 42, no. 1 (1988): 49–53; "Cancer of the Cervix: Death by Incompetence" (editorial), *Lancet*, Aug. 17, 1985, 363–364; Angela R. Raffle, B. Alden, and E. D. Mackenzie, "Detection Rates for Abnormal Cervical Screens: What Are We Screening For?" *Lancet* 345, no. 8963 (1995): 1495–1473; Ruth Etzioni and David B. Thomas, "Modeling the Effect of Screening for Cervical Cancer on the Population," *Lancet* 364, no. 9444 (2004): 224–226; Angela E. Raffle and

M. Quinn, "Harms and Benefits of Screening to Prevent Cervical Cancer," *Lancet* 364, no. 9444 (2004): 1483–1484; Michael F. Murphy and Rachel Neale, "Harms and Benefits of Screening to Prevent Cervical Cancer," *Lancet* 364, no. 9444 (2004): 1484–1485.

12. D. R. Brunet, in the name of the working group on cervical cancer screening, "Rapport au sujet de l'organization d'une campagne de dépistage du cancer du col de l'utérus en France," Ministère de la Santé, Direction Générale de la Santé, Commission du Cancer de Conseil Permanent d'Hygiène, Nov. 24, 1976, Inserm Archives, folder titled "Commission du Cancer."

13. R. Laumonier and C. Marsan, "Rapport sur la formation, les activités et les responsabilités des cytotechniciens," Ministère de la Santé, Direction Générale de la Santé, Commission du Cancer de Conseil Permanent d'Hygiène, June 1976, Inserm Archives, folder titled "Commission du Cancer."

14. A Sicard, "L'introduction des frottis cervico-vaginaux en France," *Bulletin de l'Académie de Médicine* 180, no. 5 (1996): 1109–1113; A. Garnier, C. Exbrayat, M. Bolla, et al., "Une campagne de dépistage de cancer du col avec des frottis vaginaux," *Bulletin de Cancer* 84, no. 8 (1997): 791–795.

15. Nora Liberetto, "Dépistage du cancer du col: Le point de vue des médecins généralistes" (master's thesis, Ecole des Hautes Etudes en Sciences Sociales, 2005).

16. G. Dubois, "Cytologic Screening for Cervix Cancer: Each Year or Each 3 Years," *European Journal of Obstetrics, Gynecology, and Reproductive Medicine* 65, no. 1 (1996): 57–59.

17. Institut National du Cancer, *Bilan Plan Cancer 2003–2006: Dépistage: Des avantages quantitatives et qualitatives* (Paris: Institut National du Cancer, 2006). Institut du Cancer experts found the uptake of Pap smears in France unacceptably low; however, the incidence of cervical cancer in France is not higher than in other Western European countries or in the United States.

18. George Weisz, "A Specialist Regulation in France during the First Half of the Twentieth-Century," *Social History of Medicine* 15 (2002): 457–480; George Weisz, *Divide and Conquer* (Oxford: Oxford University Press, 2005).

19. Liberetto, "Dépistage du cancer du col."

20. C. Chan Chee, M. Bessaat, and V. Kovess, "Les facteurs associés avec dépistage du cancer de col d'utérus," *Révue de l'Epidémiologie et de la Santé Publique* 53 (2005): 69–75.

21. L. Remontet, J. Estève, A. M. Bouvier, et al., "L'incidence et la mortalité du cancer en France, 1978–2000," *Révue de l'Epidémiologie et de la Santé Publique* 51 (2003): 3–30.

22. J. C. Boulanger, R. Fauvet, S. Urratiaguer, et al., "Histoire cytologique des cancers du col utérin diagnostiqués en France en 2006," *Gynécologie Obstétrique et Fertilité* 35, no. 9 (2007): 764–771.

23. J. L. Brun, R. Dachez, P. Mathé, et al., "Prévention du cancer de col d'utérus: Recommandation pour la pratique clinique," *Journées Nationals de CNGOF* (College National des Gynécologues et Obstétriciens Français), Paris, Dec. 12, 2007, 369–406. The College de Gynécologues also criticized overtreatment of low-grade cervical lesions. In 2004, 3,693 women underwent conization, and there were 139 hysterectomies for low-grade (CIN/1) cervical lesions, although the risk of progression of such lesions to invasive cancer was estimated as 0.15% to 0.26%. Ibid., 401–402.

24. C. Bergeron, J. G. Breugelmans, S. Bouée, et al., "Coût du dépistage et de la prise en charge des lésions précancéreuses du col utérin en France," *Gynécologie Obstétrique et Fertilité* 34, no. 11 (2006): 1036–1042; Patrick Arveux, Stève Bénard, Stéphane Bouée, et al., "Coût de la prise en charge du cancer invasif du col de l'utérus en France," *Bulletin du Cancer* 94, no. 2 (2007): 219–224.

25. A study of women diagnosed with invasive cervical tumors revealed that while two-thirds had not had regular Pap smears or had been underscreened, 27% had been labeled normal in the previous three years, data that point to insufficiency of screening. Boulanger, Fauvet, Urratiaguer, et al., "Histoire cytologique des cancers."

26. Laurent Greilsamer, *Le proces du sang contaminé* (Paris: La Decouverte, 1992); Pascale Robert Diard, "Hormone de croissance: Le procès d'un scandale sanitaire," *Le Monde*, Feb. 6, 2008.

27. Francis Chateauraynaud and Didier Thorny analyzed the ways "scandals" and "alerts" shaped reorganization of the French health system in the late twentieth century. F. Chateauraynaud and D. Thorny, *Les sombres précurseurs: Une sociologie pragmatique de l'alerte et de risque* (Paris: Editions EHESS, 1999).

28. For a discussion of balancing different kinds of risk when making decisions about obligatory vaccination, see, e.g., Sydney A. Halpern, *Lesser Harms: The Morality of Risk in Medical Research* (Chicago: University of Chicago Press, 2004).

29. Sanford R. K. Kimmel, "Vaccine Adverse Events: Separating Myth from Reality," *American Family Physician* 66, no. 11 (2002): 2113–2120.

30. Histories of vaccination in the United States and in Europe are different. In Europe the existence of national health systems makes obligatory vaccination rules less compelling. James Colgrove, *State of Immunity: The Politics of Vaccination in Twentieth-Century America* (Berkeley: University of California Press, 2006); see also chapters 1, 9, and 11 in this volume.

31. Jennifer Stanton and Virginia Berridge, "Vertical Ancestries and Horizontal Risks: Hepatitis B and AIDS," in *Heredity and Infection: The History of Disease Transmission*, ed. Jean Paul Gaudillière and Ilana Löwy (London: Routledge, 2001), 311–326.

32. By contrast, the (voluntary) vaccination of newborns was less successful: in 1994–97, only between 24% and 27% newborn children were vaccinated each year. Bernard Bégaud, Jan François Dartigues, Françoise Degos, et al., "Rapport de la Mission d'experts sur la politique de vaccination contre l'hépatite B en France," Feb. 15, 2002, Imprimerie de Ministère de la Santé, Paris. This report can be found at www .infectiologie.com/site/medias/_ . . . /vaccins/dartigues.pdf (accessed Dec. 23, 2009).

33. "Vaccination hépatite B," recommendation, Oct. 1, 1998, French Ministry of Health, Paris.

34. Alberto Ascherio, Shumin Zhang, Miguel Hernan, et al., "Hepatitis B Vaccination and the Risk of Multiple Sclerosis," *New England Journal of Medicine* 344 (2001): 327–332.

35. Bégaud, Dartigues, Degos, et al., "Rapport de la mission d'experts."

36. Ibid.

37. A May 2000 statement from the French Health Ministry admitted already the possibility of a weak risk of demyelination following hepatitis B vaccination, especially in "susceptible people." Communication, Secretariat d'Etat à la Santé, March 6, 2000.

The text of this message can be found at www.doctissimo.fr/html/sante/ . . . 2000/ . . . /sa_1748_hepatb.htm (accessed Dec. 23, 2009).

38. Letter sent by Lucien Abenhaim, DGS's director, to six health workers, recognizing their right for indemnification. It was published in the newspaper Parisien Libéré, May 25, 2000, under the title "Indemnisation pour la vaccination contre l'hépatite." The rationale was that the state has a direct responsibility for consequences of actions it imposes on people.

39. The majority (but not the totality) of recent publications did not find statistically meaningful connections between vaccination and neurological diseases. These publications point nevertheless to a persisting suspicion that such links may exist in "predisposed" (and, as of now, impossible-to-identify) individuals. E.g., Frank DeStephano, Thomas Verstraen, Lisa Jackson, et al., "Vaccination and Risk of Central Nervous System Demyelinating Diseases in Adults," *Archives of Neurology* 60 (2003): 504–509; Robert Nasmith and Anne Cross, "Does Hepatitis B Vaccine Cause Multiple Sclerosis?" *Neurology* 63 (2004): 772–773; Yann Mikaeloff, Guillama Caridade, Meanie Rossier, et al., "Hepatitis B Vaccination and the Risk of Childhood Onset Multiple Sclerosis," *Archives of Pediatrics and Adolescent Medicine* 161, no. 12 (2007): 1176–1182.

40. Yannick Comenge and Marc Girard, "Multiple Sclerosis and Hepatitis B Vaccination: Adding the Credibility of Molecular Biology to an Unusual Level of Clinical and Epidemiological Evidence," *Medical Hypothesis* 66 (2006): 84–86.

41. M. Brodin, president, public hearing, "Vaccination contre le virus de l'hépatite B et sclérose en plaques: L'état des lieux," INSERM and AFSSAPS, Nov. 2004. France has a recent law on "faultless compensation" of medical errors, passed as a result of the "contaminated blood" scandal, but its implementation is very uneven.

42. *Bulletin Hebrodmadaire d'Epidémiologie*, Dec. 25, 2007.

43. Marianne Enault, "Hépatite B: Le Vaccin en process," *Le Journal du Dimanche*, Jan. 31, 2008.

44. Segolene de Larquier, "Vaccin contre l'hépatite B: Les responsables des deux laboratoires mis en examin," *Le Point*, Jan. 31, 2008.

45. Ilana Löwy, "On Hybridizations, Networks, and New Disciplines: The Pasteur Institute and the Development of Microbiology in France," *Studies in the History and Philosophy of Science* 25, no. 5 (1994): 655–688.

46. Didier Hoch and Michael Watson, "Gardasil®, le vaccin qui peut prévenir le cancer du col de l'utérus a reçu une AMM europènne," Sanofi Pasteur, Sept. 28, 2006, teleconference of Sanofi Pasteur, available at www.sanofi-aventis.com/ . . . /060928_PDF_GARDASIL_slides_media_tcm29-15674.pdf (accessed Dec. 23, 2009).

47. Avis du Conseil Supérieur de l'Hygiène Publique de France (CSHPF) Dec. 5, 2006, available at the CSHPF site, www.hcsp.fr/explore.cgi/avisrapportsdomaine?ae=avisrapportsdomaine&clefdomaine (accessed Dec. 23, 2009).The mortality rates from cervical cancer in France are, nevertheless, similar to those of Western European countries with a better Pap smear coverage.

48. Avis du Conseil Supérieur de l'Hygiène Publique de France, March 9, 2007, available at the CSHPF site, www.hcsp.fr/explore.cgi/avisrapportsdomaine?ae=avisrapportsdomaine&clefdomaine (accessed Dec. 23, 2009). The majority of complementary health insurances (mutuelles) pay for all or part of the remaining 35% of Gardasil's price, 406.77 euros for three doses in 2007.

49. Avis du Comité Téchnique des Vaccinations et du Conseil Supérieur de l'Hygiène Publique de France, March 9, 2007, available at the CSHPF site, www.hcsp.fr/explore.cgi/avisrapportsdomaine?ae=avisrapportsdomaine&clefdomaine (accessed Dec. 23, 2009).

50. Haute Autorité de la Santé, "Quelle place pour le vaccin papillomavirus human (Gardasil®) dans la prévention du cancer du col?" statement, Aug. 2007, available at www.has-sante.fr/portail/jcms/c_592462/quelle-place-pour-le-vaccin-papillomavirus-humain-gardasil-dans-la-prevention-du-cancer-du-col (accessed Jan 28.2010).

51. Brun, Dachez, Mathé, et al., "Prévention du cancer de col d'utérus."

52. Pierre Bégué, Rogert Henrion, Bernard Blanc, Marc Girard, and Hélène Sancho-Garnier, "Les vaccins de papillomavirus humains: Leur place dans le prevention du cancer du col utérin," report made by commissions VI-a, VI-b, III, and two working groups of the Académie Nationale de Médecine, Dec. 11, 2008, Publications de l'Académie de Médecine, Paris. The report is also available at www.academie-medecine.fr/UserFiles/ . . . /begue_rapp_11dec_2008.doc (accessed December 23, 2009).

53. Sandrine Blanchard, "Cancer d'utérus: Interview avec le professeur Claude Béraud," *Le Monde*, July 9, 2008.

54. Winckler's blog was at http://martinwinckler.com/article.php3?id_article=908&var_recherche=gardasil. This blog was published in November 2007. Winckler's real name is Marc Zaffran. He became famous in France following the success of his novel *La maladie de Sachs*, a rather iconoclastic image of the problems of a general practitioner in the French countryside—later also a popular movie. He has since published several other novels and hosted a radio program.

55. The French-language short publicity spot (23 seconds), "Pasez l'info" (pass the information), http://fr.youtube.com/watch?v=Zjlbcj_YcRc, was coproduced by four (relatively small) associations of gynecologists and funded by Sanofi-Pasteur. This spot was not broadcast in the French media, because direct advertisement of drugs to consumers is illegal in France. Destined to an unregulated circuit, it was probably seen by only a small number of viewers. On the direct Gardasil publicity to consumers in the United States, see chapter 8.

56. The discussion appeared at http://martinwinckler.com/article.php3?id_article=911&var_recherche=Gardasil+.

57. See the DDI Web site, www.urml-reunion.net/ddi/index-ddi.html. In France, the term *liberal doctors* describes physicians who are not paid by the state; that is, in French configurations, such doctors usually do not belong to medical elites and do not work in major teaching and research hospitals.

58. Sandrine Blanchard, "Cancer d'utérus: L'age de la vaccination rémis en cause," *Le Monde*, Feb. 12, 2008.

59. Reuter, "Glaxo's Cervarix Priced on Par with Gardasil in the UK," Oct. 1, 2007; Deyna Chatzimichalaki, "GlaxoSmithKline: Cervarix Enters Potentially Lucrative Cervical Cancer Prevention Market," *Pharmaceutical Business*, Sept. 25, 2007; Ed Silverman, "Sanofi Executive: Glaxo's Cervarix Study Is a 'Gimmick,'" *Pharmalot*, Feb. 7, 2008; Elena Barton, "Sanofi Pasteur voit une poursuite de la hausse ventes de Gardasil en UE," *La Vie Financière*, Feb. 5, 2008. The sales of Gardasil in the United States were much higher.

60. In February 2008, Sanofi Pasteur reported the sale of more than 435,000 doses of Gardasil in France (corresponding to the vaccination of approximately 120,000

young women). Blanchard, "Cancer d'utérus: L'age de la vaccination rémis en cause." Blanchard states in this article that two hundred undesirable side effects of Gardasil were reported to AHSSAPS.

61. The notion of "risky girlhood" seems to be stronger in the United States than in Europe. See chapters 6 and 7.

62. The possibility is not excluded that the recent skeptical evaluations of its efficacy and the increasingly frequent emphasis on the pharmaceutical industry's interest in promoting vaccination on a large scale before the patents run out will lead to a slowdown in HPV vaccination. Charlotte J. Haug, "Human Papillomavirus Vaccination—Reasons for Caution," *New England Journal of Medicine* 359, no. 8 (2008): 861–862 ; Elizabeth Rosenthal, "Drug Makers' Push Leads to Cancer's Vaccines' Rise," *New York Times*, Aug. 20, 2008. See also chapter 3.

63. Ludwik Fleck, "Problemy naukoznawstwa," *Zycie Nauki* 1 (1946): 332–336, quote on 335–336, italics mine. Also see the English translation, Ludwik Fleck, "Problems of the Science of Science," in *Cognition and Fact: Materials on Ludwik Fleck*, ed. Robert Cohen and Thomas Schnelle (Dordrecht, Germany: Reidel, 1986), 113–128.

Individualized Risk and Public Health

Medical Perils, Political Pathways, and the Cultural Framing of Vaccination under the Shadow of Sexuality

Keith Wailoo, Julie Livingston, Steven Epstein, and Robert Aronowitz

Given the difficulty of confronting the fundamental social and environmental causes of disease, vaccines stand out as a supposedly simple solution, and they are widely acknowledged to be our best means of disease *prevention*. Modern history is replete with vaccine success stories, and vaccines have obvious appeal in a world of growing health threats. Yet, as Laura Mamo, Amber Nelson, and Aleia Clark write in chapter 7, vaccines are not neutral entities but rather sites of "cultural, social, and political contestation." As magic bullets promising intangible benefits against uncertain future perils, they inevitably provoke extreme responses: an optimism that can verge on fantasy and a skepticism that can carry over into rejection. When threats target particular segments of the population and are not perceived to be imminent, the push for vaccination may seem especially troublesome—giving rise to much cultural anxiety and sociopolitical debate. If vaccines are a loaded topic, so too are sexually transmitted infections, which have long operated as both dense signifiers and material manifestations of our complicated sexual politics. This doubly charged combination of the complexities of vaccination and those of sexuality describes the human papillomavirus debate—a case in which the larger political logic and epistemological basis for vaccination have come fully into view.

In part because of these anxieties and tensions, the global uptake of HPV vaccines has been uneven. In Scotland, March 2009 statistics showed that the "uptake of the HPV vaccine among girls in second, fifth and sixth year at school . . . [was] already 92.2 per cent for the first dose [and] 87.8 per cent for the second dose." Scotland's public health minister praised the nation's "first ever anti-cancer immunisation programme."[1] By contrast, a 2009 South African study found that Gardasil and Cervarix (although licensed for use in March 2008) were not yet available in the public health sector.[2] Meanwhile, in the United States, a September 2009 report from the Centers for Disease Control and Prevention (CDC) found that 37.2 percent of U.S. females of ages 13 to 17 had received at least one HPV vaccination, up from 25 percent the previous year.[3] Rates varied widely across states, however, with "the highest rates . . . recorded in Rhode Island (54.7 percent), New Hampshire (54.4 percent) and Massachusetts (53.3 percent); the lowest rates were seen in South Carolina (18.7 percent), Georgia (18.5 percent) and Mississippi (15.8 percent)." An earlier CDC study had also found striking disparities in vaccination levels based on race and class: "Levels were higher among non-Hispanic whites (13%) than among non-Hispanic blacks (7%) and Hispanics (6%)," with "women of higher socioeconomic status . . . more likely to be vaccinated." These were surprising findings; indeed, as the study's authors noted, the "low rates of HPV vaccination exist[ed] *despite* reportedly high awareness of both the disease and the vaccine (approximately 80%)."[4] In other words, the marketing, commercialization, and public controversy had worked (making many families knowledgeable about the details of HPV); yet, awareness had not produced uniform or truly widespread uptake.

Both the framing and the context of the U.S. HPV debates cast a shadow over uptake. Steven Epstein and April Huff explain in chapter 12 that a predicted "moral panic" about promoting promiscuity on the part of teen girls failed to emerge in a way that affected policy significantly at the federal level. Yet, attempts to promote HPV vaccination aggressively at the level of the states resurrected these concerns. A July 2009 survey of physicians in Texas, where the political controversy had flared in 2007, found that fewer than 50 percent recommended the vaccine for girls, despite expert policy statements encouraging universal vaccination. The CDC report speculated that the low immunization rate in the United States "may be related to . . . high cost of the vaccine and . . . [to the fact that] the primary target group for HPV vaccine is females aged 11–12 years old."[5] Its authors also found that Texas physicians believed that parents were hesitant—concerned about vaccine safety and influenced by "lack of education and understanding about HPV, negative media reports about the vaccine, mistrust of vaccines in general, and concern that their consent for vaccination would imply that they condoned premarital sex."[6] These findings revealed the lasting impact of how the vaccine was often framed in the American debate—as a private matter touching on the future health, and perhaps sexuality, of young girls, rather than as a public health initiative with implications for the population at large.

From the outset and still today, the HPV debate has been caught up in a swirl of commentary and debate around girls, cervical cancer, and individual risk, rather than around HPV as a sexually transmitted disease and its population impact. In September 2009, for example, ABC News reported on the case of a 17-year-old immigrant whose citizenship request was jeopardized because she refused the HPV vaccine. Reporting the case with attention to quintessentially American obsessions, ABC described the girl as "a devout Christian who has taken a virginity pledge, [and has] said she does not plan to have sex 'anytime soon.'"[7] Her mother's request for a waiver of the requirement for moral and religious reasons was rejected by the Citizenship and Immigration Services, which cited the 1996 Immigration and Naturalization Act requiring immigrants seeking citizenship to be vaccinated against specific diseases and to receive "any other vaccinations recommended by the CDC's Advisory Committee for Immunization Practices." Gardasil had been added to that list in 2008 and was only removed in December 2009 after much public outcry. The result of this intense focus on sexuality and girls has been a largely missed opportunity to talk about the larger politics of public health, as well as a missed opportunity to talk about sexual health and sexual literacy.

The black-and-white polarities of the U.S. public debate (framing the issue as parents' rights versus the power of government, or treating pharmaceutical advertising as either beneficial or coercive) mask a host of complexities that the authors of this volume have tried to expose and analyze. As they have shown, the biological certainties and uncertainties surrounding the vaccine have evolved alongside the many marketing appeals, suggesting that, at many levels, the HPV vaccine remains an experiment in progress and one impossible to assess in the kind of stark terms in which the debate has been framed. Yet, the polarized HPV vaccination debates might serve as valuable evidence of underlying tensions in our society about the management of sexuality, the role of government, the influence of industry, the place of health advocacy, and the protection of public health. Indeed, the HPV vaccine debate is, simultaneously, a biomedical and a sociocultural drama that is played out on multiple stages across the world, each with its particular history, political culture, epidemiological challenges, and cultural anxieties. As the essays in this volume insist, we need to view the vaccine debate in historical and cultural context, paying close attention to the perspectives represented, the voices excluded, and the positions of various stakeholders, including parents, the state, industry, and politically interested groups. We learn from this multidisciplinary assessment how policymakers and parents navigate this controversy and make sense of the vaccine's biological, statistical, and social uncertainties.

The Center of the Storm and the Margins of Debate

Debates over the HPV vaccine reflect the current historical moment as much as they reference the entire history of vaccination. As Robert Aronowitz argues in chapter 2, "the prototypical vaccine is directed against a specific, prevalent, serious, and communicable infectious disease . . . [and] contributes to herd immunity," but "in practice, many vaccines now and in the past have strayed from this prototype." At its heart, the HPV vaccine controversy (in the United States, the United Kingdom, the European Union, and across many parts of the global South) revolves around framing—and thereby controlling—the many uncertainties surrounding the vaccine itself. Of course, long before the HPV vaccine arrived, vaccination was already perceived as a controversial public and private health concern, often pitting individuals and families against the state and raising questions of efficacy, social control, and equality. Yet the HPV vaccine debates have shown how advocates and opponents have become increasingly sophisticated in their deployment of framing tactics, metaphors, and images to fight for their particular policy goals. Our essays illuminate and interpret these processes of *framing;* they draw particular attention to how social differences and global inequalities have been addressed and obscured.

Framing, of course, is a way of seeing and not seeing—emphasizing some concerns over others. The debate engendered by the vaccine privileged some points of view, pushing others into the background. As Lundy Braun and Ling Phoun observe in chapter 3, "men, in general, were either left out of the cycle of transmission or represented as marginal to it." Steven Epstein, in chapter 4, points to the sexual politics underpinning the debate, noting that despite the links between HPV and anal cancer, "no one is likely to advocate a targeted effort to vaccinate young MSM [men who have sex with men]. A more likely possibility is that these young men will find themselves vaccinated along the way," he notes, "as part of a broad-scale effort to vaccinate all boys." Globally, we learn, other key groups have been left out of the conversation, despite their concerns over and experiences with HPV-related cancers. The vaccine debate unfolded amid resource-rich cultures of comparative entitlement on one hand and woefully unaddressed needs on the other. As Botswana clinician Doreen Ramogola-Masire notes in chapter 5, deaths from cervical cancer have declined in the developed world, while the developing world has had "more than 500,000 new cases of cervical cancer . . . diagnosed worldwide in 2007 . . . and more than 80 percent of the deaths." Within the developed world, cervical cancer is much more prevalent among the poor and among stigmatized minorities, yet the "One Less" campaign in the United States obscured this reality with its mass marketing.

In essay after essay, we see how manufacturers, advocates on many sides of the issue, and policymakers became adept at framing the vaccine in service of their larger goals. Some issues and voices have been banished to the background, while others have been pushed to center stage. As Epstein and Huff note, one Republican state representative from Virginia who sponsored a mandatory immunization bill insisted, "This is not a prevention for a sexually transmitted disease. This is a prevention for cancer." Similarly, Merck's "One Less" campaign promoted the vaccine as a cancer concern and as a tool of empowerment for responsible young women. But, as Mamo, Nelson, and Clark argue, even as young girls and cervical cancer appear in the foreground of the vaccine debates, "the women who lack access to Pap-smear screening and other sexual health care, who are, in fact, at higher risk for cervical cancer, are made invisible and replaced with a one-size model of cancer risk that relies on a false category of girlhood and a universal girl." It is precisely the narrow interestedness of the various parties and constituencies that inflamed the debate, placing (universal) girls at the center of the rhetorical and semantic storm.

Individualizing Risk

In one ideal scenario, HPV vaccines would build herd immunity through near-universal immunization. Yet in practice to date in the United States, as Nancy Berlinger and Alison Jost (chapter 11) write, the HPV vaccine functions in public discourse "more like a car seat, protecting the individual from harm"—complementing other established methods of cancer prevention and control. Jennifer Reich observes in chapter 9 that commercial pitches have emphasized not the public good but the individual girl at risk. Heather Munro-Prescott (chapter 6) notes that "by focusing on individual attitudes and choice . . . , [the HPV vaccination campaign] overlooks larger public health issues such as socio-economic status and access to health care services." Authors Giovanna Chesler and Bree Kessler (chapter 8) similarly insist that, as a result of this focus on individual risk, safe sex has not been emphasized, nor has the necessity for routine Pap smears, since vaccination has been promoted as the safest solution. For Mamo, Nelson, and Clark, this framing of the vaccine away from public health concerns presents a critical problem. They contend that "the U.S. public health infrastructure needs to be wrestled away from capital (pharmaceutical and biotechnology companies) and returned to a multidisciplinary set of social thinkers and actors who weigh global and local STI [sexually transmitted infection] risks, preventions, and treatments . . . in the interests of public good."

From the outset, then, the HPV vaccine discussion was framed in terms of individual choices, risks, empowerment, and parental decision making. The characterization of the HPV vaccine as a question of individual risk had important implications for parents, placing before them the critical challenge of making decisions on behalf of others. As Gretchen Chapman notes in chapter 10, because vaccination decisions are highly susceptible to "biased and inaccurate risk perceptions," the ways in which risks are framed by the various interests in the controversy play a powerful role in vaccination behavior. Ideologies of parental rights and the state also came into play. For many American parents, the targeted nature of the vaccine mandate, aimed as it was at their female children, provoked many to see the public health mandate as an effort to substitute government judgment for their own, particularly in matters relating to the sexual behavior and health of their girls. From minority Washington, D.C., to conservative Colorado Springs, this broader skepticism colored parental assessments of expert claims about the safety and necessity of the vaccines. Thus, while a rhetoric of individualized risk might have encouraged some parents to vaccinate, other parents resisted vaccination precisely because it was framed as a public command by government. In this context, pressure built on governments considering vaccine mandates to include robust opportunities for parents to opt out of vaccinating their girls. Berlinger and Jost raise the important issue of whether policymakers should support the enactment of "new, customized, opt-out policies" that may reflect and encourage false beliefs about vaccines. At the same time, James Colgrove (chapter 1) emphasizes the historical significance of mandatory vaccination programs in ensuring that vaccines find their way to all sectors of society, including the poor.

Political Pathways: Public Health at a Crossroads

In many ways, the HPV debate reflects a broader challenge to public health—bringing classic public health to a political crossroads. As Julie Livingston, Keith Wailoo, and Barbara Cooper (chapter 13) observe, "in the U.S. and the African contexts, the sexual conundrums surrounding vaccination take on vastly different forms, yet they find common roots in fears about state governance." These authors portray the HPV vaccine as part of a highly varied global landscape where varieties of skepticism are evident and where the vaccine raises the "complex challenge of administrative, social, and sexual governance." Globally, people retain much skepticism about public health efforts, particularly those promoting mandatory vaccination, and it is this preexisting skepticism that gets invoked in the HPV vaccine discussion. In the United Kingdom, as Andrea Stöckl shows in chapter 14, the measles, mumps, and rubella vaccine debate of the early 2000s informed the trajectory of the HPV discussion. In France, notes Ilana Löwy (chapter 15), the HPV story "was shaped by the prolonged failure to promote a national program of screening for cervical cancer and . . . by the ill-fated attempt to introduce mass vaccination against hepatitis B . . . [during] the 1980s and 1990s." And in the United States, many of the twists and turns in the story of HPV vaccination reflect a distinct peculiarity—that drug companies there are permitted to engage in direct-to-consumer advertising. In such ways, across the world, the fate of HPV vaccination has been inseparably linked with that of other health measures and with previous controversies and developments in public health.

The HPV vaccine emerged in explicit competition with other modalities of reducing cervical cancer—and one of the competitive tensions revolved around the relative value of the vaccine versus the longstanding Pap smear. More recently, a new biotechnology product that tests directly for the presence of HPV DNA offers the possibility of an even more efficient means of detecting early-stage disease and hence a third route toward the prevention of cervical cancer morbidity and mortality. Cervical cancer rates, of course, have been declining in the developed world for decades. Interventions (the Pap smear and new modalities of treatment) and changes in socioeconomic conditions and behavior have coexisted with this decline, making the process of assessing their relative value difficult—both in the past, and in the years ahead. In this context, measuring the impact of HPV vaccines and screening tests becomes especially difficult. As Löwy notes above, the scheduled increase in numbers of cervical smears in France will surely complicate the task of interpreting the efficacy of HPV vaccine. The HPV debate cannot be understood, therefore, without looking at other interventions. For those people who lived in places where the Pap smear was fully integrated into a social system of cancer surveillance and care, the HPV vaccine emerged as a welcome next step—a chance to do more than just find and treat disease early, to prevent entirely those cases caused by the most common viral types. But as many authors in this volume have pointed out, the integration of the Pap smear into cancer surveillance is deeply uneven and highly problematic—making any shift of resources to the HPV vaccine additionally controversial. There is the possibility, for example, that the era of the HPV vaccine will privilege the haves over the have-nots once again and privilege expensive strategies over cheaper ones. It is also likely, as Aronowitz argues, that the vaccines will save more money by reducing the cost of false-positive Pap smear work-ups than by reducing the costs of HPV-related disease. The idea of vaccinating against cancer (rather than merely detecting it early and treating it) has an obvious appeal, yet in the end the discourse of HPV vaccination as one of medicine's simple solutions, or as the best tool for the job, is misleading in its simplicity.

Clearly, the meaning and long-term implications of HPV vaccination remain up for grabs. Is the HPV vaccine a vehicle for advancing private gains (the health concerns of individuals who can afford it and the economic health of Merck and GlaxoSmithKline)? Or is it a tool for advancing the public good (the health of populations and of the nations to which they belong)? Can these goals of private gains and public health work in tandem? This question has deeply marked the HPV vaccine debates on the world stage—in Europe and the United Kingdom, across North America, and to a far more limited degree in the global South, where the burden of cervical cancer is heaviest. In each context, the meanings and framing of the vaccine have been shaped by preexisting and highly politicized health debates, where these public-private issues have been central. These ongoing HPV vaccine controversies have been given personal poignancy, scientific meaning, and political life by persistent tensions among the main players (pharmaceutical companies, scientists, governments, parents, girls, other patients, and health activists). Thus they highlight a set of enduring issues that compel debate and action regarding sexuality, the cultural politics of family, trust in government, and the credibility of scientific evidence—across a range of national and international contexts. In four years, although the controversies have evolved dramatically (becoming normalized in a way that attracted far less media coverage in late 2009 than in 2007 and 2008) they have not yet disappeared. Indeed, the period since mid-2008 has seen persistent, sporadic calls for attention to the vaccines' side effects, concerns about the blurring of the line between medical advice and pharmaceutical marketing, and questions about whether public health establishments acted too far ahead of evidence in this case.[8] These debates will surely evolve further in the years ahead. Eventually, the HPV vaccine will fade from the center of the public health storm, pushed out perhaps by new vaccine and public health debates. But one lesson of this book is that the challenges embodied by the controversy—how to make decisions amid uncertainty, the consequences of gendering and individualizing risk, the costs of linking vaccination to debates over morality and good government—will not disappear. In an age of biotechnology innovation, such wrangling over medicine's simple solutions will surely continue to concern families, consumers, citizens, and states for decades to come.

NOTES

1. "High Uptake for Cervical Cancer Jag," news release, the Scottish Government, March 26, 2009, www.scotland.gov.uk/News/Releases/2009/03/26094529. Across the United Kingdom, uptake among 12–13-year-old girls was also high for the first (85.5%) and second (80.8%) doses, with data on the third dose yet to come. "HPV Vaccine Uptake February 2009," NHS Immunisation Information, www.immunisation.nhs.uk/Library/News/HPV_uptake_feb09.

2. Jane Harries, Jennifer Moodley, Mark A. Baroneb, Sumaya Malla, and Edina Sinanovic, "Preparing for HPV Vaccination in South Africa: Key Challenges and Opinions," *Vaccine* 27 (Jan. 2009): 38–44.

3. "National, State, and Local Area Vaccination Coverage among Adolescents Age 13–17 Years Old—United States, 2008," *Morbidity and Mortality Reports* 58 (2009): 997–1001; Jeannine S. Schiller and Gary L. Euler, "Vaccination Coverage Estimates from the National Health Interview Survey: United States, 2008," *NCHS Health and Stats* (July 2009): 1–7, www.cdc.gov/nchs/data/hestat/vaccine_coverage.pdf.

4. Schiller and Euler, "Vaccination Coverage Estimates," 2.

5. Ibid.

6. Jessica A. Kahn, H. Paul Cooper, Susan T. Vadaparampil, Barbara C. Pence, Armin D. Weinberg, Salvatore J. LoCoco, and Susan L. Rosenthal, "Human Papillomavi-

rus Vaccine Recommendations and Agreement with Mandated Human Papillomavirus Vaccination for 11-to-12-Year-Old Girls: A Statewide Survey of Texas Physicians," *Cancer Epidemiology, Biomarkers, and Prevention* (Aug. 2009): 2325–2332, quote on 2329.

7. Susan Donaldson James, "Girl Rejects Gardasil, Loses Path to Citizenship," ABC News, Sept. 15, 2009, http://abcnews.go.com/m/screen?id=8542051.

8. See, for example, Charlotte Haug, "The Risks and Benefits of HPV Vaccination," *Journal of the American Medical Association* 302 (Aug. 17, 2009): 795–796; and S. M. Rothman and D. J. Rothman, "Marketing HPV Vaccine: Implications for Adolescent Health and Medical Professionalism," *Journal of the American Medical Association* 302 (Aug. 17, 2009): 781–786. See also Elisabeth Rosenthal, "Researchers Question Wide Use of HPV Vaccines," *New York Times*, Aug. 21, 2008, http://query.nytimes.com/gst/fullpage.html?res=9504E3D61E39F932A1575BC0A96E9C8B63; Elisabeth Rosenthal, "The Evidence Gap: Drug Makers' Push Leads to Cancer Vaccines' Rise," *New York Times*, Aug. 19, 2008, www.nytimes.com/2008/08/20/health/policy/20vaccine.html; and Miriam Jordan, "Gardasil Requirement for Immigrants Stirs Backlash: Advocates, Experts Criticize U.S. Move to Mandate Vaccine," *Wall Street Journal*, Oct. 1, 2008, http://online.wsj.com/article/SB122282354408892791.html.

Contributors

Robert Aronowitz, Professor in the Department of History and Sociology of Science at the University of Pennsylvania, is the author of *Making Sense of Illness: Science, Society, and Disease* (Cambridge, 1998) and *Unnatural History: Breast Cancer and American Society* (Cambridge, 2007). Aronowitz was the founding director of the health and societies program at the University of Pennsylvania, and he currently codirects the university's Robert Wood Johnson Health and Society Scholars Program.

Nancy Berlinger, Deputy Director and Research Scholar at The Hastings Center, is the author of *After Harm: Medical Error and the Ethics of Forgiveness* (Johns Hopkins, 2005). Her current research includes ethics and policy issues arising from vaccination refusals and other avoidance behaviors and practices in health care.

Lundy Braun is Professor of Pathology and Laboratory Medicine and Africana Studies and a member of the Faculty Committee on Science and Technology Studies at Brown University. For fifteen years she conducted research on the role of HPVs in cervical cancer. Her current research focuses on the history of race and science, the contemporary debate over genomics, and asbestos-related diseases in South Africa. She is at work on a history of the racialization of spirometry, an instrument that measures lung capacity.

Gretchen Chapman, Professor and Chair of the Psychology Department at Rutgers University, conducts research on judgment and decision-making, comparing how people make decisions with normative models of the best or most rational method for making decisions. She has published widely on factors associated with the decision to receive a flu shot, focusing on perceived risks and benefits, anticipated emotions such as worry, social factors such as perceived norms and job role, and interpersonal factors such as altruistic benefits of vaccinating.

Giovanna Chesler, Assistant Professor in Communication Arts at Marymount Manhattan College, is a Web producer and filmmaker. Recent films include the hour-long documentary *Period: The End of Menstruation* (2006), distributed by Cinema Guild, on trends in menstrual suppression and *Bye Bi Love* (2010), a short narrative that she wrote and directed, addressing the climate of marriage, both gay and straight. Her critical theory centers on documentary and sound in film. She blogs about the intersections of media and menstruation for the Society of Menstrual Cycle Research at www.menstruationresearch.com/blog.

Aleia Clark is a doctoral student in the Department of Sociology at the University of Maryland, College Park. She received her B.A. in Sociology from Spelman College in 2005 and her M.A. in Sociology from University of Maryland in 2008. Her master's thesis uses Gardasil as a case study to explore how notions of public good and profit motives shape vaccine development.

James Colgrove is an Associate Professor in the Center for the History and Ethics of Public Health at Columbia University's Mailman School of Public Health. He is the author of *State of Immunity: The Politics of Vaccination in Twentieth-Century America* (University of California Press, 2006); coauthor, with Amy Fairchild and Ronald Bayer, of *Searching Eyes: Privacy, the State, and Disease Surveillance in America* (University of California Press, 2007); and coeditor, with David Rosner and Gerald Markowitz, of *The Contested Boundaries of American Public Health* (Rutgers University Press, 2008).

Barbara M. Cooper, Professor of History at Rutgers University, is the author of *Evangelical Christians in the Muslim Sahel* (Indiana University Press, 2006), which received the Melville J. Herskovits Prize of the African Studies Association, and *Marriage in Maradi: Gender and Culture in a Hausa Society in Niger, 1900–1989* (Heinemann, Social History of Africa Series, 1997). Cooper conducts research on the intersections between culture and political economy, focusing upon gender, religion, and family life.

Steven Epstein, John C. Shaffer Professor in the Humanities and Professor of Sociology at Northwestern University, is the author of *Impure Science: AIDS, Activism, and the Politics of Knowledge* (University of California Press 1996) and *Inclusion: The Politics of Difference in Medical Research* (University of Chicago Press 2007).

April N. Huff is a doctoral student in Sociology and Science Studies at the University of California, San Diego. Her areas of research include social movements, women's health, sexuality, and science and technology studies.

Alison Jost is a Research Assistant at the Interdisciplinary Center for Bioethics at Yale University.

Bree Kessler is a doctoral student in the Environmental Psychology Program at the Graduate Center of the City University of New York and an adjunct lecturer at Hunter College. Her interests are media, gender and the built environment, reproductive/sexual health, and participatory action research. She is a trained birthing doula and reiki practitioner and a frequent contributor to *Bitch: A Feminist Response to Pop Culture* on issues related to public health and the media.

Julie Livingston, Associate Professor of History at Rutgers University, is a scholar of African history with interdisciplinary training in public health and anthropology. She is the author of *Debility and the Moral Imagination in Botswana* (Indiana University Press, 2005) and a coeditor of *A Death Retold: Jesica Santillan, the Bungled Transplant, and Paradoxes of Medical Citizenship* (University of North Carolina Press, 2006). Her research focuses on the human body as a moral condition, including the ethical entanglements engendered by bodily vulnerability in conditions of scarce resources. She is currently writing a book-length ethnography of Botswana's only cancer ward.

Ilana Löwy, Senior Research Fellow with the Institut National de la Santé et de la Recherche Médicale (INSERM), Paris, is a historian of medicine and biomedical sciences and the author of *Between Bench and Bedside: Science, Healing, and Interleukin-2 in a Cancer Ward* (Harvard University Press, 1996), *Virus, moustiques et modernité: Science, politique et la fièvre jaune au Brésil* (Archives d'Histoire Contemporaine, 2001), and *L'emprise du genre: Masculinité, féminité, inégalité* (La Dispute, 2006).

Laura Mamo, Associate Professor at the Health Equity Institute for Research, Practice, and Policy at San Francisco State University, is the author of *Queering Reproduction: Achieving Pregnancy in the Age of Technoscience* (Duke University Press, 2007), a coauthor of *Living Green: Communities that Sustain* (New Society Press, 2009), and a coeditor of *Biomedicalization: Theorizing Health and Illness in U.S. Biomedicine* (Duke University Press, 2010). Her work has appeared in *Signs:*

Journal of Women, Culture and Society, the *American Sociological Review*, and *Sociology of Health and Illness*.

Amber Nelson is an advanced doctoral candidate in the Department of Sociology at the University of Maryland, College Park. Nelson received her master's degree in sociology in 2007 for her study of the social-scientific controversy of prescribing antidepressants to adolescents. Her doctoral research examines the rhetorical and material construction of adolescent mental health classification and practice since 1985.

Ling Phoun is a Ph.D student in the Molecular Biology and Microbiology Program at the Sackler School of Biomedical Sciences, Tufts University School of Medicine. She is interested in the genetics of breast cancer.

Heather Munro Prescott, Professor of History at Central Connecticut State University, is the author of *Student Bodies: The Impact of Student Health on American Society and Medicine* (University of Michigan Press, 2007) and of *"A Doctor of Their Own": A History of Adolescent Medicine* (Harvard University Press, 1998), Munro Prescott has also coedited Children and Youth in Sickness and Health: A Historical Handbook and Guide (Greenwood Press, 2004) and has begun a new project on the history of emergency contraception.

Doreen Ramogola-Masire, Country Director for the Botswana UPenn Partnership based in Botswana (a partnership between Botswana and the University of Pennsylvania in HIV care and support) and head of Women's Health Initiative, is an obstetrician and gynecologist with special interest in high-risk obstetrics and cervical cancer prevention. She has worked at several hospitals in southern Africa, including Chris Hani Baragwanath and Groote Schuur hospitals.

Jennifer A. Reich, Assistant Professor in the Sociology Department at the University of Denver, is the author of *Fixing Families: Parents, Power, and the Child Welfare System* (Routledge, 2005), which received the Distinguished Contribution to Scholarship Book Award from the American Sociological Association section on Race, Gender, and Class in 2007. Reich has also written numerous articles and book chapters on the intersections between family, welfare, health care, race, and public policy. Her current research explores how parents make decisions about their children's health care, particularly in terms of immunizations.

Andrea Stöckl, Lecturer in Medical Sociology at the School of Medicine, Health Policy, and Practice, University of East Anglia, has published research on how "ordinary people" learn new medical concepts and incorporate new medical technology into their everyday lives, focusing especially on SLE (systemic lupus erythematosus) and gene therapy. Her recent research and projects explore how expert knowledge and lay expertise interact in complex ways to create ever-shifting practices of dealing with dis-ease in late modernity.

Keith Wailoo is the Martin Luther King Jr. Professor of History at Rutgers University and has a joint appointment in the Institute for Health, Health Care Policy, and Aging Research. He is the author of *Dying in the City of the Blues: Sickle Cell Anemia and the Politics of Race and Health* (University of North Carolina Press, 2001) and *Drawing Blood: Technology and Disease Identity in Twentieth-Century America* (Johns Hopkins University Press, 1997) and coauthor of *The Troubled Dream of Genetic Medicine: Ethnicity and Innovation in Tay-Sachs, Cystic Fibrosis, and Sickle Cell Disease* (Johns Hopkins University Press, 2006).

Index